Key Topics in Landscape Ecology

Landscape ecology is a relatively new area of study, which aims to understand the pattern of interaction of biological and cultural communities within a landscape. This book brings together leading figures from the field to provide an up-to-date survey of recent advances, identify key research problems, and suggest a future direction for development and expansion of knowledge. Providing in-depth reviews of the principles and methods for understanding landscape patterns and changes, the book illustrates concepts with examples of innovative applications from different parts of the world. Forming a current "state-of-the-science" for the science of landscape ecology, this book forms an essential reference for graduate students, academics, professionals, and practitioners in ecology, environmental science, natural resource management, and landscape planning and design.

JIANGUO (JINGLE) WU is Professor of Ecology, Evolution, and Environmental Science at Arizona State University, Tempe, Arizona, USA. His research interests include landscape ecology, urban ecology, and sustainability science, focusing on hierarchical patch dynamics, pattern–process–scale relationships, spatial scaling, land-use change and its effects on ecosystem processes, and biodiversity and ecosystem functioning. He has published more than 120 scientific papers which involve mostly dryland ecosystems in North America and China. His professional service includes Program Chair of the 2001 Annual Symposium of the US Association of the/International Association of Landscape Ecology (US-IALE), Councillor-at-Large of US-IALE (2001–3), and Chair of the Asian Ecology Section of the Ecological Society of America (1999–2000). He is currently the editor-in-chief of the international journal *Landscape Ecology*.

RICHARD HOBBS is Professor of Environmental Science at Murdoch University, Western Australia, and has research interests in restoration ecology and landscape ecology. These focus on the conservation and management of altered landscapes, particularly the agricultural area of southwestern Australia. He is a fellow of the Australian Academy of Science and has been listed by ISI as one of the most highly cited researchers in ecology and environmental science. His professional services include President of the International Association for Landscape Ecology (1999–2003) and President of the Ecological Society of Australia (1998–1999). He is currently the editor-in-chief of the journal *Restoration Ecology*.

Cambridge Studies in Landscape Ecology

Series Editors

Professor John Wiens *Colorado State University*
Dr Peter Dennis *Macaulay Land Use Research Institute*
Dr Lenore Fahrig *Carleton University*
Dr Marie-Jose Fortin *University of Toronto*
Dr Richard Hobbs *Murdoch University, Western Australia*
Dr Bruce Milne *University of New Mexico*
Dr Joan Nassauer *University of Michigan*
Professor Paul Opdam *ALTERRA, Wageningen*

Cambridge Studies in Landscape Ecology presents synthetic and comprehensive examinations of topics that reflect the breadth of the discipline of landscape ecology. Landscape ecology deals with the development and changes in the spatial structure of landscapes and their ecological consequences. Because humans are so tightly tied to landscapes, the science explicitly includes human actions as both causes and consequences of landscape patterns. The focus is on spatial relationships at a variety of scales, in both natural and highly modified landscapes, on the factors that create landscape patterns, and on the influences of landscape structure on the functioning of ecological systems and their management. Some books in the series develop theoretical or methodological approaches to studying landscapes, while others deal more directly with the effects of landscape spatial patterns on population dynamics, community structure, or ecosystem processes. Still others examine the interplay between landscapes and human societies and cultures.

The series is aimed at advanced undergraduates, graduate students, researchers and teachers, resource and land-use managers, and practitioners in other sciences that deal with landscapes.

The series is published in collaboration with the International Association for Landscape Ecology (IALE), which has Chapters in over 50 countries. IALE aims to develop landscape ecology as the scientific basis for the analysis, planning and management of landscapes throughout the world. The organization advances international cooperation and interdisciplinary synthesis through scientific, scholarly, educational and communication activities.

EDITED BY

JIANGUO WU
ARIZONA STATE UNIVERSITY

RICHARD J. HOBBS
MURDOCH UNIVERSITY

Key Topics in Landscape Ecology

Shaftesbury Road, Cambridge CB2 8EA, United Kingdom

One Liberty Plaza, 20th Floor, New York, NY 10006, USA

477 Williamstown Road, Port Melbourne, VIC 3207, Australia

314–321, 3rd Floor, Plot 3, Splendor Forum, Jasola District Centre, New Delhi – 110025, India

103 Penang Road, #05–06/07, Visioncrest Commercial, Singapore 238467

Cambridge University Press is part of Cambridge University Press & Assessment,
a department of the University of Cambridge.

We share the University's mission to contribute to society through the pursuit of
education, learning and research at the highest international levels of excellence.

www.cambridge.org
Information on this title: www.cambridge.org/9780521616447

First published 2007
Reprinted with corrections 2009

A catalogue record for this publication is available from the British Library

ISBN   978-0-521-85094-0   Hardback
ISBN   978-0-521-61644-7   Paperback

Contents

PART III **Synthesis**

Contributors

Marc Antrop
Geography Department, Ghent University,
B9000 Ghent, Belgium

David Bowman
Key Centre for Tropical Wildlife Management,
Northern Territory University, Darwin 0909,
Australia

Jeffrey A. Cardille
Department of Zoology, University of
Wisconsin, Madison, WI 53706, USA

Lenore Fahrig
Ottawa-Carleton Institute of Biology, Carleton
University, Ottawa, Canada K1S 5B6

Curtis Flather
US Forest Service, Rocky Mountain Research
Station, Fort Collins, CO 80526, USA

James D. Forester
Department of Zoology, University of
Wisconsin, Madison, WI 53706 USA

Gary Fry
Institute of Landscape Planning, Agricultural
University of Norway, N-1432 Aas, Norway

Robert H. Gardner
University of Maryland Center for
Environmental Science, Appalachian
Laboratory, Frostburg, MD 21532, USA

Richard J. Hobbs
School of Environmental Science, Murdoch
University, Murdoch, WA 6150, Australia

John Hof
US Forest Service, Rocky Mountain Research
Station, Fort Collins, CO 80526, USA

Louis R. Iverson
USDA Forest Service, 359 Main Road,
Delaware, OH 43015, USA

Robert. G. Lesslie
Bureau of Rural Sciences, GPO Box 858,
Canberra, ACT 2601, Australia

Harbin Li
USDA Forest Service Southern Research
Station, Center for Forested Wetlands
Research, Charleston, SC 29414, USA

John A. Ludwig
Tropical Savannas Management Cooperative
Research Centre and CSIRO Sustainable
Ecosystems, Atherton, Queensland 4883,
Australia

Brendan G. Mackey
School of Resources, Environment and Society,
Faculty of Science, The Australian National
University, Canberra ACT 0200, Australia

Henry A. Nix
CRES, The Australian National University,
Canberra, ACT 0200, Australia

Paul Opdam
Wageningen University and Research Center,
Wageningen, the Netherlands

Roy E. Plotnick
Department of Earth and Environmental
Sciences, University of Illinois at Chicago,
Chicago, IL 60670, USA

Hugh P. Possingham
The Ecology Centre, Department of
Mathematics and School of Life Sciences, The
University of Queensland, St Lucia, QLD 4072,
Australia

Harry F. Recher
School of Natural Sciences, Edith Cowan
University, Joondalup, Western Australia 6027,
Australia

Rien Reijnen
Alterra Green World Research, Landscape
Centre, Wageningen, The Netherlands

Michael E. Soulé
PO Box 2010, Hotchkiss, CO 81419, USA

Eveliene Steingröver
Alterra Green World Research, Department of
Landscape Ecology, Wageningen, the
Netherlands

Bärbel Tress
Department of Geography and Environment,
University of Aberdeen, Aberdeen, AB24 3UF,
United Kingdom

Gunther Tress
Department of Geography and Environment,
University of Aberdeen, Aberdeen, AB24 3UF,
United Kingdom

Monica G. Turner
Department of Zoology, University of
Wisconsin, Madison, WI 53706, USA

Claire C. Vos
Alterra Green World Research, Department of
Landscape Ecology, Wageningen, the
Netherlands

Jann E. Williams
Centre for Sustainable Regional Communities,
La Trobe University, Bendigo, Victoria 3552,
Australia

John C. Z. Woinarski
Biodiversity Section, Natural Systems,
Department of Infrastructure, Planning and
Environment, PO Box 496, Palmerston, NT
0831, Australia

Jianguo Wu
School of Life Sciences and Global Institute of
Sustainability, Arizona State University,
Tempe, AZ 85287, USA

Preface

Landscapes are diverse, complex, beautiful, and inspirational. Spatial heterogeneity is the most salient feature that characterizes all landscapes. While the physical environment exhibits various spatial patterns on different scales, biological organisms are organized into populations and communities across landscapes. Like other biological organisms, humans live and act on landscapes, and thus have influenced, and been influenced by, landscapes. Unlike other biological organisms, however, humans represent an unparalleled force that has profoundly altered the structure and function of landscapes and even the entire biosphere. A number of worldwide environmental problems, such as land degradation, biodiversity loss, and global climate change, clearly attest to this destructive power of anthropogenic activities. Most, if not all, of the pressing ecological and environmental problems that humanity is faced with today are directly related to human alterations of landscapes. In most cases, humans strive to increase their appropriation of ecosystem goods and services from landscapes while compromising the abilities of ecosystems to perform other functionalities and resulting in serious ecological and socioeconomic consequences. Thus, landscape ecology is essential not only for understanding how Nature works in spatially heterogeneous environments, but also for providing practical guidelines and solutions for maintaining and developing sustainable landscapes.

Landscape ecology has made tremendous progress in theory and practice in recent decades. In the same time, as a rapidly developing discipline it is faced with new problems and challenges. For example, the diversification of ideas and approaches in landscape ecology, which we consider is mostly healthy and inevitable, has caused confusions among landscape ecologists as to what the identity or scientific core of this field is. Also, while all landscape ecologists seem to agree that landscape ecology should be interdisciplinary or transdisciplinary, little consensus can be found in terms of what interdisciplinarity and transdisciplinarity mean and how they should be achieved.

To address these problems and promote the further development of landscape ecology, Jianguo Wu, then Program Chair of the US Association of the International Association of Landscape Ecology (US-IALE), organized a special session entitled "Top 10 List for Landscape Ecology in the twenty-first Century" at the 16th Annual Symposium of US-IALE at Arizona State University, Tempe, Arizona in April 2001. A group of prominent landscape ecologists worldwide were invited to present their views on the most important research topics, questions, and challenges in the field. Richard Hobbs, then President of IALE, presented an overview of the outcomes of this symposium at the European Landscape Ecology Congress in Stockholm and Tartu, Estonia in July 2001. Afterwards, J. Wu and R. Hobbs developed a synthesis paper based on the diverse perspectives presented at the "Top 10 List Symposium" (Wu, J. and R. Hobbs. 2002. Key issues and research priorities in landscape ecology: An idiosyncratic synthesis. *Landscape Ecology* **17**, 355–65). While the "Top 10 List" was successful in identifying key issues and research topics, an important next step was to have in-depth discussions to examine the state-of-the-science and future directions in each subject area. This was precisely the objective of the symposium on "Key Issues and Research Priorities in Landscape Ecology" at the 2003 World Congress of IALE in Darwin, Australia in July 2003, participated by a group of well-established landscape ecologists and organized by J. Wu and R. Hobbs. This book is based on selected presentations at the Darwin symposium, with additional invited contributions.

The book focuses on the prevailing perspectives and prospects of landscape ecology across geographic and cultural boundaries. It covers the theory, methodology, and applications of landscape ecology. The chapters have in-depth discussions of the major achievements, key questions, and future directions in a series of important research topics in landscape ecology. Some of them explore holistic, interdisciplinary approaches and describe innovative applications of landscape ecology principles in conservation, management, planning, and design. We believe that identifying key research problems, synthesizing major advances, and pointing out future directions are necessary for promoting concerted development of landscape ecology and enhancing its "identity." We do not believe that any individual is in the position to dictate what landscape ecology is or direct where landscape ecology should go. Landscape ecology, as a new paradigm, has to be defined and developed by the community of landscape ecologists and practitioners. We hope that, as a whole, this book reflects the collective view of the state-of-the-science of landscape ecology.

We are most grateful to all the contributors to this book, who are not only first-rate landscape ecologists, but also the most wonderful colleagues to work with. To ensure the quality of the book, all chapters were peer-reviewed. We sincerely thank all those who participated in the review process, including: Jack Ahern, Gary Brierley, John M. Briggs, Peter Cale, Marie-Josee Fortin, G. Darrel Jenerette, Rob Jongman, Ted Lefroy, Kirk A. Moloney, Michael R. Moss, Jari Niemelä, R. Gil Pontius, Jr., Kurt Riitters, Denis Saunders, Santiago Saura, Austin Troy, Helene Wagner, James D. Wickham, and Xinyuan (Ben) Wu. Our sincere appreciation also goes to Alan Crowden at Cambridge University Press who saw the book through from concept to reality. Finally, we thank Yongfei Bai and Kaesha Neil at the Landscape Ecology and Modeling Laboratory (LEML) of Arizona State University for their assistance with reformatting the references throughout the book.

We believe that this book will be of interest to a wide audience, including graduate students, academic professionals, and practitioners in ecology, environmental science, landscape planning and design, and resource management. In addition to its value as a reference for a variety of research and application purposes, this book could be used for graduate-level courses, or a supplementary text for undergraduate-level courses, in landscape ecology and related subject areas. To help the readers to better understand the contents of the book and to stay abreast with what's going on in the forefront of landscape ecological research, a web site will be dynamically maintained to provide additional materials related to the book (e.g., color figures, chapter abstracts, and related key publications) and information on continuing discussions on the key issues in landscape ecology. The web address is http://LEML.asu.edu/Landscape-Ecology/.

This book is dedicated to the next generation of landscape ecologists, and we wish them luck with the exciting and challenging times ahead.

PART I

Introduction

1

Perspectives and prospects of landscape ecology

1.1 Introduction

Landscape ecology has rapidly established itself as an interdisciplinary research field worldwide in the past few decades. However, diversification in perspectives and approaches has apparently caused some concerns with the "identity" of the field in recent years. For example, Wiens (1999) stated that "landscape ecology continues to suffer from something of an identity crisis," while Moss (1999) warned that landscape ecology's "healthy, youthful development will be cut off before it matures if it does not recognize and develop its own distinctive core and focus." As landscape ecologists, we feel that we should not be particularly worried about the identity or the fate of the field. Its identity is to some extent self-defining through the activities that people calling themselves landscape ecologists undertake, and its fate will be determined by its utility and its ability to provide techniques, approaches, and applications which help tackle the complex environmental management challenges facing humanity. However, we do think that, after two decades of rapid developments in both theory and practice, landscape ecology can benefit from a forward-looking introspection.

For example, several questions may be asked to address some of the concerns and challenges this field now faces. What is the identity of landscape ecology that it is losing or that has never been established? Given the multidisciplinary origins and goals of the field, is it possible for landscape ecology to have "its own distinctive core and focus?" If so, what would it be? How should we solidify the interdisciplinarity or transdisciplinarity of landscape ecology? These are grand questions whose answers may be still quite elusive. Thus, this book is not intended to provide all the answers. Rather, it addresses a series of key issues

Key Topics in Landscape Ecology, ed. J. Wu and R. Hobbs.
Published by Cambridge University Press. © Cambridge University Press 2007.

and perspectives in contemporary landscape ecology identified by a group of leading scientists around the world. By closely examining these key topics, we hope that this book will contribute to the development of landscape ecology, and help resolve the grand questions posed above.

1.2   Key issues and research topics in landscape ecology

The chapters in this book were collected together to explore a set of key issues synthesized by Wu and Hobbs (2002) from a symposium which sought to draw out from leading landscape ecologists what these issues were. Many ideas from the group of 17 people was condensed to a long list of items (Table. 1.1), from which Wu and Hobbs (2002) further identified six key issues to be considered: (1) interdisciplinarity or transdisciplinarity, (2) integration between basic research and applications, (3) conceptual and theoretical development, (4) education and training, (5) international scholarly communication and collaborations, and (6) outreach and communication with the public and decision-makers.

Wu and Hobbs (2002) also identified ten key research areas dealing with these issues: (1) ecological flows in landscape mosaics, (2) causes, processes, and consequences of land use and land cover change, (3) nonlinear dynamics and landscape complexity, (4) scaling, (5) methodological development, (6) relating landscape metrics to ecological processes, (7) integrating humans and their activities into landscape ecology, (8) optimization of landscape pattern, (9) landscape conservation and sustainability, and (10) data acquisition and accuracy assessment.

The chapters in this book collectively cover most of these issues and research areas. The subject matter varies from questions regarding the collection and analysis of data for use in landscape ecological studies, through the intersection between landscape ecology, ecosystem ecology and conservation biology, to the broader application of landscape ecology in complex social–ecological systems in inter- and transdisciplinary settings. Hence this book provides a microcosm of the current state of play in landscape ecology: a lot of activity in the area of acquiring and interpreting spatial ecological data and an equivalent amount of effort in the broader aspects which interface ecology with management and planning.

There has been a lot of introspection in landscape ecology about what the subject is all about. It is apparent from the chapters in this book that this is still evident. In the subject as a whole, there seems to be something of a schism between the more biophysically oriented school and the arm that deals with the interface between science, planning and management. The first sees landscape ecology primarily as a means of dealing with spatial patterning and

TABLE 1.1. *A list of major research topics in landscape ecology based on suggestions by a group of leading landscape ecologists from around the world at the 16th Annual Symposium of the US Regional Association of the International Association for Landscape Ecology, held at Arizona State University, Tempe in April 2001[a]*

Development of theory and principles
- Landscape mosaics and ecological flows
- Land transformations
- Landscape sustainability
- Landscape complexity

Landscape metrics
- Norms or standards for metric selection, change detection, etc.
- Integration of metrics with holistic landscape properties
- Relating metrics to ecological processes
- Sensitivity to scale change

Ecological flows in landscape mosaics
- Flows of organisms, material, energy, and information
- Effects of connectivity, edges, and boundaries
- Spread of invading species
- Spatial heterogeneity and ecosystem processes
- Disturbances and patch dynamics

Optimization of landscape pattern
- Optimization of land-use pattern
- Optimal management
- Optimal design and planning
- New methods for spatial optimization

Metapopulation theory
- Integration of the view of landscape mosaics
- Integration of economic theory of land-use change and cellular automata

Scaling
- Extrapolating information across heterogeneous landscapes
- Development of scaling theory and methods
- Derivation of empirical scaling relations for landscape pattern and processes

Complexity and nonlinear dynamics of landscapes
- Landscapes as spatially extended complex systems
- Landscapes as complex adaptive systems
- Thresholds, criticality, and phase transitions
- Self-organization in landscape structure and dynamics

Land-use and land-cover change
- Biophysical and socioeconomic drivers and mechanisms
- Ecological consequences and feedbacks
- Long-term landscape changes driven by economies and climate changes

(cont.)

TABLE 1.1. (*cont.*)

Spatial heterogeneity in aquatic systems
• The relationship between spatial pattern and ecological processes in lakes, rivers, and oceans
• Terrestrial and aquatic comparisons

Landscape-scale experiments
• Experimental landscape systems
• Field manipulative studies
• Scale effects in experimental studies

New methodological developments
• Integration among observation, experimentation, and modeling
• New statistical and modeling methods for spatially explicit studies
• Interdisciplinary and transdisciplinary approaches

Data collection and accuracy assessment
• Multiple-scale landscape data
• More emphasis on collecting data on organisms and processes
• Data quality control
• Metadata and accuracy assessment

Fast changing and chaotic landscapes
• Rapidly urbanizing landscapes
• War zones
• Other highly dynamic landscapes

Landscape sustainability
• Developing operational definitions and measures that integrate ecological, social, cultural, economic, and aesthetic components
• Practical strategies for creating and maintaining landscape sustainability

Human activities in landscapes
• The role of humans in shaping landscape pattern and processes
• Effects of socioeconomic and cultural processes on landscape structure and functioning

Holistic landscape ecology
• Landscape ecology as an anticipative and prescriptive environmental science
• Development of holistic and systems approaches

a See Wu and Hobbs 2002 for more details.

heterogeneity and building this on the foundation of ecosystem and population ecology. The second sees landscape ecology primarily as the necessary scientific underpinning for spatial planning and management of landscapes, particularly in human-dominated settings. This dichotomy could simplistically be interpreted as a North American versus European divide, but that would be too simplistic since there are many European landscape ecologists working primarily on the biophysical aspects and equivalently, many North Americans dealing with the planning and management issues. In addition, there are others, such as the Australians, who perhaps take a pragmatic middle road which combines both aspects.

Is this dichotomy a problem? The obvious answer is that it should not be, since both approaches are necessary and can be highly complementary. It is only a problem if adherents of either approach fail to appreciate the value and context of the other. Clearly, landscape planning has to rely on the acquisition and analysis of complex spatial data. Similarly, to be useful, spatial data need to feed into the planning and management process. Landscape ecology's key role, therefore, is to provide an umbrella for all of these endeavors so that people with different objectives and backgrounds can interact and develop approaches which are more than the sum of the parts.

In recent years this umbrella function has succeeded in part, but has perhaps not yet achieved all it can. Landscape ecology could be accused of lacking the unifying direction of more mission-oriented sciences such as conservation biology or restoration ecology (Hobbs 1997). Landscape ecology conferences attract people who are interested in landscapes – any and all aspects of landscapes are covered, from the hard-core spatial ecology through to the more humanities-based landscape history, aesthetics, design, and so on. Often there is still a clash of cultures, with apparently little common ground between the numerical and the spiritual and aesthetic. This is perhaps inevitable, but is not necessarily a terminal problem. Its solution lies in the acceptance of the breadth of issues and approaches involved in understanding how landscapes work. It lies in greater communication among researchers and practitioners from different disciplines and backgrounds. It lies in fostering that communication through mechanisms such as workshops and meetings, joint supervision of Ph.D. students, and joint faculty appointments between ecology and landscape design departments. We have had an era of increased specialization and fragmentation of effort, which has led us to the current state of the world: the future has to be based more in integrative and transdisciplinary approaches if we wish to find effective ways of steering the world in a more sustainable direction. Landscape ecology provides much of value for those wishing to better conserve or manage the planet and its inhabitants.

1.3 Concluding remarks

Landscape ecology must, therefore, continue to develop along the lines identified in the chapters in this book. We need continued improvement in our ability to collect and interpret spatial data. We need to ensure that effective metrics are developed which aid in this interpretation. We need to develop streamlined ways of feeding complex spatial data into land-use planning and management decisions. And to do all this, we need to find ways of conducting our research in inter- and transdisciplinary settings which actually work. This set of requirements is surely enough to stimulate the field of landscape ecology to continue to develop its intellectual rigor and to mature as a science. The various chapters in this book explore the current status of endeavors in each of the areas outlined above, and we hope that they faithfully indicate the vigor and promise currently being shown within landscape ecology.

References

Hobbs, R. J. 1997. Future landscapes and the future of landscape ecology. *Landscape and Urban Planning* 37, 1–9.

Moss, M. R. 1999. Fostering academic and institutional activities in landscape ecology. Pages 138–144 in J. A. Wiens and M. R. Moss (eds.) *Issues in Landscape Ecology*. Snowmass Village: International Association for Landscape Ecology.

Wiens, J. A. 1999. Toward a unified landscape ecology. Pages 148–51 in J. A. Wiens and M. R. Moss (eds.) *Issues in Landscape Ecology*. Snowmass Village: International Association for Landscape Ecology.

Wu, J. and R. Hobbs. 2002. Key issues and research priorities in landscape ecology: an idiosyncratic synthesis. *Landscape Ecology* 17, 355–65.

PART II

Key topics and perspectives

2

Adequate data of known accuracy are critical to advancing the field of landscape ecology

2.1 Introduction

The science of landscape ecology is especially dependent on high-quality data because often it focuses on broad-scale patterns and processes and deals in the long term. Likewise, high quality data are necessary as the basis for building policy. When issues, such as climate change, can induce international political and economic consequences, it becomes clear that providing high-quality, long-term data is paramount. It is not an accident that this chapter is positioned near the front of this book. Of the priority research topics presented in this book, this is the most pervasive across other topics because the availability of high-quality data limits progress in other realms. Be it historic land-use data needed to understand the dynamics of land-use change, the independent data of varying scales needed to assess scaling phenomena or test new metrics, the socioeconomic/cultural data needed to integrate humans into landscape ecology, or the biological and population data needed to evaluate ecological flows, the quality of raw data, metadata, and derived data products is critical to the core of landscape ecology. For each of these key topics and perspectives, the availability and quality of data will affect research results and practical recommendations.

2.2 Data advances in past two decades

It has been two decades since the 1983 workshop that many say established the landscape ecology field in North America (Risser *et al.* 1984). It was attended by many who have and still contribute to the field (e.g., Barrett, Botkin, Costanza, Forman, Godron, Golley, Hoekstra, Karr, Levin, Merriam,

Key Topics in Landscape Ecology, ed. J. Wu and R. Hobbs.
Published by Cambridge University Press. © Cambridge University Press 2007.

O'Neill, Parton, Risser, Sharpe, Shugart, Steinitz, Thomas, Wiens, and also a rookie named Iverson). From a scanty list of databases available, this group identified several databases with spatial components useful in landscape ecology: aerial photos; Landsat MSS; biological sampling schemes; and statistical measures of demography. They also identified several problems requiring attention: merging data from multiple sources with various levels of precision, resolution, and timing; choosing display formats appropriate for various uses and without distortions; the need for systematic or stratified field sampling in a heterogeneous universe; and decisions about the appropriate resolution for a particular problem. Researchers still struggle with these problems.

It may be useful to remind ourselves, especially our younger readers, where we were technologically with respect to data acquisition and manipulation two decades ago. I will relay what it was like for me. I was hired by Paul Risser in late 1982 to help develop the Illinois Lands Unsuitable for Mining Program to ensure lands of particular value were deemed "unsuitable" for surface mining. Risser had the foresight to identify that the new technology called "GIS" might be appropriate to do analysis of multiple mapped features. We hired Environmental Systems Research Institute (ESRI) to help us, and we became ESRI client number 12. Risser also believed it important that the GIS technology be made available to scientists, not just computer geeks. So I and my colleagues of various scientific bents spent three weeks in Redlands, CA training with the developers (ArcInfo 2.1 at the time), and the company president, Jack Dangermond, would take us during break to the orange orchard on the property to pick a few oranges. Subsequently, Illinois was the first state with full, integrated vector GIS at 1:500K. Prior to this time, most GIS work was performed with raster processing, using paper print-outs with different symbols for different classes within the matrix. Often entire walls were plastered with these print-outs to get the overall view of the study area. Several people from the Oak Ridge National Laboratory were creating and manipulating county-level data sets for the conterminous United States (Klopatek *et al.* 1979, Olson *et al.* 1980).

ArcInfo 2.1 was vector, but the hardware and software was limited. For data, we had a statewide digitized map of pre-European settlement vegetation (Anderson 1970) and the Land Use Data Acquisition (LUDA) data from the US Geological Survey (Anderson *et al.* 1976), vintage late 1970s. With these, we could assess long-term vegetation changes (Iverson and Risser 1987) and the attributes related to these landscapes (Iverson 1988). At that time, a simple overlay process would run all night; in fact, my colleagues forbade me to run those overlay batch jobs during the day because the shared computer system (which filled a room) would slow to a crawl or crash with more than a few jobs running simultaneously. I "divided" the state into many chunks because the software could not handle so many arcs.

Other characteristics of the time include the absence of ArcView, GRID, FRAGSTATS, CDs, zip drives, disk drives bigger than 300 MB (and these occupied 1 m³). We had just advanced to 1.4 MB diskettes, and nine-track tapes were the main means of data dispersal. There was no internet and no email. With remote sensing, there was no SPOT, MODIS, radar, hyperspectral data, or any other satellite data besides Landsat MSS and the beginning, experimental phase of Landsat TM and AVHRR. I was privileged to be an early NASA principal investigator, funded to use forest plot data, TM, and AVHRR in scaling forest cover (Iverson *et al.* 1989a,b) and productivity (Cook *et al.* 1987, 1989). However, we had to use small pieces of the Landsat scenes, often only 512×512 pixels.

Civilian GPS units became available in the late 1980s. There were few satellites and few base stations so we had only a few hours of sufficient satellites and we had to do differential post-processing from a station more than 200 km away. Of course, selective availability was the norm until May 2000. There were essentially no spatial statistics or metrics for landscapes other than basic patch area/perimeter metrics. When Krummel *et al.* (1987) published on the value of the fractal, it opened the door to a flood of landscape metrics, including many by the same group in the following year (O'Neill *et al.* 1988). Gardner *et al.* (1987) also first published on neutral models to help assess landscape pattern. GIS-based habitat or suitability models had appeared earlier (e.g., Hopkins 1977, Spanner *et al.* 1983, Iverson and Perry 1985, Donovan *et al.* 1987, Risser and Iverson 1988), but spatially explicit simulation models did not begin to emerge until the later 1980s (e.g., Turner 1988, Turner *et al.* 1989, Costanza *et al.* 1990). We have, indeed, come a long way in the way we acquire and process data.

2.3   Current status

Technology and data sources have perhaps advanced at the scale of computer speed according to Moore's Law, which states that the number of transistors in computer chips will double every 18 months (Moore 1965). However, the people available to analyze these data do not double at this rate, so the workload for all landscape ecologists must necessarily nearly double every 18 months as well. (Not really, but it seems like it sometimes.) Nonetheless, data and ways to acquire data are plentiful, though not always of the nature desired, so that retrofitting with surrogate data is often necessary. A few of the recent advances in data and tools to analyze them are discussed below.

2.3.1 More powerful computers and associated technology

Moore's Law has generally held true over the past two decades, resulting in a phenomenal sustained rate of development and an increase in capacity for processing pixels. For example, Riitters *et al.* (2000, 2002) and Riitters and Wickham (2003) have assessed global patterns at 1 km and conterminous United States patterns at 30 m resolution.

2.3.2 Small data recorder technology

Small data loggers now can be attached to a plethora of devices to allow long-term data recording of various environmental attributes. For example, our group has used them to determine soil and air temperatures, by landscape position, during and in the months following prescribed fires (Iverson and Hutchinson 2002, Iverson *et al.* 2004b). With these sensors, researchers can spatially locate temperature profiles, map and analyze them across landscapes, and animate the actual fire behavior through time (e.g., see animation found at http://www.fs.fed.us/ne/delaware/4153/ffs/zaleski.burn.html). These devices are being used in more diverse and creative ways to acquire data long term and in spatially disparate locations – both very important for landscape ecology.

2.3.3 GPS/GIS on hand-held computers

With the same trend of shrinking computer components comes advances in hand-held computers. GPS and GIS software now can be used effectively on palm-sized units, thus permitting much wider access of the technology to field biologists and others who otherwise have plenty of field equipment to lug around.

2.3.4 Software in image analysis, spatial statistics, modeling, pattern metrics, GIS

Software development has been rapid and diverse as well. The field of data mining and machine learning has been rapidly developing (e.g., Breiman 1996, 2001). Spatial statistics have been a real focus for some time (e.g., Cliff and Ord 1981, Burrough 1987, Legendre and Fortin 1989, Cressie 1991). Analytical techniques not only have been developed by and for landscape ecologists (e.g., McGarigal, this volume), but also borrowed and modified from other fields.

2.3.5 Remote sensing sensors

Many sensors are orbiting that weren't a decade ago (Table 2.1). The pixel sizes have gotten considerably smaller – now often 1 m or less – and the amount of data being transmitted daily to Earth is measured in petabytes (10^{15} bytes). Several countries are involved in developing the sensors and operating the

TABLE 2.1. *Current satellites*

Satellite	Country	Launch	Best resolution (m)	Type[a]
Landsat 7	US	1999	15	Mid-Opt
EO-1	US	2000	10	Mid-Opt
SPOT-2	France	1990	10	Mid-Opt
SPOT-4	France	1998	10	Mid-Opt
SPOT-5	France	2002	2.5	Mid-Opt
CBERS-1	China/Brazil	1999	20	Mid-Opt
Ziyuan-ZY-2A	China	2000	9	Mid-Opt
Ziyuan-ZY-2B	China	2002	3	Mid-Opt
KOMPSAT-1	Korea	1999	6.6	Mid-Opt
Proba (hyperspectral)	ESA	2001	18	Mid-Opt
UoSat 12	Singapore	1999	10	Mid-Opt
DMC AlSat-1	Algeria	2002	32	Mid-Opt
ASTER	US	1999	15	Mid-Opt
ERS-2	ESA	1995	30	Mid-Rad
ENVISAT	ESA	2002	30	Mid-Rad
RadarSat 1	Canada	1995	8.5	Mid-Rad
AVHRR	US	1978	1000	Low-Opt
MODIS	US	1999	250	Low-Opt
Landsat MSS	US	1972	79	Low-Opt
IKONOS	US	1999	1	High-Opt
QuickBird-2	US	2001	0.6	High-Opt
EROS A1	Israel	2000	1.8	High-Opt
IRS TESS	India	2001	1	High-Opt
Helios-1A	France	1995	1	High-Opt
Helios-1B	France	1999	1	High-Opt

[a] Low-Mid-High = resolution class, Opt = optical sensor, Rad = Radar sensor
From: William Stoney, Mitretek Systems.

satellites. Many of the highest-resolution satellites are commercial, while the coarser sensors are publicly operated and more utilized in research. For example, the MODIS sensor, with pixels 250–1000 m, is providing numerous maps, including estimated gross primary productivity, leaf area index, and fraction of photosynthetic active radiation on a regular basis (e.g., Running 2002, Zhang *et al.* 2003).

2.3.6 Data clearing houses

Data are becoming more freely available as government and multi-government agencies and nongovernment organizations are anxious to have

TABLE 2.2. *Example data clearing houses available on the Internet*

Site	Common type of data	Organization
www.natureserve.org	Biodiversity	NatureServe
edc.usgs.gov	Environmental	US Geological Survey
www.unepwcmc.org/cis/	Biodiversity	World Conservation Monitoring Centre
www.grid.unep.ch	General	United Nations Environmental Program
gcmd.gsfc.nasa.gov/	Remotely Sensed	US National Atmospheric Space Administration
www.gbif.org	Biodiversity	Global Biodiversity Information Facility
fsgeodata.fs.fed.us	Forests, Environment	US Forest Service
geodata.gov	General	US Government
www.nbii.gov/	Biological Resources	National Biological Information Infrastructure

all data, but especially publicly supported data, available to maximize efficiency (as long as national or environmental security is not compromised). As such, several data clearing houses are on the internet to allow free download of data. Some examples are listed in Table 2.2.

2.4   What we will have soon

We should expect the recent trends in data acquisition will continue. National security reviews since September 11, 2001, have reduced the scope of high-resolution data available on the Internet, but otherwise, the trends will lead to better hardware, software, and data availability. Remote data collection via sensors attached to data recorders on the ground or satellites in the sky will pave the way for almost unimaginable sources of data on our landscapes over the long term. As an example of likely near-future data sources, William Stoney (personal communication) has compiled a list of more than 50 mid- and high-resolution sensors targeted for activation within the next few years (Table 2.3).

2.5   Issues of data quality

A better understanding of spatial data quality requires abandonment of two basic beliefs that have been the bane of GIS since the beginning: (1) information shown on maps and captured into a GIS is always correct and essentially void of uncertainty, and (2) numerical information from computers is

TABLE 2.3. *Sensors targeted for activation by 2007*[a]

Satellite	Country	Sponsor[a]	Best resolution (m)	Type[b]
OrbView 3	US	Com	1	High-Opt
IKONUS.X	US	Com	0.5	High-Opt
QuickBird.X	US	Com	0.5	High-Opt
OrbView X	US	Com	0.5	High-Opt
EROS B1	Israel	Com	0.5	High-Opt
EROS B2	Israel	Com	0.5	High-Opt
EROS B3	Israel	Com	0.5	High-Opt
EROS B4	Israel	Com	0.5	High-Opt
IRS Cartosat 2	India	Gov	1	High-Opt
Pleiades-1	France	Gov	0.7	High-Opt
Pleiades-2	France	Gov	0.7	High-Opt
Helios-2A	France	Mil	<1	High-Opt
Helios-2B	France	Mil	<1	High-Opt
IGS-01	Japan	Mil	1	High-Opt
IGS-02	Japan	Mil	1	High-Opt
Resurs DK-1	Russia	Gov	0.4	High-Opt
Resurs DK-2	Russia	Gov	0.4	High-Opt
Resurs DK-3	Russia	Gov	0.4	High-Opt
KOMPSAT-2	Korea	Gov	1	High-Opt
TerraSAR X	Germany	Gov	1	High-Rad
TerraSAR L	Germany	Gov	1	High-Rad
SAR-Lupo-1	Germany	Mil	1	High-Rad
SAR-Lupo-2	Germany	Mil	1	High-Rad
COSMO-Skymed-1	Italy	Gov	1	High-Rad
COSMO-Skymed-2	Italy	Gov	1	High-Rad
COSMO-Skymed-3	Italy	Gov	1	High-Rad
COSMO-Skymed-4	Italy	Gov	1	High-Rad
IGS-R1	Japan	Mil	1 to 3	High-Rad
IGS-R2	Japan	Mil	1 to 3	High-Rad
Resurs DK-2	Russia	Gov	1	High-Rad
Resurs DK-3	Russia	Gov	1	High-Rad
LCDM-A	US	Com	7.5	Mid-Opt
LCDM-B	US	Com	7.5	Mid-Opt
RapidEye-A	Germany	Com	6.5	Mid-Opt
RapidEye-B	Germany	Com	6.5	Mid-Opt
RapidEye-C	Germany	Com	6.5	Mid-Opt
RapidEye-D	Germany	Com	6.5	Mid-Opt
IRS ResourceSat-1	India	Gov	6	Mid-Opt

(cont.)

TABLE 2.3. (*cont.*)

Satellite	Country	Sponsor[a]	Best resolution (m)	Type[b]
IRS ResourceSat-2	India	Gov	6	Mid-Opt
CBERS-2	China/Brazil	Gov	20	Mid-Opt
DMC China DMC	China	Gov	4	Mid-Opt
CBERS-3	China/Brazil	Gov	5	Mid-Opt
CBERS-4	China/Brazil	Gov	5	Mid-Opt
RocSat2	Taiwan	Gov	2	Mid-Opt
ALOS	Japan	Gov	2.5	Mid-Opt
DMC NigeriaSat-1	Nigeria	Gov	32	Mid-Opt
DMC ThaiPhat	Thailand	Gov	36	Mid-Opt
DMC BilSat	Turkey	Gov	12	Mid-Opt
DMC UK	UK	Gov	32	Mid-Opt
TopSat	UK	Gov	2.5	Mid-Opt
DMC VinSat-1	Vietnam	Gov	4	Mid-Opt
RadarSat 2	Canada	Gov	3	Mid-Rad
ALOS	Japan	Gov	7	Mid-Rad

[a] Com = Commercial; Gov = Government; Mil = Military
[b] Low-Mid-High = resolution class, Opt = optical sensor, Rad = Radar sensor
From William Stoney, Mitretek Systems.

somehow endowed with inherent authority (Shi *et al.* 2002b). This blind acceptance of GIS data is its Achilles heel and could undermine the entire technology (Goodchild 1998). Maps present a clarified, simplified view of a world that is actually complex and confusing. People prefer this simplified view, and explicit attention to uncertainty muddles this perspective. Nonetheless, it is especially important to pay attention to uncertainty in spatial data because of its importance in decision-making. Decision-makers usually don't want to know about uncertainty and they view GIS as an attractive simplicity. However, courts are likely to hold that a GIS user should make reasonable efforts to deal with uncertainty and they are likely to take a dim view of regulations or decisions based on GIS data in which issues of uncertainty have been ignored. Therefore, avoiding the issue of uncertainty will hurt the credibility of the profession.

2.5.1 Sources of uncertainty in spatial data

Burrough and McDonnell (1998) state that most GIS procedures assume that: (1) source data are uniform, (2) digitizing is infallible, (3) map overlay is simply intersecting boundaries and reconnecting line network,

(4) boundaries can be sharply defined and drawn, (5) all algorithms operate in a fully deterministic way, and (6) class intervals defined for "natural" reasons are the best for all mapped attributes. Of course, these implications are rarely true in landscape ecological studies and must be rectified. Much of the uncertainty can be traced to the original capture and automation of the data. It is especially important to have consistency and proper error checking when a large corporate database is being developed and will be used by many people (Lund and Thomas 1995). Here are some sources of spatial data error (adapted from Stine and Hunsaker 2001):

- *Geometric error*: When data are collected on the Earth (a sphere), and transferred to a map (a plane), there are inaccuracies in projecting the locations.

- *Attribute error*: In measuring an attribute at a point, there may be bias or error in the measuring tool or the person taking the measurement. This error is especially prominent in categorical variables when interpreting class membership (e.g., which vegetation type is this?).

- *Locational/boundary uncertainty*: Positions, through a variety of reasons (e.g., digitizing errors, GPS errors, field-to-map errors), are commonly misrepresented relative to their true positions. These positional errors can matter to a greater or lesser extent depending on the attribute of interest. For example, Lewis and Hutchinson (2000) assessed the impact of positional error for estimates of slope angle and elevation, and found that small positional errors among three maps led to a highly correlated estimate for elevation ($R^2 = 0.95$–0.98) but not for slope ($R^2 = 0.18$–0.32). Boundaries of many ecological units are fuzzy, so their depictions as lines of no width in a GIS will carry significant uncertainty.

- *Physical changes of attributes over time*: Nearly all biologically relevant variables on landscapes change over time, yet most GIS systems hold data for only one time stamp. Landscape ecologists can learn much from stacking two or more time stamps and analyzing the changes, but caution is required to make sure that errors in each of the time stamps are properly handled (Walsh *et al.* 1987).

- *Data compatibility*: When combining data of different qualities, there are new errors introduced. For example, in the case of combining two dates of satellite data, if one is Landsat MSS and the other is Landsat TM, the differences in spatial and spectral resolution can be important. Or, if slope aspect is derived from two digital elevation models of different spatial resolution, the estimates are likely to be quite different.

- *Errors in interpreting and manipulating data*: This error source includes several data processes that can introduce error, such as class aggregations, changing map projections, and conversions between raster and vector data.

- *Inability to accurately detect attribute of interest*: In many cases, landscape ecologists are not able to measure the variables of interest, but instead use surrogates that hopefully are correlated to the attribute of interest. For example, in the United States, the Heinz Report on the State of the Nation's Ecosystems (Heinz Center 2002) uses 102 indicators on ecosystem status, yet only 32 percent of the indicators have adequate data for assessment. The remainder have to be estimated from surrogates or not assessed at all.

2.5.2   Considering uncertainty in landscape models

Landscape ecologists frequently are using models to better understand the system in which they work and to evaluate the influence of an altered condition (Sklar and Hunsaker 2001). Several ecological phenomena have spatially explicit characteristics important to consider in the models, including environmental gradients, migration, immigration and emigration, metapopulation dynamics, competition, fire behavior, and biogeochemical cycling (Stine and Hunsaker 2001). These models are subject to several sources of uncertainty, most of which can be traced to uncertainty in data collection, data processing, model structure, human intervention, and natural variability (Li and Wu 2006). Of these five, only model structure is unique to the development of landscape models. Within model structure, there are five places where error can influence model outputs (Sklar and Hunsaker 2001):

(1) *Inputs* – the scale of simulated events and states should match the scale of events and states of the data used by the model. For example, a habitat model is much different from a global climate model and data inputs should be matched to the questions being asked.

(2) *Initial conditions* – every model requires identification of the conditions at a particular point in time and across the entire modeled space as the model starts. Often these conditions must be estimated, with associated uncertainty, through interpolation and interpretation of point data.

(3) *Forcing functions* – these are the inputs needed to move the simulation to the next time step. Inputs collected temporally, such as temperature or precipitation, often are used as drivers in the simulation, and errors in these functions can significantly affect the outputs. The most significant uncertainty results from missing data so that, in our example, widely dispersed meteorological stations may present problems, especially for fine-scale simulations.

(4) *Calibration parameters* – the mathematical structure that defines rules, processes, statistical relationships, or state change in the model, to maximize observed and simulated resemblance. These relationships will not be perfectly modeled, so errors are imbedded in the outputs.

(5) *Verification components* – observational and simulated data again are compared, but the observational data have not been used in the model development. Again, errors are similar to those of calibration except that time increases uncertainty and error is cumulative with time in model outputs.

In general, there is a tradeoff in that the more complex the model, the more potential for learning and prediction, but the less accurate (more uncertain) the outputs. There are four categories of dynamic landscape models (Sklar and Hunsaker 2001):

(1) *Transitional probability models* – not mechanistic, but rely on maps from two or more dates to calculate historical trends, which then can be applied forward.

(2) *Gradient models* – for modeling landscapes with obvious upstream and downstream components.

(3) *Process-based mosaic models* – distributes pattern across the landscape using site-specific biogeochemical mechanisms to control energy and material flows.

(4) *Individual-based models* – focus on behavior rules for an individual or an assemblage of individuals as a function of spatial constraints and opportunities.

Sklar and Hunsaker (2001) also discuss the causes of uncertainty in each of these model types.

2.6   Needs in data acquisition and quality

In the pages following, I present 14 topics related to data acquisition and quality which I believe need additional research or effort to advance the credibility and value of the field of landscape ecology and its role in society. There is no particular order to this list. Many of these ideas have been gleaned or modified from other sources, including Mowrer and Congalton (2000), Hunsaker *et al.* (2001), Wu and Hobbs (2002), and Shi *et al.* (2002a).

2.6.1 Strengthen capacity to collect ground information

World citizens, public officials, and academic institutions need to devise a way to populate the world with many 'ologists carrying GPS units, preferably

in the context of long-term landscape monitoring programs, and to orga-
nize the acquired data into hierarchical, GIS databases. Basic biological data
on organisms and communities is still needed! Natural historians have been
diminishing in number, and when coupled with increasing information and
spatial location requirements in this spatially aware age, ground-observed
information is lacking for accurate spatial processing. These kinds of data are
critical for research on biological invasions, conservation planning and monitoring, sustainability, cause and effects of stressors, change analysis, systems
and complexity analysis, and model development and validation.

Associated with the collection of basic biological data is the nearly equally
important role of automating, managing, and serving up the data. There
are several organizations doing this to various degrees. The state of Illinois,
USA, began automating and distributing information on distributions, ecology, taxonomy, and wildlife and human interactions for more than 3200 plant
species in the 1980s (Iverson *et al.* 1997b, Iverson and Prasad 1998a, Iverson
and Prasad 1999). The National Biological Information Infrastructure, the
World Conservation Monitoring Centre, NatureServe, and the Global Biodiversity Information Facility are four other servers of this kind of information
(Table 2.2).

2.6.2 Develop key indicators of status and health of landscapes

To efficiently monitor status and trends, scientists need to identify key
indicators within various landscape types that can be readily monitored over
large areas with reasonable costs. As mentioned previously, the "The State
of the Nation's Ecosystems" report for the United States presented 102 indicators, but only a third have adequate data and many require research on
effective monitoring strategies (Heinz Center 2002). Other indicators could
be developed, especially those that may be more regional in character. Many
other projects have been conducted to assess status and trends of particular
locations or landscape components (e.g., Iverson *et al.* 1989b, Illinois Department of Energy and Natural Resources 1994, Mac *et al.* 1998, Shifley and Sullivan 2002), but all have been limited in scope and reliability by the selected
indicators and the available data.

2.6.3 Design efficient, multi-tiered sampling designs

It remains a challenge to sample across large regions in a way that provides information at multiple scales, while permitting the inference of the
effects of spatial heterogeneity. For instance, many soil and vegetation variables have substantial spatial variability within a few meters, yet we are trying

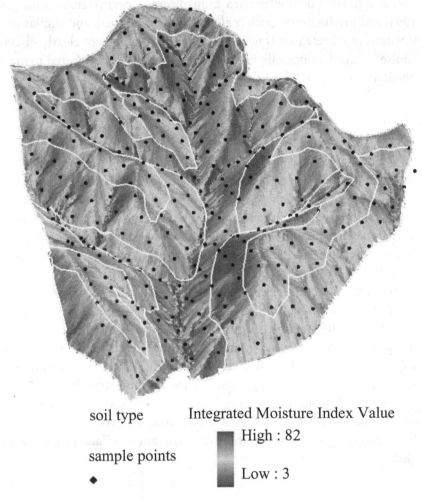

soil type   Integrated Moisture Index Value

                                          High : 82

sample points

    ◆   Low : 3

FIGURE 2.1
Integrated Moisture Index at 2 m resolution, vegetation sample points at 50 m
spacing, and soil polygons at about 100 m resolution for a study site in southern
Ohio, USA

to make repeatable, large-area assessments. For example, in Ohio a map of 1 m
digital elevation modeled for integrated moisture index shows very high local
variability (Fig. 2.1), also reported for soil nitrogen availability (Boerner *et al.*
2000). What is the best way to extend and use this type of information across
large areas? Innovative sampling methods are needed, using creative combinations of current and new methods in field sampling, experimentation, remote
sensing, statistics, and modeling. Projects like BIGFOOT (Burrows *et al.* 2002)
combine flux towers with multiple ground and remote sensing instruments to
extend detailed information across large areas. Earlier, forest plot data, Landsat

TM, and AVHRR data were used to map forest cover (Iverson *et al.* 1989a,b, 1994) and productivity (Cook *et al.* 1987, 1989). In these and similar projects, however, more research is needed to uncover methods to clearly distinguish "noise" from the fine-scale heterogeneity that can be attributed to measured phenomena.

2.6.4 Design and implement global landscape monitoring programs

Society needs to implement global monitoring programs now. The tools are currently available to begin. The incentives are high to do assessments of status and change, for these ecological processes and functions are critical to life itself! Initially, this program should be largely driven by (nearly) free satellite data, which are multiple in scale and with a time series of data. For example, we now have Landsat MSS data back to 1972, Landsat TM back to ~1982, AVHRR back to ~1978, and SPOT (*Satellite Pour l' Observation de la Terre*) back to 1986. These programs have sufficient data to establish such a program. As discussed previously, the satellite data streams are available now and are increasing dramatically. Today's hardware and software can handle the huge data sizes. The program should be interdisciplinary and be able to integrate the most appropriate methodologies from each discipline. And it should permit adaptive management so that as the science, the indicators, and public opinion evolve, so can the questions being asked of the program. In the United States, the proposed National Ecological Observatory Network (NEON) (Holsinger *et al.* 2003) is working toward this goal, but similar efforts are needed globally.

2.6.5 Develop efficient tools for strategic ground sampling

As stated in Section 2.6.1 above, there are not enough natural historians collecting data on species, etc., on the ground in a spatially organized way. There will never be enough. Therefore, strategic methods must be derived to get the most "bang for the buck" when it comes to sampling species. We need GIS tools which will better target ground sampling, so field crews will have a higher probability of encountering the species of interest. In this way, places rich in threatened and endangered species or invasive species, or biologically rich communities, could be modeled and then visited for verification. As an example, Iverson and Prasad (1998a) used a GIS model for 102 Illinois counties to predict possible plant species richness that had been under-sampled based on the richness in the well-sampled counties. Similar efforts and strategies have been presented by Palmer *et al.* (2002) and Ferrier *et al.* (2002).

2.6.6 Develop methods to share sensitive ground-specific information

Sometimes ground-specific information, or at least the specific locations of that information, is sensitive in that it cannot be freely shared without restriction. This restriction may be deemed necessary to protect national security, threatened or endangered organisms, or the rights of private landowners. It would be great if the information could be shared for research and monitoring purposes, but not cause legal or other problems. We need research into methods that might allow the ecological information to be gleaned without legal constraints. For example, the US Forest Service Forest Inventory and Analysis (FIA) program, by law, cannot release coordinates of their plots which number in the hundred thousands. This restriction greatly limits research on plant-environment studies. FIA is incorporating a "fuzz and swap" technique to fuzz locations slightly and to swap attributes with the nearest similar neighbor, which would at least allow summarizing to coarse-level polygons (Charles Scott, US Forest Service, personal communication). Somewhat related is the issue of credit versus data sharing for researchers. Too often researchers are reluctant to submit their data for meta- or regional analysis because they have not yet fully published on the data, even though the data were collected with public funds. Conversely in many instances, the collector(s) of the ground data is forgotten by the researchers doing the regional analyses.

2.6.7 Enhance and categorize methods to interpolate/extrapolate point-level data across landscapes

Because it is not possible, or at least practical, to completely sample any landscape attribute that can't be sensed remotely via satellite or aerial photograph, there always will be a need for interpolation methods to map attributes spatially across landscapes from point-sampled data. Attributes needing to be mapped include species or community distributions, fuels, basal areas, soil properties, climatic data, and air quality. There are several methods available, and the list is growing. What tools to use has been a question for a long time and has been reviewed extensively elsewhere (e.g., Lam 1983, Franklin 1995, Guisan and Zimmerman 2000, Lehmann *et al.* 2002, Leibhold 2002). Some of the methods, along with citations to case studies, follow (in no particular order):

- Regressions (general linear models, general additive models, etc.) (James and McCulloch 1990, Iverson *et al.* 1997b, Austin 1998, Franklin 1998, Cawsey *et al.* 2002, Lehmann *et al.* 2002, Moisen and Frescino 2002). Regression includes a wide array of models in which predictor variables, often in a stepwise fashion, are selected which explain variation in the response variable or variables. Often models are built by fitting lines to data that minimize the sum of the squared residuals.

- Kriging (e.g., universal, indicator) (Rossi *et al.* 1992, Leibhold *et al.* 1993, 1994, Hershey 1996, Riemann-Hershey and Reese 1999). These methods are theoretically based in multiple linear regression and use semivariograms to describe spatial structure in data, as well as predict values across nonsampled areas. Implicit is the notion that samples close together in time and/or space will be more similar than those that are farther apart. These methods preserve the spatial structure and variability inherent in the sample data but do not work well with ancillary data and usually predict a univariate response.

- Splines (e.g., thin plate splines) (Mitasova and Hofierka 1993, Hutchinson 1995, Mitasova *et al.* 1996, Price *et al.* 2000, Hofierka *et al.* 2002). These interpolation functions include tension and smoothing parameters so that a digital elevation model (DEM), for example, can be viewed as a thin plate built at a higher resolution from points, and the tension adjusted to minimize overshoots and artificial pits in the resulting DEM.

- Classification and regression trees (CART) (Breiman *et al.* 1984, Franklin 1998, Iverson and Prasad 1998b, Moisen and Frescino 2002). The model is fit using recursive partitioning rules, where data are split into left and right branches according to rules defined by the predictor variables. At the terminal node, the predicted value (regression trees) or class (classification trees) is estimated.

- Multivariate adaptive regression splines (MARS) (Freidman 1991, DeVeaux *et al.* 1993, Prasad and Iverson 2000, Moisen and Frescino 2002). MARS is related to classification and regression trees in that it is a flexible nonparametric regression method that generalizes the piecewise constant functions of CART to continuous functions by fitting (multivariate) splines.

- Computer-intensive data mining and prediction techniques (Breiman 1996, 2001, Iverson *et al.* 2004a). These advanced machine-learning techniques use multiple CART trees in determining the best predictive models, including measures of variable importance within the models. Bagging and random forests are techniques that use a bootstrap approach to identify variable importance and produce averaged models, sometimes with as many as 1000 CART trees involved.

- Inverse distance weighted methods (ESRI 1993, Price *et al.* 2000). These methods apply a simple linearly weighted combination of a set of sample points, with the weight being a function of inverse distance.

- Most-similar-neighbor methods (Moeur and Stage 1995, Ohmann and Gregory 2002). These methods provide site-specific data for nonsampled areas by choosing the most similar parcel from a set of sampled parcels to act as its surrogate. Ohmann and Gregory (2002) combined most-similar-neighbor methods with direct-gradient analysis (canonical-correspondence analysis) to produce reasonably accurate vegetation maps.
- Artificial neural networks (Ripley 1994, Cairns 2001, Moisen and Frescino 2002). With neural networks, accurate models can be built for prediction when the underlying relationships between predictor and response are unknown; the response is a transformation of a weighted combination of the predictor variables. The many coefficients and intercepts are "learned" via an optimization method. It is more of a "black box", however, in that the influences of specific variables are difficult to discern.

As a corollary to the above methods, to spread point-level information out across the landscape is also the critical, and often more important, task of determining where boundaries lie among the patches on the landscape. This is also an area of active research (e.g., Fortin 1994, Fortin and Drapeau 1995, Lopez-Blanco and Villers-Ruiz 1995, Wang and Hall 1996, Bernert *et al.* 1997, Fortin *et al.* 2000).

2.6.8 Develop techniques to best acquire and archive information on landscape history

When we learn about the history of a landscape, we can learn more about what is currently making the landscape tick. Ecological legacies are extremely important in most locations, and they can last for many decades, even centuries. Fires, clearing, grazing, wind storms, floods, hurricanes, volcanoes, and land-use changes are example legacies that can have long-lasting effects (Wallin *et al.* 1994, Foster *et al.* 1998, August *et al.* 2002, Turner *et al.* 2003).

Landscape history is also important to document so that, especially with respect to trends in deforestation, historical trends in one part of the world can be used to aid in predicting future trends in another part of the world. Then, if need be, actions can be taken to prevent history from repeating itself. One of the most distasteful, and sadly often repeated, patterns on the planet is when native peoples are "displaced" by colonists from another place (Diamond 1999). Often but not necessarily related is the subsequent rapid conversion of its lands as the new colonists settle. Deforestation patterns in the temperate

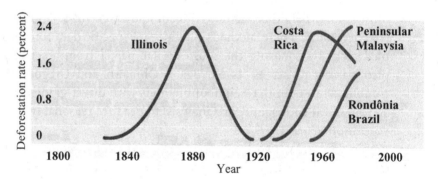

FIGURE 2.2

Rates (percent) of deforestation in Illinois, USA, followed by Costa Rica, Peninsular Malaysia, and Rondônia, Brazil (from Iverson *et al.* 1991). A rate of 2 percent a year would decimate a forest in less than a generation. History does indeed repeat itself

forests of Illinois, USA, for example, have more recently been repeated, and still continue, in the tropical forests of Costa Rica, Malaysia, and Brazil (Fig. 2.2).

We now have many years of data for data mining and evaluation of land-use histories, yet these data are being under-utilized. We have air photos since the early 1930s, Landsat MSS since ~1972, Landsat TM since ~1982, AVHRR since ~1978, and SPOT since 1986. We have sampling station data (e.g., forest plots, water quality sampling, bird census, etc.) over a very long history, but the oldest data often are not digital. Those data that are digital are yielding tremendous value, for example with respect to the breeding bird survey data, continuous since the mid 1960s (James *et al.* 1996, Sauer *et al.* 2001, Rodriguez 2002, Matthews *et al.* 2004).

Unfortunately, we also have decades of data perishing in old file cabinets and storehouses as retirements and budget issues prevent a wealth of data from being captured digitally. This is a tragic loss in these days where the evaluation of long-term trends is such a critical component of many of today's environmental issues.

2.6.9 Determine appropriate methods to merge and analyze data acquired at different scales

Often the significant biological events (e.g., rare occurrences, invasions of exotics) are happening at very fine scales, but we can't collect data everywhere at that scale. We therefore need to have suitable methods for scaling up and scaling down to obtain appropriate estimates for the scale of interest (e.g., Wiens 1989, Rastetter *et al.* 1992, Ehleringer and Field 1993, Gardner *et al.* 2001, Schneider 2001, Cushman and McGarigal 2002). This is an area of active research and discussed in separate chapters by Ludwig and Wu.

As an example from our research laboratory, we now have 1 m elevation data from LIDAR (Light detection and ranging sensor) and can calculate an integrated moisture index (Iverson *et al.* 1997a) on those data, but we cannot obtain that resolution for soils or vegetation attributes. How do we correctly merge and analyze such data so that we best understand the relationships between long-term moisture and soil and vegetation characteristics (Fig. 2.1)?

2.6.10 Efficiently handle increasing volumes of data, with minimal user pre-processing

There are petabytes of data streaming back to Earth each day. We need additional research to facilitate the pre-processing and screening of these data so landscape researchers can readily obtain and process the filtered data with less data volume and less up-front cost. As an example, the MODIS (Moderate-resolution imaging spectroradiometer) sensor has a science team that has been developing algorithms for automatic calculation for several vegetation-related metrics so that each user doesn't have to do it (Running 2002, Heinsch *et al.* 2003, Zhang *et al.* 2003).

2.6.11 New GIS technologies needed

There are at least four areas where the development of GIS technology must proceed to enhance the work of landscape ecologists and the subsequent accountability of that work. We should appeal to vendors and developers to proceed with these developments. First, we need a temporal GIS, one that allows better analysis of changes through time. Second, we need more development in three-dimensional GIS, for better analysis of volumetric and mass-flow data. Third, we need the development of an "uncertain GIS," one that allows the quantifying, display, and analysis of various forms of uncertainty (e.g., Duckham and McCreadie 2002). Fourth, we need the development of automatic metadata tracking within the GIS, so that a complete history and documentation of data generation and manipulation, including error tracking, occurs without human intervention (Beard 2001, Gan and Shi 2002).

2.6.12 Develop and test theory and methods of uncertainty analysis of landscape data

Though several books have been produced on this topic (e.g., Goodchild and Gopal 1989, Mowrer *et al.* 1996, Mowrer and Congalton 2000, Hunsaker *et al.* 2001, Shi *et al.* 2002a), there is still a lot of research needed so that every landscape ecologist and GIS user can understand the critical nature spatial uncertainty plays in their projects. Some areas needing further development include graphical visualization of uncertainty (e.g., Buttenfield 2001, Drecki 2002), error metrics calculation (e.g., Arbia *et al.* 1998), and the simulation of specific uncertainties for testing analytical procedures.

2.6.13 Devise methods so error can be evaluated and broken down into its various components (error budget)

Here I emphasize the need to be able to determine *where* the error lies in any GIS analysis – which form of error mentioned earlier in this chapter is most problematic, and therefore how might that error be trimmed? Or as an example, how can error associated with imagery classification be separated from error associated with a simulation model? Or, how does error propagate and accumulate in various spatial analyses such as overlay and buffer operations? Much of the difficulty associated with this research need is a fundamental flaw in the GIS systems that have been developed and accepted over the past 30 years. Goodchild (2002) discussed the need for a measurement-based GIS, rather than the nearly universal coordinate-based GIS, which cannot properly deal with error. Measurement-based GIS could retain details of measurements, such that error analysis is possible, and corrections to positions can be appropriately propagated through the database.

2.6.14 Devise methods to assess the effects of varying data quality and grain size on the outputs of landscape pattern analysis, model simulations, and resultant decisions

The quality of data and metadata will determine landscape ecologists' ability and effectiveness of detecting patterns and relating them to processes, and consequently affect research results, practical recommendations, and final decisions. Though some work has been done on the sensitivity of various landscape metrics from varying data quality and grain size (e.g., Wickham and Riitters 1995, Wickham *et al.* 1997, Hargis *et al.* 1998), this is an area needing more research. With respect to grain size, we need to determine with more certainty how the following processes affect uncertainty: aggregation, interpolation, transformation, and re-measurement. For many GIS applications, it is not possible to compare the outputs to an independently derived "truth"; in these cases, it is best to conduct a sensitivity analysis based on randomization of the data (Hunsaker *et al.* 2001). For example, it may be possible to use Monte Carlo simulations to determine if a decision becomes unstable because of poor data quality (Phillips and Marks 1996). Decision-support networks are needed that support error analysis and the spatial characterization of uncertainty (Eastman 2001).

2.7   Policy issues related to data acquisition and quality

In addition to the 14 research-focused issues, there are a few issues which are based primarily in policy, and so are mentioned briefly here. These issues are only presented as idea seeds, with much more effort needed to make them proposals.

First, policy-makers need to get behind the research issues to help provide the finances and exposure to make them happen. Otherwise there is no way that well-supported, globally represented, long-term monitoring programs, as an example, will come into being.

Second, mechanisms are needed to enable agencies and countries to easily cooperate, so that the best data sets possible can be derived and analyzed thoroughly and without perceived or real country-level bias.

Third, rigorous support within the policy arena is needed for adequate education and training so that the science can develop credibly in the most helpful ways for societal benefit.

Finally, the public, the decision-makers, and the researchers need to become aware of GIS/map accuracy issues and the subsequent validity of any information they use (Spear *et al.* 1996, Cornelis and Brunet 2002). For information to be used and useful in the policy arena, and not itself be the subject of debate, it must be policy relevant, technically credible, and politically legitimate (O'Malley *et al.* 2003).

2.8   Conclusions

Remote data acquisition is becoming much easier and consistent, though information obtained on the ground is still critically important, costly to acquire, and generally not achievable by remote sensing. We need to learn how to best use these data resources to monitor and manage Earth's resources. Data quality is still a major stumbling block for researchers and decision-makers, and a current critical research topic.

References

Anderson, J. R., E. E. Hardy, and J. T. Roach. 1976. *A land-use classification system for use with remote-sensor data. Circular 671.* Reston, VA: United States Geological Survey.
Anderson, R. C. 1970. Prairies in the prairie state. *Transactions of the Illinois State Academy of Science* **63**, 214–21.

Arbia, G., D. Griffith, and R. Haining. 1998. Error propagation modelling in raster GIS: overlay operations. *International Journal of Geographical Information Science* **12**, 145–67.

August, P., L. R. Iverson, and J. Nugranad. 2002. Human conversion of terrestrial landscapes. Pages 198–224 in K. Gutzwiller, (ed.). *Applying Landscape Ecology in Biological Conservation*. New York: Springer-Verlag.

Austin, M. P. 1998. An ecological perspective on biodiversity investigations: examples from Australian eucalypt forests. *Annals of the Missouri Botanical Garden* **85**, 2–17.

Beard, K. 2001. Roles of meta-information in uncertainty management. Pages 363–78 in C. T. Hunsaker, M. F. Goodchild, M. A. Friedl, and T. J. Case (eds.). *Spatial Uncertainty in Ecology*. New York: Springer-Verlag.

Bernert, J. A., J. M. Eilers, T. J. Sullivan, K. E. Freemark, and C. Ribic. 1997. A quantitative method for delineating regions: an example for the western Corn Belt plains ecoregion of the USA. *Environmental Management* **21**, 405–20.

Boerner, R. E. J., S. J. Morris, E. K. Sutherland, and T. F. Hutchinson. 2000. Spatial variability in soil nitrogen dynamics after prescribed burning in Ohio mixed-oak forests. *Landscape Ecology* **15**, 425–39.

Breiman, L. 1996. Bagging predictors. *Machine Learning* **24**, 123–40.

Breiman, L. 2001. Random forests. *Machine Learning* **45**, 5–32.

Breiman, L., J. Freidman, R. Olshen, and C. Stone. 1984. *Classification and regression trees*. Belmont, CA: Wadsworth.

Burrough, P. A. 1987. Spatial aspects of ecological data. Pages 213–57 in R. H. G. Jongman, C. J. F. ter Braak, and D. F. R. Jongeren (eds.). *Data Analysis in Community and Landscape Ecology*. Wageningen: Pudoc.

Burrough, P. A. and R. A. McDonnell. 1998. *Principles of Geographical Information Systems*. Oxford: Oxford University Press.

Burrows, S. N., S. T. Gower, M. K. Clayton, *et al.* 2002. Application of geostatistics to characterize leaf area index (LAI) from flux tower to landscape scales using a cyclic sampling design. *Ecosystems* **5**, 667–79.

Buttenfield, B. P. 2001. Mapping ecological uncertainty. Pages 115–32 in C. T. Hunsaker, M. F. Goodchild, M. A. Friedl, and T. J. Case (eds.). *Spatial Uncertainty in Ecology*. New York: Springer-Verlag.

Cairns, D. M. 2001. A comparison of methods for predicting vegetation type. *Plant Ecology* **156**, 3–18.

Cawsey, E. M., M. P. Austin, and B. L. Baker. 2002. Regional vegetation mapping in Australia: a case study in the practical use of statistical modelling. *Biodiversity and Conservation* **11**, 2239–74.

Cliff, A. D. and J. K. Ord. 1981. *Spatial Processes: Models and Applications*. London: Pion.

Cook, E. A., L. R. Iverson, and R. L. Graham (eds.). 1987. *The Relationship of Forest Productivity to Landsat Thematic Mapper Data and Supplemental Terrain Information*. Sioux Falls, SD: American Society for Photogrammetry and Remote Sensing.

Cook, E. A., L. R. Iverson, and R. L. Graham. 1989. Estimating forest productivity with thematic mapper and biogeographical data. *Remote Sensing of Environment* **28**, 131–41.

Cornelis, B. and S. Brunet. 2002. A policy-maker point of view on uncertainties in spatial decisions. Pages 168–85 in W. Shi, P. F. Fisher, and M. F. Goodchild (eds.). *Spatial Data Quality*. New York: Taylor & Francis.

Costanza, R., F. H. Sklar, and M. L. White. 1990. Modeling coastal landscape dynamics. *BioScience* **40**, 91–107.

Cressie, N. 1991. *Statistics for Spatial Data*. New York: John Wiley & Sons Inc.

Cushman, S. A. and K. McGarigal. 2002. Hierarchical, multi-scale decomposition of species-environment relationships. *Landscape Ecology* **17**, 637–46.

DeVeaux, R.D., D.C. Psichogios, and L.H. Ungar. 1993. A comparison of two nonparametric estimation schemes: MARS and neural networks. *Computations in Chemical Engineering* 8, 819–37.

Diamond, J. 1999. *Guns, Germs, and Steel: The Fates of Human Societies*. New York: W.W. Norton & Co.

Donovan, M.L., D.L. Rabe, and C.E. Olson. 1987. Use of geographic information systems to develop habitat suitability models. *Wildlife Society Bulletin* 15, 574–9.

Drecki, I. 2002. Visualisation of uncertainty in geographical data. Pages 140–59 in W. Shi, P.F. Fisher, and M.F. Goodchild (eds.). *Spatial Data Quality*. New York: Taylor & Francis.

Duckham, M. and J.E. McCreadie. 2002. Error-aware GIS development. Pages 62–75 in W. Shi, P.F. Fisher, and M.F. Goodchild (eds.). *Spatial Data Quality*. New York: Taylor & Francis.

Eastman, R. 2001. Uncertainty management in GIS: decision support tools for effective use of spatial data resources. Pages 379–90 in C.T. Hunsaker, M.F. Goodchild, M.A. Friedl, and T.J. Case (eds.). *Spatial Uncertainty in Ecology*. New York: Springer-Verlag.

Ehleringer, J.R. and C.B. Field. 1993. *Scaling Processes between Leaf and Landscape Levels*. San Diego: Academic Press.

ESRI (Environmental Systems Research Institute). 1993. *Arc/Info GRID Command Reference*. Redlands, CA: Environmental Systems Research Institute.

Ferrier, S., M. Drielsma, G. Manion, and G. Watson. 2002. Extended statistical approaches to modelling spatial pattern in biodiversity in northeast New South Wales. II. Community-level modelling. *Biodiversity and Conservation* 11, 2309–38.

Fortin, M.-J. 1994. Edge detection algorithms for two-dimensional ecological data. *Ecology* 75, 956–65.

Fortin, M.-J. and P. Drapeau. 1995. Delineation of ecological boundaries: comparison of approaches and significance tests. *Oikos* 72, 120–31.

Fortin, M.-J., R.J. Olson, S. Ferson, *et al.* 2000. Issues related to the detection of boundaries. *Landscape Ecology* 15, 453–66.

Foster, D.R., D.H. Knight, and J.F. Franklin. 1998. Landscape patterns and legacies resulting from large, infrequent forest disturbances. *Ecosystems* 1, 497–510.

Franklin, J. 1995. Predictive vegetation mapping: geographic modelling of biospatial patterns in relation to environmental gradients. *Progress in Physical Geography* 19, 494–519.

Franklin, J. 1998. Predicting the distribution of shrub species in southern California from climate and terrain-derived variables. *Journal of Vegetation Science* 9, 733–48.

Freidman, J.H. 1991. Multivariate adaptive regression splines. *Annals of Statistics* 19, 1–141.

Gan, E. and W. Shi. 2002. Error metadata management system. Pages 251–66 in W. Shi, P.F. Fisher, and M.F. Goodchild (eds.). *Spatial Data Quality*. New York: Taylor & Francis.

Gardner, R.H., W.M. Kemp, V.S. Kennedy, and J.E. Petersen. 2001. *Scaling Relations in Experimental Ecology*. New York: Columbia University Press.

Gardner, R.H., B.T. Milne, M.G. Turner, and R.V. O'Neill. 1987. Neutral models for the analysis of broad-scale landscape pattern. *Landscape Ecology* 1, 19–28.

Goodchild, M.F. 1998. Uncertainty: the Achilles heel of GIS? *Geo Info Systems* 1998 (November), 50–2.

Goodchild, M.F. 2002. Measurement-based GIS. Pages 5–17 in W. Shi, P.F. Fisher, and M.F. Goodchild (eds.). *Spatial Data Quality*. New York: Taylor & Francis.

Goodchild, M.F. and S. Gopal. 1989. *Accuracy of Spatial Databases*. Bristol, PA: Taylor & Francis.

Guisan, A. and N.E. Zimmermann. 2000. Predictive habitat distribution models in ecology. *Ecological Modelling* 135, 147–86.

Hargis, C.D., J.A. Bissonette, and J.L. David. 1998. The behavior of landscape metrics commonly used in the study of habitat fragmentation. *Landscape Ecology* 13, 167–86.

Heinsch, F. A., M. Reeves, C. F. Bowker, *et al.* 2003. *User's Guide, GPP and NPP (MOD17A2/A3) Products, NASA MODIS Land Algorithm*. Bozeman, MT: Montana State University.

Heinz Center for Science, Economics and the Environment. 2002. *The State of the Nation's Ecosystems*. Cambridge, UK: Cambridge University Press.

Hershey, R. R. 1996. Understanding the spatial distribution of tree species in Pennsylvania. Pages 73–82 in H. T. Mowrer, R. L. Czaplewski, and R. H. Hamre (eds.). *Spatial Accuracy Assessment in Natural Resources and Environmental Sciences*. General Technical Report RM-GTR-277, Fort Collins, CO: Rocky Mountain Forest and Range Experiment Station, USDA Forest Service.

Hofierka, J., J. Parajka, M. Mitasova, and L. Mitas. 2002. Multivariate interpolation of precipitation using regularized spline with tension. *Transactions in GIS* 6, 135–50.

Holsinger, K. E. and IBRCS Working Group. 2003. IBRCS White Paper. *Rationale, Blueprint, and Expectations for the National Ecological Observatory Network*. Washington, DC: American Institute of Biological Sciences.

Hopkins, L. D. 1977. Methods for generating land suitability maps: a comparative evaluation. *AIP Journal* 43, 386–400.

Hunsaker, C. T., M. F. Goodchild, M. A. Friedl, and T. J. Case. 2001. *Spatial Uncertainty in Ecology*. New York: Springer-Verlag.

Hutchinson, M. F. 1995. Interpolating mean rainfall using thin plate smoothing splines. *International Journal of Geographical Information Systems* 9, 385–403.

Illinois Department of Energy and Natural Resources. 1994. *The Changing Illinois Environment: Critical Resources*. Volume 3. Technical Report. ILENR/RE-EA-94/05. Springfield, IL: Illinois Department of Energy and Natural Resources.

Iverson, L. R. 1988. Land-use changes in Illinois, USA: the influence of landscape attributes on current and historic use. *Landscape Ecology* 2, 45–61.

Iverson, L. R., E. A. Cook, and R. L. Graham. 1989a. A technique for extrapolating and validating forest cover data across large regions: calibrating AVHRR data with TM. *International Journal of Remote Sensing* 10, 1805–12.

Iverson, L. R., E. A. Cook, and R. L. Graham. 1994. Regional forest cover estimation via remote sensing: the calibration center concept. *Landscape Ecology* 9, 159–74.

Iverson, L. R., M. E. Dale, C. T. Scott, and A. Prasad. 1997a. A GIS-derived integrated moisture index to predict forest composition and productivity in Ohio forests. *Landscape Ecology* 12, 331–48.

Iverson, L. R. and T. F. Hutchinson. 2002. Soil temperature and moisture fluctuations during and after prescribed fire in mixed-oak forests, USA. *Natural Areas Journal* 22, 296–304.

Iverson, L. R., R. L. Oliver, D. P. Tucker, *et al.* 1989b. *Forest Resources of Illinois: An Atlas and Analysis of Spatial and Temporal Trends*. Champaign, IL: Illinois Natural History Survey Special Publication 11.

Iverson, L. R. and L. G. Perry. 1985. Integration of biological pieces in the siting puzzle. Pages 99–131 in *The Siting Puzzle: Piecing Together Economic Development and Environmental Quality*. Proceedings of the 13th Annual ENR Conference. Springfield, IL: Illinois Department of Energy and Natural Resources.

Iverson, L. R. and A. Prasad. 1998a. Estimating regional plant biodiversity with GIS modeling. *Diversity and Distributions* 4, 49–61.

Iverson, L. R. and A. M. Prasad. 1998b. Predicting abundance of 80 tree species following climate change in the eastern United States. *Ecological Monographs* 68, 465–85.

Iverson, L. R. and A. M. Prasad. 1999. The Illinois Plant Information Network (database). www.fs.fed.us/ne/delaware/ilpin/ilpin.html.

Iverson, L. R., A. Prasad, and D. M. Ketzner. 1997b. A summary of the Illinois flora based on the Illinois Plant Information Network. *Transactions of the Illinois State Academy of Science* 90, 41–64.

Iverson, L. R., A. M. Prasad, and A. Liaw. 2004a. New machine learning tools for predictive vegetation mapping after climate change: bagging and random forest perform better than regression tree analysis. Pages 317–20 in R. Smithers (ed.). Proceedings, Cirencester, UK: UK-International Association for Landscape Ecology.

Iverson, L. R. and P. G. Risser. 1987. Analyzing long-term changes in vegetation with geographic information system and remotely sensed data. *Advances in Space Research* **7**, 183–94.

Iverson, L. R., D. A. Yaussy, J. Rebbeck, *et al.* 2004b. A data recording method to monitor the spatial and temporal distribution of fire behavior from prescribed fires. *International Journal of Wildland Fire* **13**, 1–12.

James, F. C. and C. E. McCulloch. 1990. Multivariate analysis in ecology and systematics: panacea or Pandora's box? *Annual Review of Ecology and Systematics* **21**, 129–66.

James, F. C., C. E. McCulloch, and D. A. Wiedenfield. 1996. New approaches to the analysis of population trends in land birds. *Ecology* **77**, 13–27.

Klopatek, J. M., R. J. Olson, C. J. Emerson, and J. L. Jones. 1979. Land-use conflicts with natural vegetation in the United States. *Environmental Conservation* **6**, 191–200.

Krummel, J. R., R. H. Gardner, G. Sugihara, R. V. O'Neill, and P. R. Coleman. 1987. Landscape patterns in a disturbed environment. *Oikos* **48**, 321–4.

Lam, N. S. 1983. Spatial interpolation methods: a review. *The American Cartographer* **10**, 129–49.

Legendre, P. and M.-J. Fortin. 1989. Spatial pattern and ecological analysis. *Vegetatio* **80**, 107–38.

Lehmann, A., J. M. Overton, and M. P. Austin. 2002. Regression models for spatial prediction: their role for biodiversity and conservation. *Biodiversity and Conservation* **11**, 2085–92.

Leibhold, A. M. 2002. Integrating the statistical analysis of spatial data in ecology. *Ecography* **25**, 553–7.

Leibhold, A. M., R. E. Rossi, and W. P. Kemp. 1993. Integrating the statistical analysis of spatial data in ecology. *Annual Review of Entomology* **38**, 303–27.

Lewis, A. and M. F. Hutchinson. 2000. From data accuracy to data quality: using spatial statistics to predict the implications of spatial error in point data. Pages 17–35 in H. T. Mowrer and R. G. Congalton (eds.). *Quantifying Spatial Uncertainty in Natural Resources*. Chelsea, MI: Ann Arbor Press.

Li, H. and J. Wu. 2006. Uncertainty analysis in ecological studies: an overview. Pages 45–66 in J. Wu, B. Jones, B. Li, and O. L. Loucks (eds.). *Scaling and Uncertainty Analysis in Ecology: Methods and Applications*. Dordrecht, the Netherlands: Springer.

Lopez-Blanco, J., and L. Villers-Ruiz. 1995. Delineating boundaries of environmental units for land management using a geomorphological approach and GIS: a study in Baja California, Mexico. *Remote Sensing and Environment* **53**, 109–17.

Lund, H. G. and C. E. Thomas. 1995. *A Primer on Evaluation and Use of Natural Resource Information for Corporate Data Bases*. US Department of Agriculture Forest Service, General Technical Report WO-62. Washington, DC: Washington Office.

Mac, M. J., P. A. Opler, C. E. Puckett Haecker, and P. D. Doran. 1998. *Status and Trends of the Nation's Biological Resources*. Reston, VA: US Geological Survey.

Matthews, S. N., R. J. O'Connor, L. R. Iverson, and A. M. Prasad. 2004. *Atlas of Current and Climate Change Distributions of Common Birds of the Eastern United States*. General Technical Report NE-318. Newtown Square, PA: USDA Forest Service, Northeastern Research Station.

Mitasova, H. and J. Hofierka. 1993. Interpolation by regularized spline with tension: II. Application to terrain modeling and surface geometry analysis. *Mathematical Geology* **25**, 657–69.

Mitasova, H., J. Hofierka, M. Zlocha, and L. R. Iverson. 1996. Modeling topographic potential for erosion and deposition using GIS. *International Journal of Geographical Information Systems* **10**, 629–41.

Moeur, M. and A. R. Stage. 1995. Most similar neighbor: an improved sampling inference procedure for natural resource planning. *Forest Science* **41**, 337–59.

Moisen, G.G. and T. Frescino. 2002. Comparing five modelling techniques for predicting forest characteristics. *Ecological Modelling* **157**, 209–25.

Moore, G.E. 1965. Cramming more components into integrated circuits. *Electronics* **38**, 1–4.

Mowrer, H.T. and R.G. Congalton. 2000. *Quantifying Spatial Uncertainty in Natural Resources: Theory and Applications for GIS and Remote Sensing*. Chelsea, MI: Ann Arbor Press.

Mowrer, H.T., R.L. Czaplewski, R.H. Hamre, and Tech coords. 1996. *Spatial Accuracy Assessment in Natural Resources and Environmental Sciences: Second International Symposium*. General Technical Report RM-GTR-277. Fort Collins, CO: USDA Forest Service, Rocky Mountain Forest and Range Experiment Station.

O'Malley, R., K. Cavender-Bares, and W.C. Clark. 2003. Providing "better" data: not as simple as it might seem. *Environmental Conservation* **45**, 8–18.

O'Neill, R.V., J.R. Krummel, R.H. Gardner, *et al.* 1988. Indices of landscape pattern. *Landscape Ecology* **1**, 153–62.

Ohmann, J.L. and M.J. Gregory. 2002. Predictive mapping of forest composition and structure with direct gradient analysis and nearest-neighbor imputation in coastal Oregon, USA. *Canadian Journal of Forest Research* **32**, 725–41.

Olson, R.J., C.J. Emerson, and M.K. Nungesser. 1980. *Geoecology: A County-Level Environmental Data Base for the Conterminous United States*. Publication No. 1537. Oak Ridge, TN: Oak Ridge National Laboratory Environmental Sciences Division.

Palmer, M.W., P.G. Earls, B.W. Hoagland, P.S. White, and T. Wohlgemuth. 2002. Quantitative tools for perfecting species lists. *Environmetrics* **13**, 121–37.

Phillips, D.L. and D.G. Marks. 1996. Spatial uncertainty analysis: propagation of interpolation errors in spatially distributed models. *Ecological Modeling* **9**, 213–30.

Prasad, A.M. and L.R. Iverson. 2000. Predictive vegetation mapping using a custom built model-chooser: comparison of regression tree analysis and multivariate adaptive regression splines. In Proceedings CD-ROM. 4th International Conference on Integrating GIS and Environmental Modeling: Problems, Prospects and Research Needs. Banff, Alberta, Canada (http://www.colorado.edu/research/cires/banff/upload/159/index.html).

Price, D.T., D.W. McKenney, I.A. Nalder, M.F. Hutchinson, and J.L. Kesteven. 2000. A comparison of two statistical methods for spatial interpolation of Canadian monthly mean climate data. *Agricultural and Forest Meteorology* **101**, 81–94.

Rastetter, E.B., A.W. King, B.J. Cosby, *et al.* 1992. Aggregating fine-scale ecological knowledge to model coarser-scale attributes of ecosystems. *Ecological Applications* **2**, 55–70.

Riemann-Hershey, R. and G. Reese. 1999. *Creating a "First-cut" Species Distribution Map for Large Areas from Forest Inventory Data*. General Technical Report NE-256. Radnor, PA: Northeastern Research Station, USDA Forest Service.

Riitters, K., J. Wickham, R. O'Neill, B. Jones, and E. Smith. 2000. Global scale patterns of forest fragmentation. *Conservation Ecology* **4**, 3 [online] URL: http://www.consecol.org/Journal/vol4/iss2/art3.

Riitters, K.H. and J.D. Wickham. 2003. How far to the nearest road. *Frontiers in Ecology and the Environment* **1**, 125–9.

Riitters, K.H., J.D. Wickham, R.V. O'Neill, *et al.* 2002. Fragmentation of continental United States forests. *Ecosystems* **5**, 815–22.

Ripley, B.D. 1994. Neural networks and related methods for classification. *Journal of the Royal Statistical Society B* **56**, 409–56.

Risser, P.G. and L.R. Iverson. 1988. Geographic information systems and natural resource issues at the state level. Pages 231–239 in D.B. Botkin, M.E. Caswell, J.E. Estes, and A.A. Orio (eds.). *Our Role in Changing the Global Environment: What We Can Do About Large Scale Environmental Issues*. New York: Academic Press.

Risser, P.G., J.R. Karr, and R.T.T. Forman. 1984. *Landscape Ecology: Directions and Approaches*. Champaign, IL: Illinois Natural History Survey Special Publication Number 2.

Rodriguez, J. P. 2002. Range contraction in declining North American bird populations. *Ecological Applications* **1**, 238–48.

Rossi, R. E., J. L. Dungan, and L. R. Beck. 1994. Kriging in the shadows: geostatistical interpolation for remote sensing. *Remote Sensing of the Environment* **49**, 32–40.

Rossi, R. E., D. J. Mulla, A. G. Journel, and E. H. Franz. 1992. Geostatistical tools for modeling and interpreting ecological spatial dependence. *Ecological Monographs* **62**, 719–35.

Running, S. W. 2002. New satellite technologies enhance study of terrestrial biosphere. *EOS, Transactions, American Geophysical Union* **83**, 458–60.

Sauer, J. R., J. E. Hines, and J. Fallon. 2001. The North American breeding bird survey, results and analysis 1966–2000. Version 2001.2. Laurel, MD: USGS Patuxent Wildlife Research Center. (http://www.mbr.nbs.gov/bbs/htm96/trn626/all.html).

Schneider, D. C. 2001. The rise of the concept of scale in ecology. *Bioscience* **51**, 545–54.

Shi, W., P. F. Fisher, and M. F. Goodchild. 2002a. *Spatial Data Quality*. New York: Taylor & Francis.

Shi, W., M. F. Goodchild, and P. F. Fisher. 2002b. Epilog: a prospective on spatial data quality. Pages 304–9 in W. Shi, P. F. Fisher, and M. F. Goodchild (eds.). *Spatial Data Quality*. New York: Taylor & Francis.

Shifley, S. R. and N. H. Sullivan. 2002. *The Status of Timber Resources in the North Central United States*. General Technical Report NC-228. St. Paul, MN: North Central Research Station, USDA Forest Service.

Sklar, F. H. and C. T. Hunsaker. 2001. The use and uncertainties of spatial data for landscape models: an overview with examples from the Florida Everglades. Pages 15–46 in C. T. Hunsaker, M. F. Goodchild, M. A. Friedl, and T. J. Case (eds.). *Spatial Uncertainty in Ecology*. New York: Springer-Verlag.

Spanner, M. A., A. M. Strahler, and J. E. Estes. 1983. Soil loss prediction in a geographic information system format. Pages 89–102 in *Seventeenth International Symposium on Remote Sensing of Environment*. Ann Arbor, MI: Environmental Research Institute of Michigan.

Spear, M., J. Hall, and R. Wadsworth. 1996. Communication of uncertainty in spatial data to policy makers. Pages 199–207 in H. T. Mowrer, R. L. Czaplewski, and R. H. Hamre (eds.). *Spatial Accuracy Assessment in Natural Resources and Environmental Sciences*. General Technical Report RM-GTR-277, Fort Collins, CO: USDA Forest Service, Rocky Mountain Forest and Range Experiment Station.

Stine, P. A. and C. T. Hunsaker. 2001. An introduction to uncertainty issues for spatial data used in ecological applications. Pages 91–107 in C. T. Hunsaker, M. F. Goodchild, M. A. Friedl, and T. J. Case (eds.). *Spatial Uncertainty in Ecology*. New York: Springer-Verlag.

Turner, M. G. 1988. A spatial simulation model of land use changes in a piedmont county in Georgia. *Applied Mathematics and Computation* **27**, 39–51.

Turner, M. G., R. Costanza, and F. H. Sklar. 1989. Methods to evaluate the performance of spatial simulation models. *Ecological Modelling* **48**, 1–18.

Turner, M. G., W. H. Romme, and D. B. Tinker. 2003. Surprises and lessons from the 1988 Yellowstone fires. *Frontiers in Ecology and the Environment* **1**, 351–8.

Wallin, D. O., F. J. Swanson, and B. Marks. 1994. Landscape pattern response to changes in pattern-generation rules: land-use legacies in forestry. *Ecological Applications* **4**, 569–80.

Walsh, S. J., D. R. Lightfoot, and D. R. Butler. 1987. Recognition and assessment of error in geographic information systems. *Photogrammetric Engineering and Remote Sensing* **53**, 1423–30.

Wang, F. and G. B. Hall. 1996. Fuzzy representation of geographical boundaries in GIS. *International Journal of Geographical Information Systems* **10**, 573–90.

Wickham, J. D., R. V. O'Neill, K. H. Riitters, T. G. Wade, and K. B. Jones. 1997. Sensitivity of selected landscape pattern metrics to misclassification and differences in land cover composition. *Photogrammetric Engineering and Remote Sensing* **63**, 397–402.

Wickham, J. D. and K. H. Riitters. 1995. Sensitivity of landscape metrics to pixel size. *International Journal of Remote Sensing* **16**, 3585–94.

Wiens, J. A. 1989. Spatial scaling in ecology. *Functional Ecology* **3**, 385–97.

Wu, J. and R. Hobbs. 2002. Key issues and research priorities in landscape ecology: an idiosyncratic synthesis. *Landscape Ecology* **17**, 355–65.

Zhang, X., M. A. Friedl, C. B. Schaaf, *et al.* 2003. Monitoring vegetation phenology using MODIS. *Remote Sensing of the Environment* **84**, 471–5.

3

Landscape pattern analysis: key issues and challenges

3.1 Introduction

Landscape pattern analysis (LPA) has been a major part of landscape ecological research for the last two decades (Romme 1982, O'Neill *et al.* 1988,
Turner 1989, 1990, Turner and Gardner 1991, Pickett and Cadenasso 1995,
Gustafson 1998, Wu and Hobbs 2002). The ultimate goal of LPA is to link spatial patterns to ecological processes at different scales. The importance of LPA
lies in the needs to: (1) monitor, quantify, and project the change of a given
landscape; (2) compare and contrast patterns between different landscapes;
and (3) help understand processes underlying observed patterns, so that landscape dynamics may be better understood and predicted (Turner *et al.* 2001,
Wu 2004). Thus, appropriate and effective use of LPA methods is vital to the
development of landscape ecology.

After two decades of rapid development, landscape ecology has begun to
mature. However, many problems still persist in the application of LPA (Li and
Reynolds 1995, Tischendorf 2001, Fortin *et al.* 2003, Li and Wu 2004). Li and
Wu (2004) have called for a serious rethinking of why and how landscape pattern analysis should be used, with an intent to discourage the rampant and
blind use of LPA methods. They argued that theoretical guidance should be
sought in the practice of LPA. Fortin *et al.* (2003) stated that methodological
developments often undergo four phases: (1) the introduction phase with key
papers describing a new methodology, (2) the testing phase with many papers
applying the new methodology, (3) the critical review phase with limitations
of the methodology identified and with rethinking of its fundamental purposes, assumptions, and formulations, and (4) the standardization phase with
the most effective methods being selected as the norm. They suggested that

Key Topics in Landscape Ecology, ed. J. Wu and R. Hobbs.
Published by Cambridge University Press. © Cambridge University Press 2007.

landscape pattern analysis was in the third phase, suffering from some growing pains. However, one may also argue that critical review should always be part of any scientific process, and the four phases actually form a cycle around which methods are continuously being developed, tested, and improved (or otherwise discarded).

The objectives of this chapter are: (1) to critically assess the current status of LPA, (2) to review the assumptions, behaviors, and limitations of commonly used LPA methods, (3) to discuss major challenges in LPA, and (4) to develop guidelines for the effective application of LPA. We will focus on the proper selection, use, and interpretation of various methods, based on key spatial pattern attributes, basic assumptions, general behaviors, and major limitations. For more in-depth reviews of the specific LPA methods and their equations, the reader is referred to Li (1989), Turner and Gardner (1991), McGarigal and Marks (1995), Riitters *et al.* (1995), Dale *et al.* (2002), and Fortin *et al.* (2002).

3.2   General classification of LPA methods

The numerous methods used in LPA may be classified based on different criteria such as research objectives, data types, and mathematical properties of the methods (Li and Reynolds 1995, Gustafson 1998, Wu *et al.* 2000, Dale *et al.* 2002, Fortin *et al.* 2002). Most LPA studies deal with spatially continuous data with known locations and complete coverage in the form of maps. Such map data used in LPA and landscape modeling are frequently obtained from remote sensing sources (e.g., satellite imagery, aerial photos) and usually stored in geographic information systems, or GIS (e.g., vegetation, soil, land-use maps). Discrete data like point samples will not be considered here, but good reviews exist (Upton and Fingleton 1985, Dale 1999). In this review, we concentrate on the measurement types of required data and classify these methods into two groups: spatial statistics for numerical (i.e., ratio and interval) data and landscape metrics for categorical (i.e., rank and nominal) data. This is because data types influence what methods may be used, what spatial pattern may be revealed, and what research questions may be addressed (Li and Reynolds 1995, Fortin *et al.* 2002).

Many LPA methods may be applied to one data type only, but some can be used to analyze both data types (Dale *et al.* 2002). Commonly used for numerical maps are several methods of spatial statistics (including geostatistics), such as semivariogram analysis and autocorrelation indices like Moran's I, and Geary's C (see Cliff and Ord 1981, Robertson 1987, Cressie 1991, Rossi *et al.* 1992, Dale *et al.* 2002). As a fundamental property of spatial statistics, spatial autocorrelation means that things closer in space are more related than things farther apart. Thus, strong spatial autocorrelation signifies a strong influence of processes at work. The methods for categorical maps are primarily indices of landscape mosaics (Romme 1982, O'Neill *et al.* 1988, Li 1989, Li and Reynolds 1993, Riitters *et al.* 1995). Landscape indices quantify the spatial structure of landscape mosaics, and such quantitative information can then be used to study effects of pattern on process. While most of these methods are borrowed from geosciences and other branches of ecology, landscape ecology has created some of its own in the last two decades (Romme 1982, O'Neill *et al.* 1988, Li 1989, Li and Reynolds 1993, Riitters *et al.* 1995).

To ensure that the methods of LPA are applied appropriately and effectively, it is essential to understand what exactly each method does (Li and Reynolds 1994, 1995, Riitters *et al.* 1995, Dale *et al.* 2002, Fortin *et al.* 2002, Li and Wu 2004, Wu 2004). In addition to learning how to calculate them, landscape ecologists must understand key components of spatial pattern as well as the assumptions, behaviors, and limitations of the methods.

3.3 Key components of spatial pattern in relation to LPA

Different methods of LPA reveal different aspects of spatial pattern in landscapes. Thus, it is critical to recognize the different components of spatial patterns in data as well as the different attributes of methods in interpreting results. Generally speaking, spatial pattern in landscapes is defined by the complexity or variability of a system property of interest in space and time (Li and Reynolds 1995). Complexity refers to qualitative or categorical descriptors of this system property, while variability refers to quantitative or numerical descriptors of the property. Each data type has its own characteristic complexity and variability that may serve as linkages of spatial pattern to ecological processes and provide ways to interpret their relationships. Even though we emphasize the structural characteristics in data that can be observed and analyzed, their functional linkages to ecological processes (e.g., population dynamics, nesting or foraging behavior, biogeochemical cycling) are critical to the interpretation of LPA results and must be considered (Li and Reynolds 1994, 1995, Fortin *et al.* 2003, Li and Wu 2004).

In landscape pattern analysis with numerical maps, the key quantifiable components of spatial pattern are domain variation, autocorrelated variation, random variation, and anisotropy of landscape elements (Table 3.1; Burrough 1987, Li and Reynolds 1995, Fortin *et al.* 2002). These components display a continuum of variability (Li and Reynolds 1994, 1995). Domain variation is long-range, deterministic and structured, and is represented by trends and magnitude of change of the mean or variance in space (e.g., those defined by trend surface analysis). Autocorrelated variation is medium-range, random but spatially correlated, and is characterized by the scale, within which

TABLE 3.1. *Data types, components and attributes of spatial pattern, and examples of methods to analyze them[a]. SV stands for semivariogram, and SH% is the degree of spatial autocorrelation (see text for more explanation)*

Data Type	Component of Spatial Pattern	Attribute	Method	Reference
Categorical Map	Composition	• Number of patch types • Proportion of patch types	• Contagion index • Diversity index	• Li and Reynolds 1993 • Pielou 1975
	Configuration	• Spatial arrangement • Patch shape • Contrast between neighboring patches • Connectivity of patches	• Fractal dimension • Shape index • Patchiness index	• O' Neill *et al.* 1988 • Forman and Godron 1986 • Romme 1982 • Tischendorf and Fahrig 2000
	Anisotropy	• Directional difference in spatial arrangement	• Directional analysis	• Wu *et al.* 2002 • Wu 2004
Numerical Map	Domain variation	• Trend • Magnitude of mean or variance	• Trend surface analysis	• Sokal and Rohlf 1981
	Autocorrelated variation	• Degree • Intensity • Scale	• SH% of SV • Fractal dimension of SV • Range of SV	• Li and Reynolds 1995 • Cressie 1991 • Rossi *et al.* 1992
	Random variation	• Degree at short range • Degree at long range	• Nugget of SV • Sill of SV	• Burrough 1987 • Li and Reynolds 1995
	Anisotropy	• Directional difference in domain and autocorrelated variation	• Directional analysis	• Cressie 1991 • Rossi *et al.* 1992 • Burrough 1983

[a] Modified from Li and Reynolds 1995.

autocorrelation exists, and the degree and intensity (i.e., rate of change) of spatial autocorrelation. Random variation is short-range or long-range, independent and not structured, and is marked by random noise or total variance. Anisotropy is the variation of trend and autocorrelation in different directions. For example, the autocorrelation structure of a system property may be quantified by parameters (i.e., range, nugget, sill) calculated from semivariogram analysis (Li and Reynolds 1995; Table 3.1). The short-range random noise is represented by nugget and the long-range random variation by sill. The scale of spatial autocorrelation is defined by the range of semivariograms, whereas the degree of spatial autocorrelation may be characterized by the percentage of structured variance, which is represented by the index of SH%, i.e., SH% = (sill − nugget)/sill. Notice that SH% corresponds to the proportion of spatial autocorrelation in the total variation of the system property because sill represents the maximum (total) variation and nugget the random variation. The intensity of spatial autocorrelation may be quantified by the slope of logarithm-transformed semivariograms because the slope expresses the rate of change in semivariance with respect to lag distance. Thus, the steeper the slope, the higher the intensity of autocorrelation. The slope can be represented by the fractal dimension of the semivariogram (Burrough 1983).

In landscape pattern analysis with categorical maps, the key quantifiable components of spatial pattern are the composition and configuration of a landscape mosaic (Table 3.1; Li and Reynolds 1994, 1995). Composition is nonspatial and includes the number and proportions of patch types, while configuration is spatial and includes spatial arrangement of patches, patch shape, contrast between neighboring patches, connectivity among patches of the same type, and anisotropy (i.e., variations of other attributes in different directions). These seven attributes in Table 3.1 have been widely recognized as factors of spatial pattern (e.g., Pielou 1977, Romme 1982, Forman and Godron 1986, Ludwig and Reynolds 1988, O'Neill et al. 1988, Wiens et al. 1993, Li and Reynolds 1994, 1995, Tischendorf and Fahrig 2000). The attributes are self-explanatory and easy to understand even though some of them (e.g., spatial arrangement, connectivity) are difficult to quantify. Use of these attributes in LPA can be illustrated by the following example. Suppose that we have a habitat map for a species, and then each pattern attribute of the map may be linked to functional responses of the species (Li and Reynolds 1994, 1995). For example, the number of habitat patch types may indicate the level of resource diversity; the proportion may determine the dominance (or lack) of critical resources; spatial distribution of resources may affect species dispersal and foraging efficiency; irregular habitat patch shape may signify great edge effects; changes in the neighboring contrast may determine the magnitude of edge effects and the permeability of boundaries to flows of matter and biological organisms;

connectivity of habitat patches may influence the success rate of species dispersal and the spread of diseases; and anisotropy in spatial pattern may be related to influences of topographic or edaphic factors, or differences in underlying processes in different directions.

3.4   Statistical and ecological assumptions of LPA methods

The assumptions of LPA methods may be general or specific, and explicitly defined or implied. Recognizing the assumptions of a method is an important first step to apply it properly. Below we discuss some major statistical and ecological assumptions behind the commonly used LPA methods.

3.4.1 Statistical assumptions

Spatial statistics (e.g., semivariogram, Moran's I, Geary's C) assume stationarity in the system property of interest. The stationarity assumption dictates that the mean (i.e., the first-order stationarity) and the variance (i.e., the second-order stationarity) of the system property is constant over sampling space (Webster 1985, Cressie 1991, Fortin et al. 2002). A weak, first-order stationarity can often be assumed for ecological data when some neighborhood effects or processes are at work. As Fortin et al. (2002) pointed out, stationarity is a property of the process rather than the observation. This means that this assumption may not be validated by the data used in analysis alone. The observation is just one realization of the pattern generated by the corresponding process. Fortunately, minor departures from the stationarity assumption should not preclude the use of these spatial statistics. What is important is to ensure that there is no change in the process or its dominance within the study area and that there is no change of support or grain size (Webster 1985).

Some LPA studies assume that a landscape map (numerical or categorical) is a sample from a population of maps. This assumption makes it possible to compare results of LPA of two (or more) maps if statistical characteristics (e.g., mean, variance) of the populations are known (Fortin et al. 2003). However, when the statistical characteristics of the populations are not known, no statistical tests may be made about the differences between two maps from different times or locations because each map represents only one single data point (Li and Wu 2004). We will revisit the consequences of this assumption later.

3.4.2 Relationship between pattern and process

A fundamental principle in landscape ecology is that spatial patterns of landscapes not only affect but also are affected by ecological processes (Turner

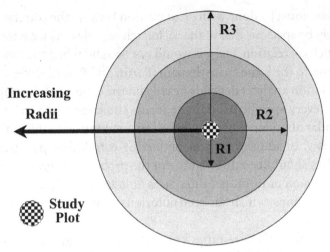

FIGURE 3.1
Schematic demonstration of a problematic research approach to establishing
relationships of wildlife abundance to habitat sites at increasing radii. The study
plot is the only site within which wildlife information is available. As a result, a
potential flaw exists because of a scale mismatch (i.e., the two variants used in
correlation analysis not belonging to the same space domain) and because of an
invalid assumption (i.e., similar abundance of the species being supported at every
specific site of habitat inside the radius of R1, R2 and R3)

1989, Pickett and Cadenasso 1995). The principal message here is that spatial pattern matters in our understanding of the structure, functioning, and
dynamics of landscapes. However, this pattern and process principle requires
critical scrutiny because a conceptual flaw may occur in LPA if pattern and process are not interactive (Li and Wu 2004). Although it is true in general that any
spatial pattern is a result of some process or processes at some point in time,
it is not appropriate to assume pattern and process always exhibit reciprocal
effects on ecological scales. The interaction between pattern and process can
only occur when they both operate at commensurate temporal scales and in the
same spatial domain (Wu and Loucks 1995, Li and Wu 2004). In addition, when
a process is not involved in the formation of a particular spatial pattern, there
should not be any inherent relationship between the pattern and the process;
when a process is nonspatial, spatial pattern becomes nonconsequential. Thus,
whether spatial pattern affects a process of interest cannot be assumed a priori;
rather, it needs to be verified with empirical evidence.

For example, in the search for relationships between forest fragmentation
and wildlife population dynamics (e.g., abundance, predation), many attempts
have been made to relate landscape metrics at large scales with increasing radii
to wildlife population quantities observed in study plots (Fig. 3.1; Robinson
et al. 1995). However, such practice of relating broad-scale landscape metrics to

fine-scale processes must be done with great caution because the correlational relationships may be specious due to the mismatch of scales. As we have suggested earlier, such correlation analysis would not be valid when the two variants do not belong to the same space domain (Fortin and Payette 2002). Also, this sort of correlation analysis depends heavily on the validity of the implicit assumption that every specific habitat site inside the larger radii must support or have similar abundance of the species, an assumption that is unlikely to be met in reality. In addition, when conducting correlation or regression analysis, ecologists should be acutely aware of the problem of "ecological fallacy," the phenomenon of improper inferences of lower-level attributes from higher-level relationships, which has been notorious in the social sciences (Wu, Chapter 7, this volume).

Establishing causal relations requires knowledge of mechanisms of pattern generation. This is often difficult because different processes may generate the same pattern (Pielou 1977, Cale et al. 1989). Without identification of responsible mechanisms for the pattern, observed correlation may be misleading because it is possible that a critical factor in the pattern generation may be unmeasured and omitted from analysis (Fig. 3.2; Fortin and Payette 2002). In addition, spatial pattern is not a simple factor to wildlife populations because it may often affect population processes indirectly through altering the distribution of other abiotic and biotic factors. As a result, predictions based on relationships between spatial pattern and wildlife population dynamics may generally tend to be weak for many species.

3.4.3 Ecological relevance of categorical data and landscape metrics

LPA assumes the ecological relevance of landscape data used. Ecologically relevant map data should contain information closely related to the ecological process under study. However, this simple, but implicit, assumption is often overlooked when categorical maps are used (Li and Wu 2004). The main reason for such oversight is that map data used for LPA are often collected for other purposes (e.g., pre-existing maps of vegetation, soil, land-use). Consequently, whether available maps are appropriate for addressing the specific objectives of a study is not always adequately discussed in landscape ecological studies. Of course, if the ecological relevance of map data is questionable, the results of LPA can be of little use. Therefore, to ensure the ecological relevance of data, one should explicitly address the question in the research planning, data collection, and publication processes. This is mainly a problem with categorical maps because numerical maps directly present variables of ecological interest. Li and Wu (2004) provided an example of and a solution to this problem. They argued that, in habitat suitability analysis, a vegetation map

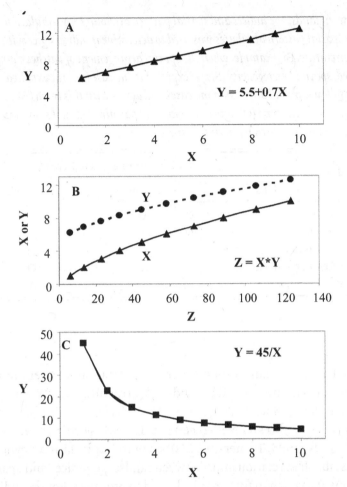

FIGURE 3.2
Example of misleading relationships when a key factor is omitted in correlation analysis. X and Y seem to have a positive linear relationship (A). However, if another variable, Z, affects both X and Y (B), then X and Y could in reality have a negative nonlinear relationship (C). Note that A is generated with the X and Y data in B and that C is generated by keeping Z as a constant (= 45; thus, the missing variable) and obtaining Y with each value of X

should not be used directly and blindly because the same landscape may be viewed by wildlife species differently. Instead, the vegetation map should be converted into a habitat rank map based on the unique habitat requirements of the species of interest (Harris and Sanderson 2000, Li *et al.* 2000), and then LPA is performed on the habitat rank map. Such data transformations (reclassification) relate species requirements to landscape attributes, thus making them ecologically relevant.

LPA also assumes ecological relevance of landscape metrics used. Ecologically relevant metrics should functionally link ecological processes (e.g., flows

TABLE 3.2. *Relationships between landscape indices and spatial pattern attributes.[a] A simulation experiment with full control of the spatial pattern attributes was used to determine if indices agree with the expected trends obtained from the literature. For example, when the spatial arrangement of patches changes from uniform to random to clumped, spatial heterogeneity (SH) is expected to increase (i.e., the expected trend), while the indices behave as follows: fractal dimension increases and agrees with this trend (A), evenness decreases and thus shows an opposite trend (O), and contagion and patchiness do not show any trend (N). Note that only patchiness can be used to characterize contrast*

Spatial Pattern Attribute	Expected Trend Low SH ← – – – → High SH	Agreement to Expected Trend			
		Fractal	Contagion	Evenness	Patchiness
No. of Patch Types	Small < Large	O	A	N	O
Proportion	Uneven < Even	O	O	A	A
Spatial Arrangement	Uniform < Random < Clumped	A	N	O	N
Patch Shape	Square < Regular < Irregular	A	N	A	O
Neighboring Contrast	Low < Mixed < High	…	…	…	A

[a] Modified from Li and Reynolds 1995.

of materials and energy in landscapes: water transportation, nutrient movement, species dispersal, fire spread) to landscape structure (e.g., Wiens *et al.* 1993, Li and Wu 2004). Such linkages have been discussed with regard to many commonly used indices (O'Neill *et al.* 1988, Li 1989, Turner 1989, Li and Reynolds 1993, 1994, 1995, Turner *et al.* 2001). For example, Li and Reynolds (1994, 1995) established relationships between landscape indices and spatial pattern components, which in turn may be linked to processes. For the indices examined (i.e., evenness, contagion, fractal dimension, patchiness), they found that the relationships of the four indices to different spatial patterns were not always consistent with the common expectations (Table 3.2; see Li and Reynolds 1994). Their findings highlighted the need that ecological relevance of landscape indices should be established with empirical evidence. Li and Wu (2004) warned that, in the absence of such evidence, landscape indices may simply be mathematical constructs without any ecological meaning. It must be pointed out that, even for indices that are effective measures of spatial pattern, their effects on ecological processes (e.g., edge effects, corridor usage) should not be taken for granted. It is essential, therefore, to ensure ecological relevance of indices in LPA by developing indices that have close associations with ecological processes. In the case of wildlife habitat evaluation, for example, LPA is in principle more effective if indices used are related to food (e.g., foraging efficiency), cover (e.g., shelters for energy conservation), reproduction (e.g., nesting sites), and other population processes (e.g., dispersal) (Li and Wu 2004).

In LPA, the scale of map data is often changed before analysis (i.e., data rescaling; Gardner *et al.* 1987, Turner *et al.* 1989, Wu *et al.* 2000, Saura and Martinez-Millan 2001, Wu 2004). Rescaling is necessary for analysis or modeling at multiple scales because of the difficulties in collecting such data by direct observations. While such data transformation usually assumes that the rescaled data are equivalent to those directly observed at the new scales, patterns revealed with rescaled data may be distorted (Li and Wu 2004, Wu 2004). As argued by Li and Reynolds (1995), one needs to distinguish between the scale of observation, at which the natural world is translated into data, and the scale of analysis, at which patterns are revealed from the data. Rescaling changes the scale of analysis, but its results are still strongly affected by the scale of original observations that determines the finest grain size of the data set. Studies have shown that the accuracy of rescaled data using different methods (e.g., majority rules, systematic selection, random selection) can be quite adequate for certain purposes, but generally varies with specific landscapes and metrics of interest (Justice *et al.* 1989, Benson and Mackenzie 1995, Bian 1999, Saura 2004, Wu 2004). The bottom line is that the validity of rescaled data must be assured for the intended analysis because data are the basis for deriving relationships and testing hypotheses.

3.5   Behavior of LPA methods

In addition to the underlying assumptions, the behavioral characteristics of LPA methods also need to be known to assure their appropriate applications. The behavior of a method is a function of its mathematical formulation and the data used in calculation. Several recent studies have systematically analyzed the behaviors of a number of LPA methods, including correspondence between landscape measures and spatial pattern attributes (Li and Reynolds 1994, 1995, Wickham *et al.* 1997, Hargis *et al.* 1998, Neel *et al.* 2004), relationships among landscape indices and among spatial statistical methods (Riitters *et al.* 1995, Dale *et al.* 2002), and sensitivity of landscape metrics to scale (Turner *et al.* 1989, Wickham and Riitters 1995, Jelinski and Wu 1996, Wu 2004).

3.5.1 Correspondence between landscape measures and pattern attributes

Recognizing the different attributes of spatial pattern is important to successful application of LPA methods and interpretation of their results. In Table 3.1 we described the correspondences of some major landscape indices and the semivariogram to various attributes of spatial pattern (Li and Reynolds 1994, 1995). These linkages can help relate quantitative information on spatial pattern to dynamics of ecological processes and phenomena. For numerical map data, relationships of spatial statistical methods (e.g., semivariograms) to spatial pattern attributes are usually straightforward. For example, the parameters calculated from semivariograms (or other similar methods) are by design all positively correlated to the spatial pattern attributes (Table 3.1): (1) the autocorrelated variation increases with SH% (degree), fractal dimension (intensity), and range (scale), (2) the random variation increases with nugget (degree at short range) and sill (degree at long range), and (3) domain variation and anisotropy can be determined to exist by trend surface analysis and directional analysis, respectively.

For categorical map data, relationships of landscape indices to spatial pattern attributes are complicated because most of these indices have complex formulas that are not based directly on spatial pattern attributes (Table 3.1). Through a simulation experiment, Li and Reynolds (1994, 1995) showed that the selected indices (i.e., fractal dimension, contagion, evenness, patchiness) corresponded well to the expected trends of spatial pattern in most cases, but exceptions and reverse patterns did occur (Table 3.2). The expected trends of how spatial heterogeneity should change with the spatial pattern attributes were obtained from the literature. Usually, fractal dimension, contagion, evenness, and patchiness are all assumed to be positively related to spatial heterogeneity. However, when different spatial pattern attributes are considered, these commonly used metrics do not always show such simple behavior. For example, fractal dimension increased and agreed with the expected trends as the spatial arrangement changed from uniform to random to clumped and as the patch shape changed from square to regular to irregular. But it decreased and thus showed the opposite to the expected trend as the number of patch types changed from small to large and as the proportions of patch types changed from uneven to even (Table 3.2). Thus, Li and Reynolds (1994, 1995) proposed that each of the spatial pattern attributes should be represented by a simple landscape index. For example, the proportion of patch types may be represented by evenness, the spatial arrangement of patches by fractal dimension, the patch shape by the shape index, the contrast between neighboring patches by patchiness index, the patch size by mean patch size, and the amount of edge by edge density. As for the number of patch types, it should be used directly as a measure of patch richness.

3.5.2 Relationships among LPA methods

Landscape indices can be highly correlated because they usually use the same five basic measurements of landscape mosaics: the number of patch types, proportion, perimeter–area relation, distance, and contrast (Li and Reynolds 1994, 1995, Riitters et al. 1995). High correlation among indices indicates great redundancy in information content; thus, a criterion for selecting indices is to avoid using indices with high correlation. Working with landscape maps from across the United States, Riitters *et al.* (1995) established the correlation structure of 55 landscape metrics based on a factor analysis. They found that more than half of the indices examined were closely correlated with at least one other index with a correlation coefficient exceeding 0.9 (including negative correlation). They classified these landscape metrics into five relatively independent groups: (1) patch compaction, (2) texture, (3) patch shape, (4) perimeter–area scaling, and (5) number of patch types. Most of the groupings by Riitters *et al.* (1995) can be explained simply by the formulations of the landscape metrics based on the five basic pattern attributes. These results were similar to the findings by Li and Reynolds (1994; unpublished data) that grouped landscape indices by their relations to spatial pattern attributes based on a different perspective (see discussion above; Table 3.2). The correlation coefficients among indices by Riitters *et al.* (1995) should be useful in selecting a group of indices for any given study to maximize the information content of results and improve the understanding and interpretation of the observed spatial pattern of landscapes.

Many spatial statistical methods also have close mathematical or conceptual relations because of the similarities in their mathematical formula and data used. In a recent review, Dale *et al.* (2002) performed an analysis of the relationships among diverse groups of spatial statistics based on their mathematical foundations, empirical calculations, and conceptual linkages. Similar to the correlation study by Riitters *et al.* (1995), Dale *et al.* (2002) used an ordination technique to show relative positions or groupings of the methods in the space defined by the first two ordination axes. Among those groups commonly used in landscape ecology are autocorrelation-based methods (e.g., semivariogram, Moran's I, Geary's C), wavelet-based methods (e.g., various wavelet approaches, spectral analysis), moving-window-based methods (e.g., fractal dimension, lacunarity), variance-to-mean-ratio-based methods (e.g., Ripley's K, Morisita's index), Mantel test, and join-count methods. The methods in each group show great similarity and, thus, need not to be used together.

3.5.3   Changes of landscape measures with respect to scale

Scaling relations as measured by changes in landscape measures over a range of scales (grain size and extent) provide critical information for the understanding and interpretation of landscape structure and functioning, as well as for identifying the characteristic scales and quantifying multiple-scale or hierarchical structures in landscapes (Wu *et al.* 2000, Dale *et al.* 2002, Wu 2004). This is because landscape pattern is often scale-dependent and its multiscale characteristics may be explicitly quantified by scaling functions. Most of the spatial statistics (e.g., semivariogram, Moran's I, Geary's C, fractal dimension based on the box method) are formulated (or can be extended) to provide outputs in the form of scalograms (Dale *et al.* 2002, Wu *et al.* 2002). As a result, they offer a straightforward way of investigating the changes of landscape measures in response to changing the scale of analysis.

However, scaling relations for landscape metrics require in-depth and often labor-intensive analyses to establish (Turner *et al.* 1989, Wu *et al.* 2000, 2002, Saura 2004, Wu 2004). Wu *et al.* (2002) and Wu (2004) performed systematic analyses of the scaling relations of a number of landscape metrics with map data of real landscapes. They found that the responses of landscape metrics to changing scale at the landscape level fell into three categories: simple (e.g., power-law), unpredictable, and staircase scaling functions. When the analysis was done at the class level (i.e., individual land-cover types), only the first two categories were observed. Effects of changing grain size were generally more predictable than those of changing extent. Such knowledge on how different metrics may behave with changing scale (i.e., scalograms) can guide applications of LPA in scaling of spatial pattern because landscape metrics with simple scaling relations reflect those landscape features whose extrapolation across spatial scales can be readily achieved and, in contrast, metrics with unpredictable scaling relations represent landscape features whose extrapolation is difficult (Wu 2004).

3.6   Limitations and challenges of LPA

Effective applications of LPA require that landscape ecologists understand the limitations of LPA methods, which hinge on their purposes, assumptions, and behaviors. LPA is limited by what a method can do. For example, using evenness or coefficient of variation to quantify spatial arrangement of patches is a futile effort because these indices do not contain spatial information. LPA is also restricted by what assumptions a particular method must meet. For example, the implicit assumption of ecological relevancy of categorical map data is required for all landscape indices, and should not be overlooked. Finally, choosing the right methods for a particular study is of great importance, and can be achieved by closely examining the behaviors of potential methods in terms of their linkages to spatial pattern attributes, relationships among themselves, and changes with scale.

Below, we discuss four major challenges in relation to LPA: (1) how to interpret results of LPA to address specific research questions, (2) how to effectively establish relationships between landscape pattern and process such that the relationship can be helpful in understanding and projecting landscape change,

(3) how to use known spatial heterogeneity to improve predictions of landscape change (or scaling), and (4) how to assess significance in comparing two landscapes such that one can determine with some certainty whether or not significant changes have occurred (Turner *et al.* 2001, Wu and Hobbs 2002, Fortin *et al.* 2003, Li and Wu 2004). Resolving these challenges will greatly enhance the progress of landscape ecology and the successful applications of LPA.

3.6.1 Difficulties in interpreting indices

Interpreting results from LPA can be difficult, especially with complex landscape indices (Li and Wu 2004). The reasons for this difficulty include: (1) uncertainty about what an index really measures (e.g., multiple formulas of contagion and fractal dimension (Riitters *et al.* 1996, Dale 1999)), (2) problems of using landscape indices in correlation analysis (Li and Wu 2004, Wu, Chapter 7, this volume), (3) lack of specific thresholds to identify significant changes in landscape structure and function (Turner *et al.* 2001), and (4) lack of a well-tested group of indices that can directly measure rates of ecological processes (O'Neill *et al.* 1999, Ludwig *et al.* 2000) or are closely related to spatial pattern attributes (see Tables 3.1 and 3.2). Successful interpretation of LPA results requires that landscape ecologists understand not only the methods of spatial pattern analysis, but also the concepts on which the methods are based (Dale 1999, Li and Wu 2004). To do so, one must understand the key spatial pattern attributes of landscapes as well as the assumptions and behaviors of LPA methods as discussed above.

3.6.2 Establishing relationships between pattern and process

Establishing relationships between landscape pattern and ecological processes is a main goal of landscape ecology. A commonly used approach is to perform correlation analysis with landscape indices and process data (e.g., nutrient loading in streams, species abundance, dispersal success). This approach is effective especially when mechanisms for the relationships can be developed for causal explanation and testing (Pickett *et al.* 1994). For example, Ludwig *et al.* (2000) developed an empirical scaling equation (i.e., correlation) for savanna landscapes in Australia that relates soil nitrogen content to the size of vegetation patches. They attributed the observed relationships to the redistribution of resources (nutrients, water) by run-off/run-on processes that form resource islands (e.g., Schlesinger *et al.* 1990).

The most difficult challenge in establishing relationships between landscape pattern and process is the lack of large-scale data. The lack of full coverage of process data in landscapes is due to difficulties in measuring processes at large scales. The consequence of the problem was demonstrated in the wildlife example in Fig. 3.1. Effective applications of remote sensing need to be explored to provide a solution to the problem. The lack of replications for landscape analysis is due to difficulties in sampling large areas and to the problem of pseudoreplication (Li and Wu 2004). Meta-analysis with a large number of case studies may be used to tackle the problem given that the same protocols of methodology and similarity in landscapes involved can be assured. The lack of large-scale experimentation to test relationships or hypotheses in landscape ecology is often due to difficulties with experimental control and logistic constraints. Simulation experimentation with models will continue to provide some remedies to this problem because it allows for landscape dynamics to be studied with different scenarios and sufficient replications (e.g., Li and Reynolds 1994). However, simulation modeling alone cannot solve the problem simply because simulated results are not real-world observations. So, field-manipulative experiments should be encouraged in landscape ecology whenever feasible.

3.6.3 Improving prediction based on known spatial heterogeneity

Spatial heterogeneity is one of the most fundamental characteristics of ecological systems at all scales and exerts significant influences on sampling, analysis, and modeling (Risser *et al.* 1984; Wiens 1989; Li and Reynolds 1995; Turner *et al.* 2001). However, few studies have explored ways of improving predictions of landscape dynamics with quantitative information on landscape heterogeneity from LPA. A common practice in modeling processes at large scales (or scaling up) is to run a local-scale model repeatedly for all patches (or pixels) in the landscape of study (known as the direct integration method; see Wu, Chapter 7, this volume). Usually, landscape heterogeneity is handled with a spatial data set that defines model input and parameters across the entire study area in a spatially explicit fashion. In this case, the uncertainty of model predictions may increase with the degree of within-patch or subpixel spatial heterogeneity. For example, models for regional assessments are often run on patches of one to several km^2, within which distinct land-cover types may be mixed. As a result, model parameters for a patch are usually determined by the major land-cover type or an average of all types present. This results in model uncertainty similar to the well-known problem of pixel-mixing in remote sensing.

To improve model predictions, one has to deal with two issues: error reduction with known within-patch spatial heterogeneity, and uncertainty assessment with incomplete coverage of parameter values in space. When information on spatial heterogeneity at a sub-patch scale is available, one can incorporate it directly into models. For example, Asner *et al.* (1997) used a spectral mixture analysis (SMA) to incorporate subpixel spatial heterogeneity from fractional coverage of two vegetation types (tree and grass) into the calculation of LAI for a savanna landscape. SMA characterizes subpixel heterogeneity by using a linear combination of the reflectance spectra of ground components to define reflectance at each pixel. Fractional coverage of vegetation types at a subpixel scale (30 m from Landsat TM) was used to reduce error in estimating LAI at a coarser resolution (1 km from AVHRR). When spatial coverage of model parameters is incomplete (i.e., unknown parameter values at some locations), one may use Monte Carlo techniques to quantify the level of uncertainty and to determine if variability of the parameters of concern significantly affects model output.

3.6.4 Determining the significance of differences between two landscapes

Landscape ecologists are frequently confronted with the following questions: Does the spatial configuration of two landscapes differ significantly? Have significant changes in landscape structure occurred during a certain period of time? Such questions are not always easy to answer because field studies usually deal only with one or a few landscapes so that no simple test is available to make statistical inferences (Fortin and Payette 2002). The inability to address such basic questions statistically is a critical limitation of LPA. This problem is not unique to landscape ecology, but common to many other kinds of large-scale studies (e.g., regional assessment, climate change) where the number of replicates is limited. In these situations, visual comparisons with spatial statistics or landscape indices are often used to draw conclusions about differences or similarities between landscape maps. In general, landscape indices are descriptive and do not allow for statistical tests, whereas spatial statistics like semivariograms may allow for statistical tests locally where there are a large number of sample points, but not globally where the whole landscape represents only one data point. Hence, solutions to this problem of LPA are urgently needed.

One such solution is Monte Carlo simulation. For example, Fortin *et al.* (2003) demonstrated such an approach by comparing two landscapes. The basic idea is that a landscape at any given moment is one realization of a stochastic process, and thus the population characteristics can be generated by Monte Carlo simulation as long as a stochastic model of the process can be defined. With two sets of model parameters estimated to represent two different landscapes, a large number of realizations of maps can be produced for each landscape through Monte Carlo simulation. Thus, the resultant population characteristics of the two sets of simulated maps can be used in an inferential test to define significant differences between the two landscapes (e.g., less than 5 percent overlap of the probability distribution; Fortin *et al.* 2003). In this approach, the lack of observations is dealt with by making the stationarity assumption, i.e., the model parameters are spatially invariant. This approach is promising only if the spatial process can be correctly identified and only if a mechanistic model can be constructed. In addition, the approach may not work well for categorical variables when the main processes that generate landscape mosaics are human activities (e.g., management regimes, land-use policies) that often defy statistical characterization by stochastic processes.

When the stochastic process is unknown, however, an alternative, complementary approach may be used as demonstrated by Li and Reynolds (1995). They treated spatial pattern as an observable phenomenon that was defined stochastically by spatial pattern components discussed above for either numerical or categorical data (Table 3.1). Li and Reynolds (1994, 1995) developed two spatial pattern simulation models: SHAPN and SHAPC. SHAPN is an autoregressive-moving-average (ARMA) model (e.g., Bras and Rodriguez-Iturbe 1985) that generates numerical transects based on variability in magnitude, trend surface, and scale and degree of autocorrelation. SHAPC is a landscape mosaic simulator that generates categorical maps with characteristic spatial pattern based on complexity in number of patch types, proportion of each type, patch distribution, patch shape, and patch-size distribution. The key characteristics of SHAPC and SHAPN are that: (1) they do not have to refer to any underlying stochastic processes given the difficulties involved in identifying which process is at work in most situations, and (2) they generate spatial data with controllable parameters that closely represent spatial pattern attributes (Li and Reynolds 1994, 1995). In a pilot study, Li (unpublished manuscript) used this Monte Carlo approach to simulate a large number of categorical maps based on chosen model parameters (which can be estimated from real landscapes), and to determine how much change in the values of indices must occur to indicate significant change in landscape structure (Fig. 3.3). In this experiment, the significance tests were based on the population characteristics (i.e., mean and variance) of each pair of groupings, which were composed of landscape maps with the same settings of the selected factors (e.g., all maps with 4 patch types vs. those with 5). The preliminary results indicated that this approach could determine the thresholds of index change that were required for two maps to be statistically different (Fig. 3.3). Such threshold values of indices may help landscape ecologists make better inferences about observed changes in landscapes. Both of these approaches need to be tested with real landscapes.

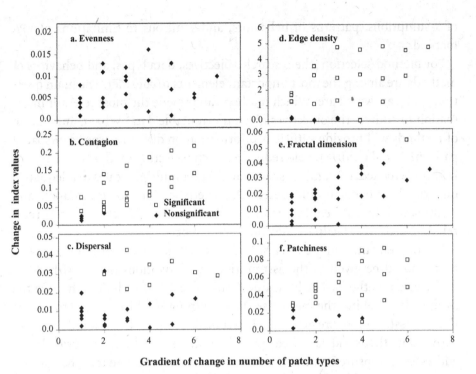

FIGURE 3.3

Changes in landscape metrics along the gradient of the number of patch types. The x-axis represents absolute change in number of patch types between the data groups, whereas the y-axis represents absolute change in a given index. The landscape maps used in the analysis were generated in a factorial experiment (with 864 treatments and five replicates) by SHAPC under controlled conditions as defined by five factors of spatial pattern: (1) number of patch types, (2) proportion, (3) patch size, (4) patch shape, and (5) spatial arrangement. The response variables in the analysis were evenness, contagion, dispersal, edge density, fractal dimension, and patchiness. The open squares represent significant differences based on the Tukey comparison with z-scores at the 90 percent confidence level. The solid diamonds indicate nonsignificance. Threshold values may be derived for four of the six indices; e.g., a change of 0.05 in contagion may result in a significant difference.

3.7   Concluding remarks

In this chapter we have provided an overview of landscape pattern analysis in terms of its usefulness, classification, basic assumptions, key characteristics, and major limitations and challenges. These are all fundamental elements of LPA. Methodology is the cornerstone of science, and LPA is one of the primary methodological foundations of landscape ecology. Therefore, the effective application of LPA is of paramount importance to the advancement of landscape ecology. Thus, general guidelines for the proper selection, implementation, and interpretation of various methods in LPA are needed. Here we provide several such guidelines based on the directions of theory, consequences of assumptions, patterns of behaviors, and solutions to limitations of LPA methods.

For method selection, the research objectives, data types, and behaviors of methods are among the most important elements to consider. The main questions to be answered are: Which method can provide the most relevant information to address the objectives given the available data? What combination of methods will provide sufficient information on different aspects of spatial pattern? One should: (1) avoid redundancies by selecting methods that are not highly correlated, (2) conduct spatial analysis at multiple scales to adequately quantify heterogeneity and detect characteristic scales of landscapes, and (3) use simple metrics (e.g., patch size, edge, shape, inter-patch distance) in correlation analysis to facilitate meaningful inferences.

For implementation (i.e., mathematical operations), important considerations should be given to the assumptions and equations (or algorithms) of the selected methods. The basic questions to be asked include: What are the assumptions behind the methods? Does anything need to be done to ensure that the assumptions are met (e.g., data transformation)? One should: (1) make sure that methods and data used are ecologically relevant to the question being addressed, (2) ensure that the support (or grain size) and extent of the data are consistent throughout the analysis, (3) avoid scale mismatch between variables in analyses such as correlation and regression, and (4) compute uncertainty measures whenever possible.

For interpretation of results, the central elements to take into account include the basic behaviors and limitations of the methods, the key pattern attributes involved, and the data used in the analysis. The important questions to ask are: What does the observed pattern mean, ecologically and statistically? How do the quantitative measures of landscape structure relate to ecological processes of interest? One should: (1) pursue empirical evidence for the existence of a pattern–process relationship, instead of assuming it a priori, (2) identify possible mechanisms for observed correlation and avoid omitting a critical factor in the pattern generation, (3) use different pattern attributes as effective linkages that can help relate landscape structure to ecological processes, (4) interpret results from rescaled data with extra caution, and (5) be aware that a landscape map may represent only a single data point in many LPA studies.

In conclusion, to guarantee effective use of any method in science one must understand both the method itself and the data used in its implementation. This is the overarching theme of this chapter. These guidelines will, we hope, facilitate successful applications of LPA and continue to be refined as landscape ecology progresses. In addition, even though we discuss the selection, implementation, and interpretation of methods as separate issues of LPA, they must be considered together in practice. It is certain that, with emerging new ideas and theories of spatially extended systems and rapidly developing technologies in computing and remote sensing, LPA will continue to make progress and play a pivotal role in landscape ecology.

Acknowledgments

We thank Santiago Saura, James Wickham, and Xinyuan Wu for their comments. JW's research in landscape ecology has been supported by U.S. Environmental Protection Agency, U.S. National Science Foundation, and National Natural Science Foundation of China.

References

Asner, G.P., C.A.Wessman, and J.L. Privette. 1997. Unmixing the directional reflectances of AVHRR sub-pixel landcovers. *IEEE Transactions in Geoscience and Remote Sensing* **35**, 868–78.

Benson, B.J. and M.D. Mackenzie. 1995. Effects of sensor spatial resolution on landscape structure parameters. *Landscape Ecology* **10**, 113–20.

Bian, L. 1999. Comparing effects of aggregation methods on statistical and spatial properties of simulated spatial data. *Photogrammetric Engineering and Remote Sensing* **65**, 73–84.

Bras, R.L. and I. Rodriguez-Iturbe. 1985. *Random Functions and Hydrology*. Reading, Massachusetts: Addison-Wesley.

Burrough, P.A. 1983. Multiscale sources of spatial variation in soil. I. The application of fractal concepts to nested levels of soil variation. *Journal of Soil Science* **34**, 577–97.

Burrough, P.A. 1987. Spatial aspects of ecological data. Pages 89–125 in R.H. Jongman, C.J.F. ter Braak, and O.F.R. van Tongeren (eds.). *Data Analysis in Community and Landscape Ecology*. The Netherlands: Pudoc Wageningen.

Cale, W.G., G.M. Henebry, and J.A. Yeakley. 1989. Inferring process from pattern in natural communities. *BioScience* **39**, 600–5.

Cliff, A.D. and J.K. Ord. 1981. *Spatial Processes. Models and Applications*. London: Pion.

Cressie, N.A.C. 1991. *Statistics for Spatial Data*. New York: John Wiley & Sons, Inc.

Dale, M.R.T. 1999. *Spatial Pattern Analysis in Plant Ecology*. Cambridge: Cambridge University Press.

Dale, M.R.T., P. Dixon, M.-J. Fortin, *et al.* 2002. Conceptual and mathematical relationships among methods for spatial analysis. *Ecography* **25**, 558–77.

Forman, R.T.T. and M. Godron. 1986. *Landscape Ecology*. New York: John Wiley & Sons, Inc.

Fortin, M.-J., B. Boots, F. Csillag, and T.K. Remmel. 2003. On the role of spatial stochastic models in understanding landscape indices in ecology. *Oikos* **102**, 203–12.

Fortin, M.-J., M.R.T. Dale, and J. ver Heof. 2002. Statistical analysis in ecology. Pages 2051–2058 in A.H. El-Shaarawi and W.W. Piegorsch (eds.). *Encyclopedia of Environment*. Chichester: John Wiley & Sons Ltd.

Fortin, M.-J. and S. Payette. 2002. How to test the significance of the relation between spatially autocorrelated data at the landscape scale: a case study using fire and forest maps. *Ecoscience* **9**, 213–18.

Gardner, R.H., B.T. Milne, M.G. Turner, and R.V. O'Neill. 1987. Neutral models for the analysis of broad-scale landscape pattern. *Landscape Ecology* **1**, 19–28.

Gustafson, E.J. 1998. Quantifying landscape spatial pattern: What is the state of the art? *Ecosystems* **1**, 143–56.

Hargis, C.D., J.A. Bissonette, and J.L. David. 1998. The behavior of landscape metrics commonly used in the study of habitat fragmentation. *Landscape Ecology* **13**, 167–86.

Harris, L.D. and J. Sanderson. 2000. The re-membered landscape. Pages 91–112 in J. Sanderson and L.D. Harris (eds.). *Landscape Ecology: A Top-down Approach*. Boca Raton, LA: Lewis Publishers.

Jelinski, D.E. and J. Wu. 1996. The modifiable areal unit problem and implications for landscape ecology. *Landscape Ecology* **11(3)**, 129–40.

Justice, C.O., B.L. Markham, J.R.G. Townshend, and R.L. Kennard. 1989. Spatial degradation of satellite data. *International Journal of Remote Sensing* **10**, 1539–61.

Li, H. 1989. Spatio-temporal pattern analysis of managed forest landscapes: a simulation approach. Ph.D. Dissertation, Corvallis, Oregon: Oregon State University.

Li, H., D. Gartner, P. Mou, and C.C. Trettin. 2000. A landscape model (LEEMATH) to evaluate effects of management impacts on timber and wildlife habitat. *Computers and Electronics in Agriculture* **27**, 263–92.

Li, H. and J.F. Reynolds. 1993. A new contagion index to quantify spatial patterns of landscapes. *Landscape Ecology* **8**, 155–62.

Li, H. and J.F. Reynolds. 1994. A simulation experiment to quantify spatial heterogeneity in categorical maps. *Ecology* **75**, 2446–55.

Li, H. and J.F. Reynolds. 1995. On definition and quantification of heterogeneity. *Oikos* **73**, 280–4.

Li, H. and J. Wu. 2004. Use and misuse of landscape indices. *Landscape Ecology* **19**, 389–99.

Ludwig, J.A. and J.F. Reynolds. 1988. *Statistical Ecology*. New York: John Wiley & Sons, Inc.

Ludwig, J.A., J.A. Wiens, and D.J. Tongway. 2000. A scaling rule for landscape patches and how it applies to conserving soil resources in savannas. *Ecosystems* **3**, 84–97.

McGarigal, K. and B.J. Marks. 1995. FRAGSTATS: Spatial pattern analysis program for quantifying landscape structure. General Technical Report PNW-GTR-351. Portland, OR: Pacific Northwest Research Station.

Neel, M.C., K. McGarigal, and S. Cushman. 2004. Behavior of class-level landscape metrics across gradients of class aggregation and area. *Landscape Ecology* **19**, 435–55.

O'Neill, R.V., J.R. Krummel, R.H. Gardner, G. Sugihara, B. Jackson, D.L. DeAngelis, B.T. Milne, M.G. Turner, B. Zygmunt, S.W. Christensen, V.H. Dale, and R.L. Graham. 1988. Indices of landscape pattern. *Landscape Ecology* **1**, 153–62.

O'Neill, R.V., K.H. Riitters, J.D. Wickham, and K.B. Jones. 1999. Landscape pattern metrics and regional assessment. *Ecosystem Health* **5**, 225–33.

Pickett, S.T.A. and M.L. Cadenasso. 1995. Landscape ecology: spatial heterogeneity in ecological systems. *Science* **269**, 331–4.

Pickett, S.T.A., J. Kolasa, and C.G. Jones. 1994. *Ecological Understanding*. San Diego: Academic Press.

Pielou, E.C. 1975. *Ecological diversity*. New York: John Wiley & Sons, Inc.

Pielou, E.C. 1977. *Mathematical Ecology*. New York: John Wiley & Sons, Inc.

Riitters, K.H., R.V. O'Neill, C.T. Hunsaker, *et al.* 1995. A factor analysis of landscape pattern and structure metrics. *Landscape Ecology* **10**, 23–39.

Riitters, K.H., R.V. O'Neill, J.D. Wickham, and K.B. Jones. 1996. A note on contagion indices for landscape analysis. *Landscape Ecology* **11**, 197–202.

Risser, P.G., J.R. Karr, and R.T.T. Forman. 1984. *Landscape Ecology: Directions and Approaches*. Special Publ. 2, Champaign: Illinois Natural History Survey.

Robertson, G.P. 1987. Geostatistics in ecology: interpolating with known variance. *Ecology* **68**, 744–8.

Robinson, S.K., F.R. Thompson, T.M. Donovan, D.R. Whitehead, and J. Faaborg. 1995. Regional forest fragmentation and the nesting success of migratory birds. *Science* **267**, 1987–9.

Romme, W.H. 1982. Fire and landscape diversity in subalpine forests of Yellowstone National Park. *Ecological Monograph* **52**, 199–221.

Rossi, R.E., D.J. Mulla, A.G. Journel, and E.H. Franz. 1992. Geostatistical tools for modeling and interpreting ecological spatial dependence. *Ecological Monographs* **62**, 277–314.

Saura, S. 2004. Effects of remote sensor spatial resolution and data aggregation on selected fragmentation indices. *Landscape Ecology* **19**, 197–209.

Saura, S. and J. Martinez-Millan. 2001. Sensitivity of landscape pattern metrics to map spatial extent. *Photogrammetric Engineering and Remote Sensing* **67**, 1027–36.

Schlesinger, W. H., R. F. Reynolds, G. I. Cunningham, *et al.* 1990. Biological feedbacks in global desertification. *Science* **247**, 1043–8.

Sokal, R. R. and F. J. Rohlf. 1981. *Biometry: The Principles and Practice of Statistics in Biological Research*, 2nd edn. San Francisco, CA: Freeman.

Tischendorf, L. 2001. Can landscape indices predict ecological processes consistently? *Landscape Ecology* **16**, 235–54.

Tischendorf, L. and L. Fahrig. 2000. How should we measure landscape connectivity? *Landscape Ecology* **15**, 633–41.

Turner, M. G. 1989. Landscape ecology: the effect of pattern on process. *Annual Review of Ecology and Systematics* **20**, 171–97.

Turner, M. G. 1990. Spatial and temporal analysis of landscape pattern. *Landscape Ecology* **4**, 21–30.

Turner, M. G. and R. H. Gardner. 1991. *Quantitative Methods in Landscape Ecology: The Analysis and Interpretation of Landscape Heterogeneity*. New York: Springer-Verlag.

Turner, M. G., R. H. Gardner, and R. V. O'Neill. 2001. *Landscape Ecology in Theory and Practice: Pattern and Process*. New York: Springer.

Turner, M. G., R. V. O'Neill, R. H. Gardner, and B. T. Milne. 1989. Effects of changing spatial scale on the analysis of landscape pattern. *Landscape Ecology* **3**, 153–62.

Upton, G. J. G. and B. Fingleton. 1985. *Spatial Data Analysis by Example*, Volume 1: *Point Pattern and Quantitative Data*. New York: John Wiley & Sons, Inc.

Webster, R. 1985. Quantitative spatial analysis of soil in the field. *Advances in Soil Science* **3**, 1–70.

Wickham, J. D. and K. H. Riitters. 1995. Sensitivity of landscape metrics to pixel size. *International Journal of Remote Sensing* **16**, 3585–95.

Wickham, J. D., R. V. O'Neill, K. H. Riitters, T. G. Wade, and K. B. Jones. 1997. Sensitivity of landscape metrics to land cover misclassification and differences in land cover composition. *Photogrammetric Engineering and Remote Sensing* **63**, 397–402.

Wiens, J. A. 1989. Spatial scaling in ecology. *Functional Ecology* **3**, 385–97.

Wiens, J. A., N. C. Stenseth, B. V. Horne, and R. A. Ims. 1993. Ecological mechanisms and landscape ecology. *Oikos* **66**, 369–80.

Wu, J. 2004. Effects of changing scale on landscape pattern analysis: scaling relations. *Landscape Ecology* **19**, 125–38.

Wu, J. and R. Hobbs. 2002. Key issues and research priorities in landscape ecology: An idiosyncratic synthesis. *Landscape Ecology* **17**, 355–65.

Wu, J., D. E. Jelinski, M. Luck, and P. T. Tueller. 2000. Multiscale analysis of landscape heterogeneity. *Geographic Information Sciences* **6**, 6–19.

Wu, J. and O. L. Loucks. 1995. From balance-of-nature to hierarchical patch dynamics: a paradigm shift in ecology. *Quarter Review of Biology* **70**, 439–66.

Wu, J., W. Shen, W. Sun, and P. T. Tueller. 2002. Empirical patterns of the effects of changing scale on landscape metrics. *Landscape Ecology* **17**, 761–82.

4

Spatial heterogeneity and ecosystem processes

4.1 Introduction

Understanding the patterns, causes, and consequences of spatial heterogeneity for ecosystem function is a research frontier in both landscape ecology and ecosystem ecology (Turner *et al.* 2001, Chapin *et al.* 2002, Wu and Hobbs 2002, Lovett *et al.* 2005). Landscape ecology research has contributed to tremendous gains in understanding the causes and consequences of spatial heterogeneity, how relationships between patterns and processes change with scale, and the management of both natural and human-dominated landscapes. There are now many studies in widely varied landscapes that elucidate, for example, the conditions under which organisms may respond to landscape composition or configuration, or disturbance spread may be constrained or enhanced by landscape pattern. The inclusion of a spatial component is now pro forma in many ecological studies, and tools developed by landscape ecologists for spatial analysis and modeling now enjoy widespread use (e.g., Baskent and Jordan 1995, McGarigal and Marks 1995, Gustafson 1998, Gergel and Turner 2002). Landscape ecological approaches are not limited only to "land" but are also applied in aquatic and marine ecosystems (e.g., Fonseca and Bell 1998; Bell *et al.* 1999; Garrabou *et al.* 2002; Teixido *et al.* 2002; Ward *et al.* 2002). However, with a few exceptions, the consideration of ecosystem function has lagged behind progress in understanding the causes and consequences of spatial heterogeneity for other ecological processes.

Ecosystem ecology focuses on the flow of energy and matter through organisms and their surroundings, seeking to understand pools, fluxes, and regulating factors. From its initial descriptions of how different ecosystems (e.g., forests, grasslands, lakes and streams) vary in structure and function, ecosystem ecology moved toward quite sophisticated analyses of function – e.g., food web analyses, biogeochemistry, regulation of productivity, and so on (Golley 1993, Pace and Groffman 1998, Chapin *et al.* 2002). However, ecosystem ecology has typically emphasized understanding changes through time within a single ecosystem rather than understanding variation across space (but see Ryszkowski *et al.* 1999). Recent studies suggest that spatial variability in some ecosystem processes may be of similar magnitude to temporal variation (e.g., Burrows *et al.* 2002, Turner *et al.* 2004), and efforts to explain and predict such variation are increasing. The importance of transfers among patches, representing losses from donor ecosystems and subsidies to recipient ecosystems, for the long-term sustainability of ecosystems is also now acknowledged explicitly (Naiman 1996, Carpenter *et al.* 1999, Chapin *et al.* 2002). Ecosystem studies have elucidated the mechanisms underlying temporal dynamics of many processes, but there has been comparatively little explicit treatment of spatial heterogeneity.

Progress at the interface of ecosystem and landscape ecology has been relatively slow, despite a tradition in Eastern Europe (e.g., Ryszkowski and Kedziora 1993, Ryszkowski *et al.* 1999) and stronger connections during the early development of landscape ecology in North America (e.g., Risser *et al.* 1984, Gosz 1986). Integration of the understanding gained from ecosystem and landscape ecology would likely enhance progress in both disciplines while generating new insights into how landscapes function. Indeed, gaining a more functional understanding of landscapes is a goal shared by ecosystem and landscape ecology. In this chapter, we identify key questions that could guide a research agenda in spatial heterogeneity and ecosystem function, focusing on four key research areas in which significant progress can be made: (1) understanding spatial heterogeneity of process rates, (2) land-use legacies, (3) lateral fluxes in landscape mosaics, and (4) linking species and ecosystems.

4.2   Understanding the spatial heterogeneity of process rates

Understanding variability in the rates of key ecosystem processes is a major focus of ecosystem ecology (Chapin *et al.* 2002). "Point processes" are those that can be well represented by rates measured at a particular location in space and time (Turner and Chapin 2005), and for these processes, spatial variation among replicate measurements is often averaged to estimate a mean value. For example, net primary production, net ecosystem production, denitrification, and nitrogen mineralization are processes understood in many systems using methods of analysis focused on spatially independent measurements. Most ecosystem ecologists have focused on understanding the mean rates and their temporal dynamics, in spite of the "noise" owing to spatial variation.

However, the basic causes of spatial heterogeneity in point-process rates have been well known for a long time (Jenny 1941). Heterogeneity is derived from the abiotic template, including factors such as climate, topography, and substrate. In addition, ecosystem process rates may vary with the biotic assemblage, disturbance events (including long-term legacies), and the activities of humans (Chapin *et al.* 1996, Amundson and Jenny 1997). Despite the acknowledgement of sources of spatial heterogeneity, there has been relatively little empirical work designed to characterize the spatial variation of process rates, the spatial scales over which variation is manifest, and the factors that control such variation.

Recent studies have demonstrated that understanding temporal behavior in mean rates may not be adequate; understanding spatial variance in process rates may lead to new insights into the mechanisms governing ecosystem dynamics and new approaches for predicting landscape function (van Dokkum *et al.* 2002, Benedetti-Cecchi 2003). Understanding the locations and direct and indirect effects of the spatial and temporal variation in process rates across landscapes could help reveal the relative importance of abiotic, biotic, and human factors, which interact across potentially different scales of time and space to both constrain and produce observed spatial pattern.

Studies linking disturbance, succession, and ecosystem processes in Yellowstone National Park illustrate how new insights and predictive power can be gained from understanding variation in process rates (Turner *et al.* 1994, 1997, 2004). The 1988 Yellowstone fires created a landscape mosaic in which post-fire lodgepole pine densities varied from 0 to >500000 stems ha^{-1}. This spatial heterogeneity in sapling density resulted largely from contingencies such as the spatial variation in fire severity and in pre-fire serotiny within the stand, rather than from the abiotic template. The tremendous variation in stand density in turn generated substantial heterogeneity in aboveground net primary production (ANPP), which ranged from 1 to 15 Mg ha^{-1} yr^{-1} ten years after the fires (Turner *et al.* 2004). Analyses of how spatial variation in stand density and growth rate (basal area increment, an index of ANPP) changes with stand age revealed that effects of the initial post-fire mosaic persists for at least a century (Kashian 2002, Kashian *et al.* 2005). Had only mean ANPP been studied and the spatial variability in ANPP ignored, it is likely that key factors influencing the process would not have been identified.

Understanding the spatial patterns of ecosystem process rates is also fundamental to spatial extrapolation over large areas. Obtaining field measurements of many ecosystem process rates across large areas is costly, and relatively few spatially extensive empirical data sets exist. Remote sensing methods and platforms offer promise for some variables on land (Groffman and Turner 1995, Martin and Aber 1997, Serrano *et al.* 2002), in wetlands (Urban *et al.* 1993), and in the open ocean (Karl 2002). Extrapolation of process rates across heterogeneous landscapes using empirical data or simulation models combined with GIS data layers can be used to test hypotheses about the influence of independent variables (Miller *et al.* 2004). Running *et al.* (1989) were among the first to integrate biophysical data obtained from many sources and combine these data with an ecosystem simulation model to predict evapotranspiration, leaf-area index and net photosynthesis across a large landscape. Their estimates demonstrated the power of these new integrative methods for producing spatially explicit projections of variation in ecosystem processes and offered insights into interactions among the controls on these processes (Running *et al.* 1989). Empirical extrapolations combined with GIS data have been used to predict rates of denitrification in southern Michigan (Groffman *et al.* 1992), net nitrogen mineralization within forests of the Midwestern Great Lakes region (Fan *et al.* 1998), and aboveground net primary production in western Yellowstone National Park (Hansen *et al.* 2000). Using spatial extrapolation in a hypothesis-testing mode represents a powerful approach that could be used much more widely in studies of spatial heterogeneity of point-process rates (Miller *et al.* 2004).

A first research priority for linking ecosystem and landscape ecology should focus on understanding the spatial structure of variation in rates at multiple scales, the factors that produce the spatial variation, and the consequences of that variation for other ecological phenomena (Table 4.1). Methods from landscape ecology that consider both continuous and discrete representations of spatial data should be integrated with studies of ecosystem processes to build understanding of landscape function.

4.3   Influence of land-use legacies

Landscape ecology has made important contributions to our understanding of land-use change, including the natural and socioeconomic drivers of land-use change, how it affects landscape structure, and how organisms may respond. Recent studies have documented the importance of historical land use in explaining contemporary ecosystems and landscapes (Foster *et al.* 2003). For example, historical land use influences current vegetation composition in New England forests (Currie and Nadelhoffer 2002, Foster 2002, Hall *et al.* 2002, Eberhardt *et al.* 2003). Comparisons of formerly cultivated forests with reference forests in North America and Europe suggest that agricultural practices can alter soil nutrient content and net nitrification rates for at least a century after abandonment (Koerner *et al.* 1999, Compton and Boone 2000, Goodale and Aber 2001, Dupouey *et al.* 2002, Jussy *et al.* 2002). Thus, historical patterns of land use may be important drivers of the pattern and variability in current

TABLE 4.1. *Suggested general questions that could guide research in each of four areas in which progress is both needed and possible at the interface of landscape ecology and ecosystem ecology*

Topic area	Research questions
Spatial heterogeneity of process rates	• How spatially heterogeneous are ecosystem process rates? • What causes variation in ecosystem process rates? • What are the consequences of variation in process rates on key ecological phenomena?
Influence of land-use legacies	• What is the role of land-use legacies in explaining the state of contemporary ecosystems? • How persistent are the effects of historical land use?
Lateral fluxes in landscape mosaics	• In landscape mosaics, how does spatial configuration influence pools and lateral fluxes of matter, energy and information? • What are the relative roles of spatial variation in initial conditions, local process rates, and lateral connections for pools and fluxes? • How do effects of spatial heterogeneity differ in one-way networks and mosaics with multidirectional flows?
Linking species and ecosystems	• How do trophic cascades influence vegetation mosaics and rates of ecosystem processes? • How do the spatial movements of organisms respond to and create spatial heterogeneity in ecosystem process rates?

rates of ecosystem processes. Landscape ecologists have often conducted studies that quantify how landscape patterns have changed through time; however, in few cases have linkages been made between historical landscapes, their trajectories, and the ecosystem processes.

Understanding the functional role of land-use legacies could be addressed by combining the spatial analysis methods of landscape ecology with the process-based approach of ecosystem ecology. Landscape ecology offers sophisticated methods to quantify land-use patterns as they change through time. This information could be used to stratify field sampling locations by historical land use and other appropriate variables, such as factors that relate to the abiotic template (e.g., elevation, slope, aspect, substrate). Pools and process rates for key functional variables can be measured using traditional methods from ecosystem ecology (e.g., Sala *et al.* 2000). Spatial extrapolation can again serve as a means of testing the predictive power of current understanding, and also of identifying locations in a landscape where some pool or flux of interest may be especially high or low.

The use of new quantitative methods that consider the magnitude and scale of spatial variability in ecosystem response variables may also yield important new insights. Ecosystem processes are usually measured as continuous rather than categorical variables, and methods derived from spatial statistics are ideally suited for studying spatial variation in continuous data. Many of these methods (e.g., semivariograms, correlograms, kriging) are similar to time-series analyses that identify temporal periodicities in a data set. Spatial statistics also provide guidance for efficient sampling schemes to assess the spatial structure of continuous data (e.g., Burrows *et al.* 2002, Fraterrigo *et al.* 2005).

Understanding the influence of historical land-use patterns on vegetation and soils in the Southern Appalachian Mountains illustrates how a blending of landscape and ecosystem ecology can be used to understand the effects of historical land use on landscape function. Spatial-pattern analyses of these landscapes have identified topographic positions and forest communities that have been influenced by land-use changes to a greater or lesser degree (e.g., Wear and Bolstad 1998, Turner *et al.* 2003). Historical land use has strongly affected mesic forest communities and the occurrence and abundance of herbaceous plants within these forests (Duffy and Meier 1992; Pearson *et al.* 1998, Mitchell *et al.* 2002, Turner *et al.* 2003). However, the long-term (> 50 yr) impacts of land use on the spatial heterogeneity of soil nutrients are poorly understood. Fraterrigo *et al.* (in review) examined patterns of nutrient heterogeneity in the mineral soil (0–15 cm depth) of 13 southern Appalachian forest stands in western North Carolina > 60 yr after abandonment from pasture or timber harvest using a cyclic sampling design derived from spatial statistics. Mean concentrations rarely indicated an enduring effect of historical land use on nutrient pools, but the spatial heterogeneity of nutrient pools differed substantially with past land use. Nutrient pools were most variable in reference stands, and this variability was greatest at fine scales. In contrast, formerly pastured and logged stands generally exhibited less variability, and soil nutrients were relatively more variable at coarse spatial scales. Geostatistical analysis of fine-scale patterns further revealed that spatial structure of soil cations was more closely linked to former land use than observed for other soil nutrients. These results suggest that land use has persistent effects on the spatial heterogeneity of soil resources, which may not be detectable when values are averaged across sites (Fraterrigo *et al.* in review). These insights were only possible by combining the spatial approaches of landscape ecology with the analytical methods of ecosystem ecology.

All landscapes exist and change in a framework of both natural and cultural legacies. Historical natural disturbances such as fire, floods, and storms appear to strongly influence contemporary systems, and analysis of cultural history of contemporary landscapes has assumed greater importance in recent decades (Foster *et al.* 2003). Yet studies of the impact of prior historical conditions of a landscape are relatively few. Landscape ecology can contribute by linking a temporally extended understanding of landscape spatial dynamics with functional measurements and the application of methods for analyzing continuous data. We suggest a second research priority for linking landscape and ecosystem ecology directed toward understanding the relative importance of historical landscape conditions for explaining contemporary ecosystem dynamics, along with quantifying the persistence time of legacy effects on different ecosystem characteristics and processes (Table 4.1).

4.4   Lateral fluxes in landscape mosaics

Lateral fluxes of matter, energy or information in spatially heterogeneous systems have been recognized as key foci within landscape ecology in particular (Risser *et al.* 1984, Wiens *et al.* 1985, Turner *et al.* 1989, Shaver *et al.* 1991) and ecology in general (e.g., Reiners and Driese 2001). Broad conceptual frameworks have considered the conditions under which spatial pattern, or particular aspects of spatial pattern, should influence a lateral flux. For example, Wiens *et al.* (1985) proposed a framework for considering fluxes across boundaries that included the factors determining the location of boundaries between patch types in a landscape mosaic, how boundaries affect ecological processes and the movement of materials over an area, and how imbalances in these transfers in space can affect landscape configuration. Weller *et al.* (1998) explored how and why different riparian buffer configurations would vary in their ability to intercept nutrient fluxes moving from a source ecosystem to an aquatic system. Simulation models ranging from simple representations (e.g., Gardner *et al.* 1989, Turner *et al.* 1989, Gardner *et al.* 1992) to complex, process-based spatial models (e.g., Costanza *et al.* 1990, Sklar and Costanza 1990, Fitz *et al.* 1996) have also been employed to identify the aspects of spatial configuration that could enhance or retard a lateral flux. However, a general understanding of lateral fluxes in landscape mosaics has remained elusive, despite promising conceptual frameworks developed for semi-arid systems (e.g., Tongway and Ludwig 2001).

Many empirical studies have taken a comparative approach using integrative measurements, such as nutrient concentrations in aquatic ecosystems, as indicators of how spatial heterogeneity influences the end result of lateral fluxes (Correll *et al.* 1992, Hunsaker and Levine 1995). Most of these studies focus on nutrients, such as nitrogen or phosphorus, related to eutrophication of surface waters (e.g., Lowrance *et al.* 1984, Peterjohn and Correll 1984, Soranno *et al.* 1996, Jordan *et al.* 1997, Bennett *et al.* 1999). For example, in a recent study of the US Mid-Atlantic region, landscape heterogeneity explained from 65–86 percent of the variation in nitrogen yields to streams (Jones *et al.* 2001). Variation in topography, the amount of impervious surfaces (e.g., pavement), and the extent of agricultural and urban land uses have all been related to the concentration or loading of nutrients in waters. However, the particular aspects of spatial heterogeneity that are significant or the spatial scales over which that influence is most important have varied among studies (Gergel *et al.* 2002). The lack of consistency among the comparative studies may arise, in part, from the absence of mechanistic understanding about how materials actually flow horizontally across heterogeneous landscapes.

The insights to be gained by focusing on the pathways of lateral fluxes are exemplified by studies of nitrogen retention in Sycamore Creek, Arizona focusing on hydrologic flowpaths as functional integrators of spatial heterogeneity in streams (Fisher and Welter 2005). Building upon a long history of research on this desert stream, Fisher and Welter found that nitrogen retention of the whole system could not be predicted simply by summing the rates observed in system components; rather, the lateral transfers through spatially heterogeneous space had to be understood explicitly. In particular, the geometry of different patches, such as sand bars, that influenced nitrogen processing was critical to understanding nitrogen transport and retention.

Understanding surface- and groundwater fluxes among lake chains in northern Wisconsin has demonstrated the importance of lateral fluxes for lakes. A lake's landscape position is described by its hydrologic position within the local to regional flow system and the relative spatial placement of neighboring lakes within a landscape (Webster *et al.* 1996, Kratz *et al.* 1997, Riera *et al.* 2000). Many hydrologic and biological properties of a lake are determined directly by landscape position, which reflects the relative contributions of surface- and groundwater to the lake (Kratz *et al.* 1997, Soranno *et al.* 1999, Riera *et al.* 2000). Yet across large areas (e.g., an entire lake district containing thousands of lakes), surface- and groundwater connections among lakes are not well understood, making it difficult to predict the function of individual lakes that have not been intensively studied or of the integrated land–water mosaic.

Approaches from landscape ecology could contribute to general understanding of the influence of spatial structure on stocks and fluxes across space. For example, measures of composition and configuration could be adapted to the node-and-link structure of systems with lateral fluxes. Spatial models that track the movement of organisms or propagules might be considered for applicability to matter and energy. Furthermore, only a small subset of the lateral transfers of matter, energy, and information across landscape mosaics has been studied. There is a tremendous opportunity to seek a general understanding of lateral transfers in heterogeneous landscapes. We suggest that landscape ecologists extend their frameworks and approaches for the reciprocal interactions between pattern and process to the realm of fluxes of matter, energy, and information. Priorities should focus on understanding the importance of spatial configuration of fluxes, the relative importance of controlling factors, differences between uni- and multidirectional flows, and the role of disturbance (Table 4.1).

4.5 Linking species and ecosystems

Strengthening the ties between species and ecosystems, between population ecology and ecosystem ecology, has been recognized as an important disciplinary bridge within ecology (e.g., Jones and Lawton 1995). Organisms exist in heterogeneous space; they also use, transform, and transport matter and energy. The importance of herbivores in redistributing nutrients across landscapes has been recognized for some time. For example, grazers can enhance mineral availability by increasing nutrient cycling in patches of their waste (McNaughton et al. 1988, Day and Detling 1990, Holland et al. 1992). The cascading influence of herbivores on nutrient cycling through their modification of plant community composition has also been recognized (e.g., McInnes et al. 1992, Pastor et al. 1997). Recent studies have also identified the role of piscivores in transporting nutrients derived from aquatic ecosystems to terrestrial ecosystems through their foraging patterns (e.g., Willson et al. 1998, Helfield and Naiman 2002, Naiman et al. 2002). Considering habitat use and movement patterns of species in a spatial context provides a wealth of opportunities to enhance the linkage between species and ecosystems and again enhance functional understanding of landscape mosaics.

Recent studies have identified the importance for vegetation patterns of spatial heterogeneity in trophic cascades. For example, in the western US, extirpation of wolves in the twentieth century has been linked to increased ungulate population sizes and high rates of herbivory on woody plants such as aspen (*Populus tremuloides*) and willow (*Salix* spp.) (e.g., Romme et al. 1995, Ripple and Larsen 2000, Berger et al. 2001, Beschta 2003). With predator restoration in some North American national parks, numerical or behavioral responses of ungulates to predators may lead to spatial heterogeneity in browsing and possibly the recovery of woody vegetation in some locations on the land-scape (White *et al.* 1998, Ripple *et al.* 2001, National Research Council 2002, Ripple and Beschta 2003). Such trophic cascades, when played out spatially in dynamic landscapes, may have important implications for dynamics of the vegetation mosaic. In tropical forest fragments, predator elimination has also been associated with increased herbivore abundance and a severe reduction in seedlings and saplings of canopy tree species (Terborgh *et al.* 2001).

Large herbivores are known to respond to spatial heterogeneity in the dis-tribution of forage resources, but how important herbivores are in creating those spatial patterns, how their influence may be scale dependent, and how herbivore-induced patterns affect ecosystem processes remain unclear (Augus-tine and Frank 2001). Herbivore-mediated changes in forest composition have been shown to have important implications for patterns of nutrient cycling (Pastor *et al.* 1998, 1999). In Isle Royale National Park, selective browsing by moose (*Alces alces*) altered forest community composition which, in turn, changed nutrient cycling rates in the soil. Augustine and Frank (2001) demon-strated an influence of grazers on the distribution of soil N properties at every spatial scale from individual plants to landscapes. These studies suggest that much may be learned through integrative studies of population dynamics and ecosystem processes.

Taking a landscape perspective in which the linkages between species and ecosystems play out in space offers an unprecedented opportunity to enhance the linkages between these traditionally separate sub-disciplines within ecol-ogy. Populations both respond to and create heterogeneity in their environ-ments; ecosystem processes, similarly, can both influence species' patterns of occurrence and behaviors and also respond to biota. Population/community and ecosystem ecologists have historically asked quite different research ques-tions. We suggest that the landscape ecology may provide the conceptual framework through its emphasis on spatially explicit studies to integrate pop-ulations and ecosystems much more effectively (Table 4.1).

4.6   Concluding comments

The successful integration of ecosystem ecology and landscape ecol-ogy should produce a much more complete understanding of landscape func-tion than has been developed to date. We have identified four areas in which progress is both important and possible: understanding the causes and con-sequences of spatial heterogeneity in ecosystem process rates; the influence of land-use legacies on current ecosystem condition; horizontal flows of mat-ter and energy in landscape mosaics; and the linkage between species and ecosystems.

Achieving this integration will require progress in several areas. First, continuous and categorical conceptualizations of space must be used in much more complimentary ways (Gustafson 1998). Discrete or patch-based representations of spatial heterogeneity dominate in landscape ecology, yet ecosystem ecology is often characterized by continuous variation in pools or fluxes. Second, models and empirical studies both must be brought to bear on questions of how spatially heterogeneous landscapes both create and respond to fluxes of matter, energy, and information. Studies that encompass broad spatial extents remain logistically difficult; while this is stating the obvious, it is important to recognize that studying ecosystem processes in large and heterogenous areas remains a nontrivial challenge. Third, landscape and ecosystem ecologists should collaborate to explore new technologies that may facilitate spatially extensive measurements. Landscape-ecosystem ecologists should be proactive, describing the measurements that are highly desirable but not yet technologically feasible at particular spatial–temporal scales. Fourth, collaborative research should be the rule rather than the exception. Most scientists do not have the training in all aspects of the science required to address the research questions we have identified – e.g., understanding spatial analysis, landscape patterns, and their change through time; knowing all the field and analytical procedures for ecosystem process measurements; spatial statistics; microbial ecology; and modeling. Effective collaborations may be requisite for progress.

Understanding the implications of the dynamic landscape mosaic for ecosystem processes remains a frontier in ecosystem and landscape ecology. The potential benefits of integrating landscape and ecosystem ecology are important for landscape management and ecological restoration. Maintenance of ecosystem services in changing landscapes has been identified as a key priority for resource management from local to global scales (e.g., Daly 1997, Naiman and Turner 2000, Amundson *et al.* 2003, Loreau *et al.* 2003, Schmitz *et al.* 2003). Clearly, achieving this goal requires a much greater functional understanding of landscapes than is currently available. Landscape ecology offers tremendous promise for providing a conceptual framework to understand reciprocal interactions between spatial heterogeneity and ecosystem processes. We challenge landscape ecologists to embrace the functional complexity of ecosystem ecology, and ecosystem ecologists to similarly embrace the spatial complexity of their systems.

Acknowledgments

We thank Jianguo Wu and Richard Hobbs for the invitation to participate in the symposium at the IALE World Congress in Darwin, Australia, and for partial support for travel to that conference. We also acknowledge funding for this work from the National Science Foundation and the Andrew W. Mellon Foundation.

References

Amundson, R., Y. Guo, and P. Gong. 2003. Soil diversity and land use in the United States. *Ecosystems* **6**, 470–82.

Amundson, R.H. and H. Jenny. 1997. On a state factor model of ecosystems. *BioScience* **47**, 536–43.

Augustine, D.J. and D.A. Frank. 2001. Effects of migratory grazers on spatial heterogeneity of soil nitrogen properties in a grassland ecosystem. *Ecology* **82**, 3149–62.

Baskent, E.Z. and G.A. Jordan. 1995. Characterizing spatial structure of forest landscapes. *Canadian Journal of Forest Research* **25**, 1830–49.

Bell, S.S., B.D. Robbins, and S.L. Jensen. 1999. Gap dynamics in a seagrass landscape. *Ecosystems* **2**, 493–504.

Benedetti-Cecchi, L. 2003. The importance of the variance around the mean effect size of ecological processes. *Ecology* **84**, 2335–46.

Bennett, E.M., T. Reed-Andersen, J.N. Houser, J.R. Gabriel, and S.R. Carpenter. 1999. A phosphorus budget for the Lake Mendota watershed. *Ecosystems* **2**, 69–75.

Berger, J., P.B. Stacey, L. Bellis, and M.P. Johnson. 2001. A mammalian predator–prey imbalance: grizzly bear and wolf extinction affect avian neotropical migrants. *Ecological Applications* **11**, 947–60.

Beschta, R.L. 2003. Cottonwoods, elk, and wolves in the Lamar Valley of Yellowstone National Park. *Ecological Applications* **13**, 1295–309.

Burrows, S.N., S.T. Gower, M.K. Clayton, *et al.* 2002. Application of geostatistics to characterize leaf area index (LAI) from flux tower to landscape scales using a cyclic sampling design. *Ecosystems* **5**, 667–79.

Carpenter, S.R., D. Ludwig, and W.A. Brock. 1999. Management of eutrophication for lakes subject to potentially irreversible change. *Ecological Applications* **9**, 751–71.

Chapin III, F.S., P.A. Matson, and H.A. Mooney. 2002. *Principles of Terrestrial Ecosystem Ecology*. New York: Springer-Verlag.

Chapin III, F.S., M.S. Torn, and M. Tateno. 1996. Principles of ecosystem sustainability. *American Naturalist* **148**, 1016–37.

Compton, J.E. and R.D. Boone. 2000. Long-term impacts of agriculture on soil carbon and nitrogen in New England forests. *Ecology* **81**, 2314–30.

Correll, D.L., T.E. Jordan, and D.E. Weller. 1992. Nutrient flux in a landscape: effects of coastal land use and terrestrial community mosaic on nutrient transport to coastal waters. *Estuaries* **15**, 431–42.

Costanza, R., F.H. Sklar, and M.L. White. 1990. Modeling coastal landscape dynamics: process-based dynamic spatial ecosystem simulation can examine long-term natural changes and human impacts. *BioScience* **40**, 91–107.

Currie, W.S. and K.J. Nadelhoffer. 2002. The imprint of land-use history: patterns of carbon and nitrogen in downed woody debris at the Harvard Forest. *Ecosystems* **5**, 446–60.

Daly, G.C. 1997. *Nature's Services*. Washington, DC: Island Press.

Day, T.A. and J.K. Detling. 1990. Grassland patch dynamics and herbivore grazing preference following urine deposition. *Ecology* **71**, 180–8.

Duffy, D.C. and A.J. Meier. 1992. Do Appalachian understories ever recover from clearcutting? *Conservation Biology* **6**, 196–201.

Dupouey, J.L., E. Dambrine, J.D. Laffite, and C. Moares. 2002. Irreversible impact of past land use on forest soils and biodiversity. *Ecology* **83**, 2978–84.

Eberhardt, R. W., D. R. Foster, G. Motzkin, and B. Hall. 2003. Conservation of changing landscapes: vegetation and land-use history of Cape Cod national seashore. *Ecological Applications* **13**, 68–84.

Fan, W., J. C. Randolph, and J. L. Ehman. 1998. Regional estimation of nitrogen mineralization in forest ecosystems using Geographic Information Systems. *Ecological Applications* **8**, 734–47.

Fisher, S. G. and J. R. Welter. 2005. Flowpaths as integrators of heterogeneity in streams and landscapes. Pages 311–328 in G. Lovett, C. Jones, M. G. Turner, and K. C. Weathers (eds.). *Ecosystem Function in Heterogeneous Landscapes*. New York: Springer-Verlag.

Fitz, H. C., E. B. DeBellevue, R. Costanza, *et al.* 1996. Development of a general ecosystem model for a range of scales and ecosystems. *Ecological Modeling* **88**, 263–95.

Fonseca, M. S., and S. S. Bell. 1998. Influence of physical setting on seagrass landscapes near Beaufort, North Carolina, USA. *Marine Ecology-Progress Series* **171**, 109–21.

Foster, D. R. 2002. Insights from historical geography to ecology and conservation: lessons from the New England landscape. *Journal of Biogeography* **29**, 1269–75.

Foster, D. R., F. Swanson, J. Aber, I. Burke, *et al.* 2003. The importance of land-use legacies to ecology and conservation. *BioScience* **53**, 77–88.

Fraterrigo, J. M., M. G. Turner, S. M. Pearson, and P. Dixon. 2005. Effects of past land use on spatial heterogeneity of soil nutrients in Southern Appalachian forests. *Ecological Monographs* **75**, 215–30.

Gardner, R. H., V. H. Dale, R. V. O'Neill, and M. G. Turner. 1992. A percolation model of ecological flows. Pages 259–69 in A. J. Hansen and F. Di Castri (eds.). *Landscape Boundaries: Consequences for Biotic Diversity and Ecological Flow*. New York: Springer-Verlag.

Gardner, R. H., R. V. O'Neill, M. G. Turner, and V. H. Dale. 1989. Quantifying scale-dependent effects with simple percolation models. *Landscape Ecology* **3**, 217–27.

Garrabou, J., E. Ballesteros, and M. Zabala. 2002. Structure and dynamics of northwestern Mediterranean rocky benthic communities along a depth gradient. *Estuarine Coastal and Shelf Science* **55**, 493–508.

Gergel, S. E. and M. G. Turner. 2002. *Learning Landscape Ecology: a Practical Guide to Concepts and Techniques*. New York: Springer-Verlag.

Gergel, S. E., M. G. Turner, J. R. Miller, J. M. Melack, and E. H. Stanley. 2002. Landscape indicators of human impacts to river-floodplain systems. *Aquatic Sciences* **64**, 118–28.

Golley, F. B. 1993. *A History of the Ecosystem Concept in Ecology: More than the Sum of the Parts*. New Haven, Connecticut: Yale University Press.

Goodale, C. L. and J. D. Aber. 2001. The long-term effects of land-use history on nitrogen cycling in northern hardwood forests. *Ecological Applications* **11**, 253–67.

Gosz, J. R. 1986. Biogeochemistry research needs: observations from the ecosystem studies program at the National Science Foundation. *Biogeochemistry* **2**, 101–12.

Groffman, P. M., T. M. Tiedje, D. L. Mokma, and S. Simkins. 1992. Regional-scale analysis of denitrification in north temperate forest soils. *Landscape Ecology* **7**, 45–54.

Groffman, P. M., and C. L. Turner. 1995. Plant productivity and nitrogen gas fluxes in a tallgrass prairie landscape. *Landscape Ecology* **10**, 255–66.

Gustafson, E. J. 1998. Quantifying landscape spatial pattern: what is the state of the art? *Ecosystems* **1**, 143–56.

Hall, B., G. Motzkin, D. R. Foster, M. Syfert, and J. Burk. 2002. Three hundred years of forest and land-use change in Massachusetts, USA. *Journal of Biogeography* **29**, 1319–35.

Hansen, A. J., J. J. Rotella, M. P. V. Kraska, and D. Brown. 2000. Spatial patterns of primary productivity in the Greater Yellowstone Ecosystem. *Landscape Ecology* **15**, 505–22.

Helfield, J. M. and R. J. Naiman. 2002. Salmon and alder as nitrogen sources to riparian forests in a boreal Alaskan watershed. *Oecologia* **133**, 573–82.

Holland, E. A., W. J. Parton, J. K. Detling, and D. L. Coppock. 1992. Physiological responses of plant populations to herbivory and other consequences of ecosystem nutrient flow. *American Naturalist* **140**, 685–706.

Hunsaker, C.T. and D.A. Levine. 1995. Hierarchical approaches to the study of water quality in rivers. *BioScience* **45**, 193–203.

Jenny, H. 1941. *Factors of Soil Formation*. New York: McGraw-Hill.

Jones, C.G. and J.H. Lawton. 1995. *Linking Species and Ecosystems*. New York: Chapman and Hall.

Jones, K.B., A.C. Neale, M.S. Nash, *et al.* 2001. Predicting nutrient and sediment loadings to streams from landscape metrics: a multiple watershed study from the United States Mid-Atlantic Region. *Landscape Ecology* **16**, 301–12.

Jordan, T.E., D.L. Correll, and D.E. Weller. 1997. Relating nutrient discharges from watersheds to land use and streamflow variability. *Water Resources Research* **33**, 2579–90.

Jussy, J.H., W. Koerner, E. Dambrine, J.L. Dupouey, and M. Benoit. 2002. Influence of former agricultural land use on net nitrate production in forest soils. *European Journal of Soil Science* **53**, 367–74.

Karl, D.M. 2002. Nutrient dynamics in the deep blue sea. *Trends in Microbiology* **10**, 410–18.

Kashian, D.M. 2002. *Landscape Variability and Convergence in Forest Structure and Function Following Large Fires in Yellowstone National Park*. Ph.D. Dissertation. Madison: University of Wisconsin.

Kashian, D.M., M.G. Turner, and W.H. Romme. 2005. Variability and convergence in stand structure with forest development on a fire-dominated landscape. *Ecology* **86**, 643–54.

Koerner, W., E. Dambrine, J.L. Dupouey, and M. Benoit. 1999. Delta N–15 of forest soil and understorey vegetation reflect the former agricultural land use. *Oecologia* **121**, 421–5.

Kratz, T.K., K.E. Webster, C.J. Bowser, J.J. Magnuson, and B.J. Benson. 1997. The influence of landscape position on lakes in northern Wisconsin. *Freshwater Biology* **37**, 209–17.

Loreau, M., N. Mouquet, and R.D. Holt. 2003. Meta-ecosystems: a theoretical framework for a spatial ecosystem ecology. *Ecology Letters* **6**, 673–9.

Lovett, G., C. Jones, M.G. Turner, and K.C. Weathers (eds.). 2005. *Ecosystem Function in Heterogeneous Landscapes*. New York: Springer-Verlag.

Lowrance, R., R. Todd, J. Fail, O. Hendrickson, and R. Leonard. 1984. Riparian forests as nutrient filters in agricultural watersheds. *BioScience* **34**, 374–7.

Martin, M.E. and J.D. Aber. 1997. High spectral resolution remote sensing of forest canopy lignin, nitrogen, and ecosystem processes. *Ecological Applications* **7**, 431–43.

McGarigal, K. and B.J. Marks. 1995. *FRAGSTATS. Spatial Analysis Program for Quantifying Landscape Structure*. USDA Forest Service General Technical Report PNW-GTR-351. Portland, OR: US Dept. of Agriculture Forest Service, Pacific Northwest Research Station.

McInnes, P.F., R.J. Naiman, J. Pastor, and Y. Cohen. 1992. Effects of moose browsing on vegetation and litter of the boreal forest, Isle Royale, Michigan, USA. *Ecology* **75**, 478–88.

McNaughton, S.J., R.W. Reuss, and S.W. Seagle. 1988. Large mammals and process dynamics in African ecosystems. *BioScience* **38**, 794–800.

Miller, J.R., M.G. Turner, E.H. Stanley, L.C. Dent, and E.A.H. Smithwick. 2004. Extrapolation: the science of predicting ecological patterns and processes. *BioScience* **54**, 310–20.

Mitchell, C.E., M.G. Turner, and S.M. Pearson. 2002. Effects of historical land use and forest patch size on myrmecochores and ant communities. *Ecological Applications* **12**, 1364–77.

Naiman, R.J. 1996. Water, society and landscape ecology. *Landscape Ecology* **11**, 193–6.

Naiman, R.J., R.E. Bilby, D.E. Schindler, and J.M. Helfield. 2002. Pacific salmon, nutrients, and the dynamics of freshwater and riparian ecosystems. *Ecosystems* **5**, 399–417.

Naiman, R.J. and M.G. Turner. 2000. A future perspective on North America's freshwater ecosystems. *Ecological Applications* **10**, 958–70.

National Research Council. 2002. *Ecological Dynamics on Yellowstone's Northern Range*. Washington, DC: National Academy Press.

Pace, M.L. and P.M. Groffman. 1998. *Successes, Limitations and Frontiers in Ecosystem Science*. New York: Springer-Verlag.

Pastor, J., Y. Cohen, and R. Moen. 1999. Generation of spatial patterns in boreal forest landscapes. *Ecosystems* **2**, 439–50.

Pastor, J., B. Dewey, R. Moen, *et al.* 1998. Spatial patterns in the moose–forest–soil ecosystem on Isle Royale, Michigan, USA. *Ecological Applications* **8**, 411–24.

Pastor, J., R. Moen, and Y. Cohen. 1997. Spatial heterogeneities, carrying capacity, and feedbacks in animal–landscape interactions. *Journal of Mammalogy* **78**, 1040–52.

Pearson, S. M., A. B. Smith, and M. G. Turner. 1998. Forest fragmentation, land use, and cove-forest herbs in the French Broad River Basin. *Castanea* **63**, 382–95.

Peterjohn, W. T. and D. L. Correll. 1984. Nutrient dynamics in an agricultural watershed: observations on the role of a riparian forest. *Ecology* **65**, 1466–75.

Reiners, W. A. and K. L. Driese. 2001. The propagation of ecological influences through heterogeneous environmental space. *BioScience* **51**, 939–50.

Riera, J. L., J. J. Magnuson, T. K. Kratz, and K. E. Webster. 2000. A geomorphic template for the analysis of lake districts applied to the Northern Highland Lake District, Wisconsin, USA. *Freshwater Biology* **43**, 301–18.

Ripple, W. J. and R. L. Beschta. 2003. Wolf reintroduction, predation risk, and cottonwood recovery in Yellowstone National Park. *Forest Ecology and Management* **184**, 299–313.

Ripple, W. J. and E. J. Larsen. 2000. Historic aspen recruitment, elk, and wolves in northern Yellowstone National Park, USA. *Biological Conservation* **95**, 361–70.

Ripple, W. J., E. J. Larsen, R. A. Renkin, and D. W. Smith. 2001. Trophic cascades among wolves, elk, and aspen on Yellowstone National Park's northern range. *Biological Conservation* **102**, 227–34.

Risser, P. G., J. R. Karr, and R. T. T. Forman. 1984. *Landscape Ecology: Directions and Approaches.* Special Publication Number 2. Champaign, IL: Illinois Natural History Survey.

Romme, W. H., M. G. Turner, L. L. Wallace, and J. Walker. 1995. Aspen, elk and fire in northern Yellowstone National Park. *Ecology* **76**, 2097–106.

Running, S. W., R. R. Nemani, D. L. Peterson, *et al.* 1989. Mapping regional forest evapotranspiration and photosynthesis by coupling satellite data with ecosystem simulation. *Ecology* **70**, 1090–101.

Ryszkowski, L., A. Bartoszewicz, and A. Kedziora. 1999. Management of matter fluxes by biogeochemical barriers at the agricultural landscape level. *Landscape Ecology* **14**, 479–92.

Ryszkowski, L. and A. Kedziora. 1993. Energy control of matter fluxes through land–water ecotones in an agricultural landscape. *Hydrobiologia* **251**, 239–48.

Sala, O. E., R. B. Jackson, H. A. Mooney, and R. W. Howarth. 2000. *Methods in Ecosystem Science.* New York: Springer-Verlag.

Schmitz, O. J., E. Post, C. E. Burns, and K. M. Johnston. 2003. Ecosystem responses to global climate change: moving beyond color mapping. *BioScience* **53**, 1199–205.

Serrano, L., J. Penuelas, and S. L. Ustin. 2002. Remote sensing of nitrogen and lignin in Mediterranean vegetation from AVIRIS data: decomposing biochemical from structural signals. *Remote Sensing of Environment* **81**, 355–64.

Shaver, G. R., K. J. Knadelhoffer, and A. E. Giblin. 1991. Biogeochemical diversity and element transport in a heterogeneous landscape, the north slope of Alaska. Pages 105–125 in M. G. Turner and R. H. Gardner (eds.). *Quantitative Methods in Landscape Ecology.* New York: Springer-Verlag.

Sklar, F. H. and R. Costanza. 1990. The development of dynamic spatial models for landscape ecology: a review and prognosis. Pages 239–88 in M. G. Turner and R. H. Gardner (eds.). *Quantitative Methods in Landscape Ecology.* New York: Springer-Verlag.

Soranno, P. A., S. L. Hubler, S. R. Carpenter, and R. C. Lathrop. 1996. Phosphorus loads to surface waters: a simple model to account for spatial pattern of land use. *Ecological Applications* **6**, 865–78.

Soranno, P. A., K. E. Webster, J. L. Riera, *et al.* 1999. Spatial variation among lakes within landscapes: ecological organization along lake chains. *Ecosystems* **2**, 395–410.

Teixido, N., J. Garrabou, and W. E. Arntz. 2002. Spatial pattern quantification of Antarctic benthic communities using landscape indices. *Marine Ecology – Progress Series* **242**, 1–14.

Terborgh, J., L. Lopez, P. Nunez, *et al.* 2001. Ecological meltdown in predator-free forest fragments. *Science* **294**, 1923–6.

Tongway, D.J. and J.A. Ludwig. 2001. Theories on the origins, maintenance, dynamics and functioning of banded landscapes. Pages 20–31 in D. Tongway, C. Valentin, and J. Seghieri (eds.). *Banded Vegetation Patterning in Arid and Semiarid Environments: Ecological Processes and Consequences for Management.* New York: Springer-Verlag.

Turner, M.G. and F.S. Chapin III. 2005. Causes and consequences of spatial heterogeneity in ecosystem function. In G. Lovett, C. Jones, M.G. Turner, and K.C. Weathers (eds.). *Ecosystem Function in Heterogeneous Landscapes.* New York: Springer-Verlag.

Turner, M.G., R.H. Gardner, V.H. Dale, and R.V. O'Neill. 1989. Predicting the spread of disturbance across heterogeneous landscapes. *Oikos* **55**, 121–9.

Turner, M.G., R.H. Gardner, and R.V. O'Neill. 2001. *Landscape Ecology in Theory and Practice.* New York: Springer-Verlag.

Turner, M.G., W.H. Hargrove, R.H. Gardner, and W.H. Romme. 1994. Effects of fire on landscape heterogeneity in Yellowstone National Park, Wyoming. *Journal of Vegetation Science* **5**, 731–42.

Turner, M.G., S.M. Pearson, P. Bolstad, and D.N. Wear. 2003. Effects of land-cover change on spatial pattern of forest communities in the southern Appalachian Mountains (USA). *Landscape Ecology* **18**, 449–64.

Turner, M.G., W.H. Romme, R.H. Gardner, and W.W. Hargrove. 1997. Effects of fire size and pattern on early succession in Yellowstone National Park. *Ecological Monographs* **67**, 411–33.

Turner, M.G., W.H. Romme, D.B. Tinker, D.M. Kashian, and C.M. Litton. 2004. Landscape patterns of sapling density, leaf area, and aboveground net primary production in postfire lodgepole pine forests, Yellowstone National Park (USA). *Ecosystems* **7**, 751–75.

Urban, N.H., S.M. Davis, and N.G. Aumen. 1993. Fluctuations in sawgrass and cattail densities in Everglades-Water-Conservation-Area-2a under varying nutrient, hydrologic and fire regimes. *Aquatic Botany* **46**, 203–23.

van Dokkum, H.P., D.M.E. Slijkerman, L. Rossi, and M.L. Costantini. 2002. Variation in the decomposition of *Phragmites australis* litter in a monomictic lake: the role of gammarids. *Hydrobiologia* **482**, 69–77.

Ward, J.V., F. Malard, and K. Tockner. 2002. Landscape ecology: a framework for integrating pattern and process in river corridors. *Landscape Ecology* **17**, S35–45.

Wear, D.N. and P. Bolstad. 1998. Land-use changes in southern Appalachian landscapes: spatial analysis and forecast evaluation. *Ecosystems* **1**, 575–94.

Webster, K.E., T.K. Kratz, C.J. Bowser, J.J. Magnuson, and W.J. Rose. 1996. The influence of landscape position on lake chemical responses to drought in northern Wisconsin. *Limnology and Oceanography* **41**, 977–84.

Weller, D.E., T.E. Jordan, and D.L. Correll. 1998. Heuristic models for material discharge from landscapes with riparian buffers. *Ecological Applications* **8**, 1156–69.

White, C.A., C.E. Olmsted, and C.E. Kay. 1998. Aspen, elk and fire in the Rocky Mountain national parks of North America. *Wildlife Society Bulletin* **26**, 449–62.

Wiens, J.A., C.S. Crawford, and J.R. Gosz. 1985. Boundary dynamics – a conceptual framework for studying landscape ecosystems. *Oikos* **45**, 421–7.

Willson, M.F., S.M. Gende, and B.H. Marston. 1998. Fishes and the forest. *BioScience* **48**, 455–62.

Wu, J. and R.J. Hobbs. 2002. Key issues and research priorities in landscape ecology: an idiosyncratic synthesis. *Landscape Ecology* **17**, 355–65.

5

Landscape heterogeneity and metapopulation dynamics

5.1 Introduction

Landscape ecologists became interested in how landscape structure affects ecological responses during the mid-1980s (Risser *et al.* 1984). One ecological response of interest to landscape ecologists is population dynamics. In the mid-1980s, metapopulation ecology, the study of habitat spatial structure in population dynamics, had already been in existence for 14 years (Levins 1970). It was therefore natural for landscape ecologists with an interest in population dynamics to take the metapopulation ecology perspective as a starting point in developing a landscape-scale population ecology.

In this chapter I review the original metapopulation model and describe how the spatial structure incorporated in metapopulation models has changed over the past 35 years. I then discuss limitations of the classical metapopulation framework for predicting population dynamics in heterogeneous landscapes, and I argue for continued development of landscape population models.

5.2 Levins' metapopulation model

Levins' metapopulation model is arguably the first model of population dynamics devised for the "study of population processes in a heterogeneous environment" (Levins 1969). This model represents a population existing in T patches (called "sites" by Levins.) The number of these patches that is occupied by the species is N, and the rate of change of occupied patches is

$$\frac{dN}{dt} = mN\left(1 - \frac{N}{T}\right) - EN$$

Key Topics in Landscape Ecology, ed. J. Wu and R. Hobbs.
Published by Cambridge University Press. © Cambridge University Press 2007.

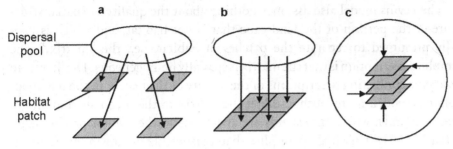

FIGURE 5.1
Three equally valid illustrations of spatial heterogeneity as represented in the
Levins (1969) model. All patches have the same colonization probability. (a) patches
are disjunct; (b) patches are contiguous (one patch); (c) patches are superimposed

where m is the rate of colonization of empty patches (called "migration" by
Levins) and E is the rate of extinction of occupied patches. The model is analyzed to give the number of occupied patches at equilibrium: $T(1 - E/m)$.
Therefore, the number of occupied patches increases with increasing colonization rate and with decreasing extinction rate. Note that the local populations
within patches are not represented in the model; there is no explicit consideration of births, deaths, emigration, or immigration. Instead, patches are simply either occupied or not occupied, and the only two processes considered
are establishment of new populations (colonization) and extinction of existing populations. The original metapopulation model and its derivatives have
therefore also been called "patch occupancy" models (Higgins and Cain 2002,
Ovaskainen and Hanski 2003) or "presence–absence" models (Baguette and
Schtickzelle 2003) or "extinction–colonization" models (Fahrig 2002).

The Levins metapopulation model includes landscape spatial structure in
the sense that the habitat in the model is assumed to be divided into T pieces.
However, since the colonization and extinction rates are the same for all
patches, the model implicitly assumes that all patches are identical in every
sense. Of particular importance is that, since the pieces of habitat are all equally
likely to be colonized, the model does not include spatial relationships among
the habitat pieces. All pieces of habitat are assumed to be in the same location relative to potential colonists. This is sometimes envisioned as a "dispersal pool" in which dispersing individuals mix and then are randomly redis-
tributed among the patches (Fig. 5.1). However, there is in fact nothing in
the model that requires the "patches" to be spatially disjunct from each other
(Fig. 5.1). In some ways the most realistic way of viewing the model is to think
of all the patches as being in the same location (Fig. 5.1). Therefore, although
the Levins model subdivides the environment into T pieces, there is no explicit
spatial structure to the habitat.

The Levins model also assumes nothing about the quality or spatial structure of the portion of the landscape that is not habitat, and which is usually presumed to separate the patches of habitat, i.e., the "matrix." The implicit assumption is that the matrix is spatially homogeneous. The literature presents conflicting descriptions of the quality of the matrix in the metapopulation model. Many authors liken the matrix to the ocean surrounding a set of islands, where the islands are analogous to the habitat patches (e.g., Hanski 1994). This analogy implies that a dispersing organism that lands in the matrix will inevitably "drown;" the matrix is therefore viewed as a hostile environment and dispersal mortality is implicitly high. However, the original metapopulation model and its derivatives do not actually include the processes of emigration from patches or dispersal mortality. The potential effects of these processes on population dynamics are not obtainable from the models. Therefore, it may be more accurate to describe the matrix in metapopulation models as being "sufficiently benign to allow passage of dispersing organisms" (Vandermeer and Carvajal 2001).

5.3   Spatially realistic metapopulation models

In the 35 years since Levins first introduced his model, hundreds of papers have analyzed and expanded on its basic structure. Current metapopulation models represent additional spatial structure beyond that represented in the Levins model, in two important respects: patches are assumed to vary in size and in location relative to each other. Metapopulation models that include patch sizes and relative locations have been termed "spatially realistic" metapopulation models (e.g., Wahlberg *et al.* 1996) and are reviewed in Hanski and Ovaskainen (2003). There are many different possible ways of formulating such models. As a particular example, in the metapopulation model presented by Drechsler *et al.* (2003), colonization of an empty patch i is both: (1) a decreasing function of the distances from i to the occupied patches in the metapopulation, on the assumption that immigration increases with decreasing distance, and (2) an increasing function of the sizes of the occupied patches, on the assumption that larger occupied patches produce more potential colonists (Fig. 5.2). The probability of extinction of occupied patch i is assumed to be both: (1) a decreasing function of the size of i, on the assumption that larger occupied patches contain larger populations, which have lower extinction probabilities, and (2) a decreasing function of the colonization probability of i, on the assumption that colonization probability is correlated to immigration rate, and increasing immigration rate should decrease extinction probability through the rescue effect. Since colonization probability is a function of patch

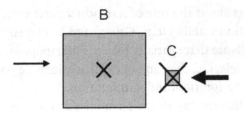

FIGURE 5.2

Illustration of the relationships between patch size and interpatch distance and extinction and colonization probabilities in a "spatially realistic" metapopulation model (Drechsler *et al*. 2003). The size of the × over each patch represents the probability of location extinction when the patch is occupied. The thickness of the arrow entering each patch represents the probability of colonization when the patch is unoccupied. Patch A has the lowest colonization rate because of the large distance to potentially occupied patches such as patch B. Patch A also has the highest extinction probability because it is a small patch (with a presumed small population) that is far from other occupied patches, thus reducing the chance of rescue. Patch B also has a low colonization rate; however, it is higher than the colonization rate of patch A, because patch B is close to patch C which is likely to be occupied (due to its proximity to a large patch, B). Patch B has the lowest extinction probability because it is very large (which implies a large population). Patch C has the highest colonization probability because it is close to a patch that is highly likely to be occupied and to produce many potential colonists because of its large size (patch B). Patch C has an intermediate extinction probability; its extinction probability based on only its patch size would be high, but it should be frequently rescued from extinction by immigration from patch B

size and interpatch distance, extinction probability is therefore also a function of patch size and interpatch distance in this model (Fig. 5.2).

Spatially realistic metapopulation models are typically analyzed for persistence probability of the metapopulation. Persistence probability increases with increasing colonization rates and decreasing extinction probabilities, and with increasing variance in patch sizes and interpatch distances. Increasing variance in patch sizes implies some large patches which have very low probabilities of extinction, and increasing variance in isolation values implies spatial contagion of patches, i.e., groups of patches within the landscape that are close together and therefore have high colonization rates (Ovaskainen *et al.* 2002, Ovaskainen and Hanski 2003). These models can also be used to study questions about the role of individual patches or groups of patches in overall population viability (e.g., Cabeza and Moilanen 2003, Ovaskainen and Hanski 2003). Note that spatially realistic metapopulation models are patch-occupancy models, i.e., they do not explicitly include population processes of births, deaths, emigration, and immigration.

Spatially realistic metapopulation models do include more landscape spatial heterogeneity than the Levins model, because they include variation among patches in patch sizes and relative spatial locations. However, they do not include any consideration of the quality or heterogeneity of the nonhabitat (matrix) portion of the landscape. As in the Levins model, they implicitly assume that the matrix is homogeneous and, since the models do not explic-itly include emigration or dispersal mortality, they implicitly assume that the matrix is benign, i.e., that dispersal mortality is not important to population dynamics.

5.4   PVA tools based on the metapopulation framework

Most applied ecologists who deal with real-world conservation prob-lems encounter metapopulation theory indirectly, through tools for popula-tion viability analysis (PVA) such as "ALEX", "RAMAS-space" and "VORTEX" (reviewed in Lindenmayer *et al.* 1995). These models are different from the clas-sical metapopulation theory discussed above in that the population dynam-ics within patches are included in the models. This is an important distinc-tion; several authors have shown that by collapsing the population processes of births, deaths, emigration, and immigration into the two processes of local col-onizations and extinctions, classical metapopulation models can lead to large errors in prediction (Amaresekare and Nisbet 2001, Higgins and Cain 2002, Léon-Cortés *et al.* 2003).

On the other hand, PVA metapopulation models do adhere to the assump-tions of classical metapopulation theory in their representation of habitat and landscape structure. Specifically, these models are habitat-patch based; each local population is assumed to occur within a habitat patch. Similar to the spa-tially realistic metapopulation models (above), patch sizes and interpatch dis-tances are included in the PVA metapopulation models. Like other metapop-ulation models, the PVA metapopulation tools do not model the movement of organisms in the matrix, and they do not include dispersal mortality. The num-ber of individuals moving from patch A to patch B is a function of the size of the population in patch A and the distance from A to B. There is no accounting for individuals that emigrate from patches but fail to reach other patches, i.e., dispersal mortality. Therefore, the PVA metapopulation models, like classical metapopulation theory, assume a benign, homogeneous matrix.

5.5   Landscape population models

Like spatially realistic metapopulation models and PVA metapopulation tools, landscape population models incorporate the effects of habitat-patch size and relative patch locations on population dynamics. However, landscape population models represent landscape structure in a more complete way than do metapopulation models. While the metapopulation models consider only the distribution of habitat, landscape population models explicitly include the quality and pattern of the matrix. In landscape population models the locations of all individuals (or portions of populations) are simulated on the entire landscape, including in the habitat and in the matrix (e.g., dispersing individuals). Landscape population models can be either general in that they are not meant to simulate a particular landscape or species (e.g., Fahrig 1998, Flather and Bevers 2002), or they may be designed to simulate the response of a particular species to landscape structure (e.g., Topping and Sunderland 1994, Henein *et al.* 1998). Inclusion of the effects of matrix quality and heterogeneity on population dynamics can have important effects on model predictions. In fact, landscape population models can produce very different model predictions than one would get using a metapopulation model, as discussed in the following two sections.

5.5.1 Matrix quality

As discussed above, metapopulation models do not explicitly include the matrix. There is no effect of dispersal mortality on population persistence in metapopulation models. This is an important omission; in reality not all emigrants from a patch will successfully find a new patch; some proportion of them will die. This means that emigration can reduce overall population persistence because it adds to mortality. This mortality will be balanced to some extent by the positive colonization and rescue effects of successful emigrants (i.e., immigrants) on overall population persistence. However, metapopulation models only include the positive effects of immigration on population persistence and neglect the possible negative effects of emigration, i.e., dispersal mortality.

Landscape population models explicitly include emigration, dispersal mortality, and immigration (Fig. 5.3). In these models, population persistence is generally found to be a declining function of emigration rate, except at low emigration rates (Fig. 5.4). At very low emigration rates, an increase in emigration rate causes an increase in persistence, due to rescue and recolonization of local populations. However, at higher emigration rates, further increases in emigration rate result in decreasing persistence probability of the

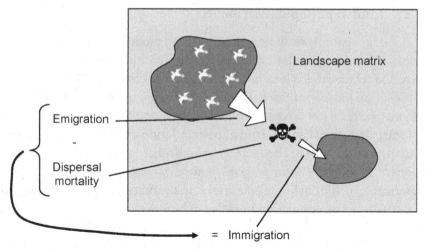

FIGURE 5.3
Illustration of the effect of matrix quality on immigration rate. Immigration is the result of emigration minus dispersal mortality. The lower the matrix quality, the higher the dispersal mortality. The net effect of an increase in emigration rate on overall population persistence in the landscape depends on the balance between the negative effect of dispersal mortality and the positive effects of immigration (i.e., colonization and rescue)

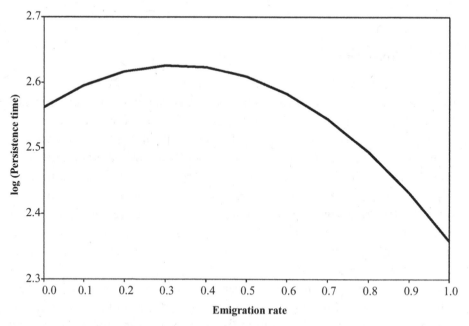

FIGURE 5.4
Relationship between emigration rate and log (population persistence time), based on simulations using a landscape population model (Fahrig 1998). Note that the location of the maximum and the steepness of the curve change with changing model parameters (e.g., reproductive rate, disturbance probability)

population due to the added dispersal mortality (Fahrig 1990, Casagrandi and Gatto 1999).

The negative effect of emigration on population persistence in landscape population models leads to conclusions that are opposite to those normally drawn from a metapopulation analysis. For example, based on their landscape population model of a rare butterfly species, León-Cortés et al. (2003) concluded that "contrary to most metapopulation model predictions, system persistence declined with increasing migration rate, suggesting that the mortality of migrating individuals in fragmented landscapes may pose significant risks to system-wide persistence." Similarly, Gibbs (1998) and Carr and Fahrig (2001) found in empirical studies that more mobile amphibian species are more strongly negatively affected by human-caused landscape changes than are less mobile species. Gibbs (2001) points out that this is in contrast to the "widely held notion" that more dispersive species should perform better in human-modified landscapes. This notion is taken from the metapopulation prediction that higher colonization rates lead to higher population persistence, which has been incorrectly interpreted to mean that increasing dispersal (emigration) always has a positive effect on population persistence. Landscape population models, which explicitly include the matrix, do not lead to this erroneous prediction.

Elsewhere I have also argued that the lack of explicit consideration of the matrix in metapopulation models has led to an over-estimate of the effect of habitat subdivision or fragmentation per se relative to the effect of habitat loss on population persistence (Fahrig 2002). In metapopulation models habitat loss reduces population persistence by an assumed reduction in colonization or immigration rate with decreasing habitat amount. In landscape population models, loss of habitat increases the proportion of the population that spends time in the matrix, where reproduction is not possible and where mortality rate is usually assumed to be higher than in breeding habitat. Habitat loss therefore decreases the overall reproduction rate and increases the overall mortality rate in landscape population models. I have argued that this imposes a constraint on the potential for reduced habitat fragmentation to mitigate effects of habitat loss in landscape population models (Fahrig 2002).

In fact, the critical role of dispersal mortality in population persistence was anticipated over 20 years ago in theoretical studies of the evolution of optimal emigration rate, using evolutionary stable strategy (ESS) models (Comins et al. 1980, Levin et al. 1984, Klinkhamer et al. 1987; Fig. 5.5). Optimal emigration rate was shown to be a decreasing function of dispersal mortality rate. Therefore, as matrix quality decreases (i.e., dispersal mortality rate increases), the optimal emigration rate should decrease. This means that, in the face of human alterations to the landscape that reduce matrix quality, such as addition

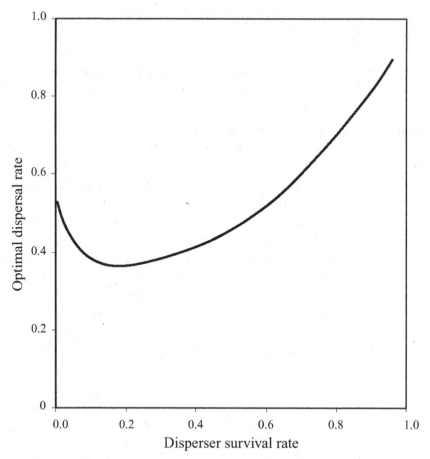

FIGURE 5.5
Optimal dispersal rate as a function of disperser survival rate, based on an evolutionary stable strategy (ESS) analysis of a stochastic spatially implicit patchy population model with random local extinctions (Comins *et al.* 1980). The curve is redrawn from Figure 5.4 in Comins *et al.*, where local extinction probability was 0.1

of roads or pesticide-laden crop fields, species with low emigration rates are more likely to persist than species with high emigration rates, despite the fact that, in the short term, they will have lower rates of colonization of empty patches. The negative effect of emigration is due to an overall increase in mortality rate of the population, which reduces overall population size. This reduction in population size eventually also reduces the probability of recolonization of local extinctions, leading to a downward spiral to extinction (Venier and Fahrig 1996).

5.5.2 Matrix heterogeneity

In addition to overall matrix quality (affecting dispersal mortality), some landscape population models include different types of landcover in the matrix. Since metapopulation models (including spatially realistic metapopulation models and PVA metapopulation tools) do not include the matrix, they also do not include matrix heterogeneity. Does matrix heterogeneity alter metapopulation predictions of population persistence? Theoretical work has not yet directly addressed this question. However, simulation studies (Gustafson and Gardner 1996, Tischendorf *et al.* 2003) have shown that patch size and isolation are good predictors of patch immigration rates only when the matrix is homogeneous. Bender and Fahrig (2005) conducted spatially explicit simulations and a field study of small mammal movement. They found that when the matrix was homogeneous, patch size and isolation accounted for up to 75 percent of the variation in patch immigration rate in the simulation study, and for 61 percent of the variation in patch immigration rate in the field study. However, when the matrix was heterogeneous, the amount of variation explained by patch size and isolation dropped to as little as 35 percent in the simulation study and to 17 percent in the field study. In an empirical study, Walker *et al.* (2003) found that patch sizes and interpatch distances did not adequately predict the distribution of a rock-dwelling rodent; presence of movement barriers in the landscape (rivers) needed to be included for the model to successfully predict distribution. Similarly, Cronin (2003) found that interpatch movement of an insect parasitoid depended on the type of matrix between the two patches. Therefore, metapopulation predictions, which assume that patch colonization rates are a function of interpatch distances, are likely to be poor when the matrix is heterogeneous. A landscape population model is needed in this situation.

5.5.3   When should population models include matrix quality and heterogeneity?

The more spatial structure that is incorporated into population models, the less feasible they are to parameterize for real species. Therefore, it is important to delineate the situations in which information on landscape structure is needed and when it is not needed. Due to the large potential effect of dispersal mortality on population persistence (Fahrig 2001), information on overall matrix quality is almost certainly always necessary.

This leaves the question: when does the heterogeneity of the matrix (independent of its average quality) affect population persistence? There are two situations in which matrix heterogeneity should matter. First, it seems obvious that information on matrix heterogeneity will be needed if the risk of mortality differs among different types of cover in the matrix. For example, predators may favour certain matrix-cover types, which will result in higher risk of mortality for prey when they travel through them than when they travel through

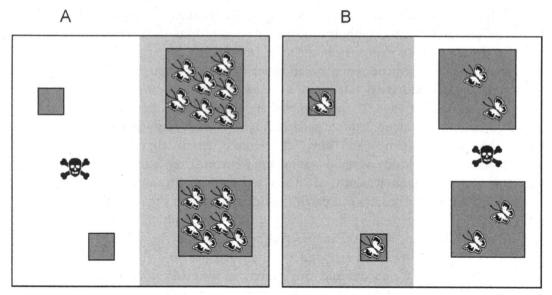

FIGURE 5.6
Illustration of the effect of matrix pattern relative to the habitat pattern,
on population size and persistence. Dispersal mortality is high in the
white matrix cover type and low in the grey matrix cover type. A and B have the
same average matrix quality (averaged over the landscape). However, the overall
population size and persistence probability is higher in A than in B because in
A most dispersing individuals (i.e., those from the large patches) encounter
high-quality matrix, whereas in B most dispersing individuals encounter
low-quality matrix. The latter situation results in a higher overall mortality rate
for the population

other cover types. In this situation the rate of movement between patches will
depend on the cover type(s) that separates them (Fig. 5.6).

The second situation in which matrix heterogeneity will affect population
persistence is when the species shows different affinities for different matrix
cover types. Landscape population models can incorporate this by using different boundary crossing probabilities for different cover types, such that the
probability of a disperser crossing into a benign cover type is high and out of
a benign cover type is low, relative to the same probabilities for a more risky
matrix-cover type. This type of movement behavior was included in the simulation models of Tischendorf *et al.* (2003) and Bender and Fahrig (2005), and led
to a large predicted effect of matrix heterogeneity on interpatch movement. In
contrast, Goodwin and Fahrig (2002) simulated a species that showed different
movement behaviors within different matrix-cover types, but no difference in
mortality among the matrix-cover types and no differential boundary crossing
probabilities among matrix-cover types. In this model, matrix heterogeneity
had very little effect on interpatch movement.

5.6   Conclusions

Levins' model was an important development in population ecology because it represented a transition from a spatially homogeneous to a heterogeneous representation of habitat. Major changes in metapopulation models over the past 35 years include: (1) the development of spatially realistic metapopulation models, which incorporate the effects of habitat patch sizes and relative locations on extinction and colonization rates, and (2) the development of PVA metapopulation tools which incorporate local population dynamics into a realistic metapopulation modeling framework.

These metapopulation models are useful in some situations. However, they are likely to fail in situations where: (1) the landscape matrix is not benign, i.e., dispersal mortality is potentially important to population dynamics, and (2) the matrix is heterogeneous, resulting in low predictability of colonization from habitat structure (i.e., patch sizes and locations) alone. For many organisms, human alterations to the landscape (e.g., urban and agricultural development) increase the probability of dispersal mortality, thus reducing matrix quality. In addition, these alterations create a heterogeneous landscape matrix from the perspective of dispersing organisms. Therefore, the conditions that compromise the predictive ability of metapopulation models are likely to occur for species of conservation concern in human-dominated landscapes. In these situations, further development of landscape population models will be needed to improve predictions of the effects of landscape structure on population dynamics. Application of landscape population models to species conservation problems will require collection of information that is not currently available in the literature for most species, including rates of emigration from habitat, and movement rates and mortality rates in various matrix-cover types.

Acknowledgments

I thank members of the Landscape Ecology Laboratory at Carleton for comments on and discussion of an earlier draft of this chapter. Two anonymous reviewers provided helpful comments. This work was supported by the Natural Sciences and Engineering Research Council of Canada.

References

Amarasekare, P. and R. M. Nisbet. 2001. Spatial heterogeneity, source-sink dynamics, and the local coexistence of competing species. *American Naturalist* **158**, 572–84.

Baguette, M. and N. Schtickzelle. 2003. Local population dynamics are important to the conservation of metapopulations in highly fragmented landscapes. *Journal of Animal Ecology* **40**, 404–12.

Bender, D. J. and L. Fahrig. 2005. Matrix heterogeneity can obscure the relationship between inter-patch movement and patch size and isolation. *Ecology* **86**, 1023–33.

Cabeza, M. and A. Moilanen. 2003. Site-selection algorithms and habitat loss. *Conservation Biology* **17**, 1402–13.

Carr, L. W. and L. Fahrig. 2001. Impact of road traffic on two amphibian species of differing vagility. *Conservation Biology* **15**, 1071–8.

Casagrandi, R. and M. Gatto. 1999. A mesoscale approach to extinction risk in fragmented habitats. *Nature* **400**, 560–2.

Comins, H. N., W. D. Hamilton, and R. M. May. 1980. Evolutionary stable dispersal strategies. *Journal of Theoretical Biology* **82**, 205–30.

Cronin, J. T. 2003. Matrix heterogeneity and host-parasitoid interactions in space. *Ecology* **84**, 1506–16.

Drechsler, M., K. Frank, I. Hanski, R. B. O'Hara, and C. Wissel. 2003. Ranking metapopulation extinction risk: from patterns in data to conservation management decisions. *Ecological Applications* **13**, 990–8.

Fahrig, L. 1990. Interacting effects of disturbance and dispersal on individual selection and population stability. *Comments on Theoretical Biology* **1**, 275–97.

Fahrig, L. 1998. When does fragmentation of breeding habitat affect population survival? *Ecological Modelling* **105**, 273–92.

Fahrig, L. 2001. How much habitat is enough? *Biological Conservation* **100**, 65–74.

Fahrig, L. 2002. Effect of habitat fragmentation on the extinction threshold: a synthesis. *Ecological Applications* **12**, 346–53.

Flather, C. H. and M. Bevers. 2002. Patchy reaction-diffusion and population abundance: the relative importance of habitat amount and arrangement. *American Naturalist* **159**, 40–56.

Gibbs, J. P. 1998. Distribution of woodland amphibians along a forest fragmentation gradient. *Landscape Ecology* **13**, 263–8.

Gibbs, J. P. 2001. Demography versus habitat fragmentation as determinants of genetic variation in wild populations. *Biological Conservation* **100**, 15–20.

Goodwin, B. J. and L. Fahrig. 2002. How does landscape structure influence landscape connectivity? *Oikos* **99**, 552–70.

Gustafson, E. J. and R. H. Gardner. 1996. The effect of landscape heterogeneity on the probability of patch colonization. *Ecology* **77**, 94–107.

Hanski, I. 1994. Patch-occupancy dynamics in fragmented landscapes. *Trends in Ecology and Evolution* **9**, 131–5.

Hanksi, I. and O. Ovaskainen. 2003. Metapopulation theory for fragmented landscapes. *Theoretical Population Biology* **64**, 119–27.

Henein, K., J. Wegner, and G. Merriam. 1998. Population effects of landscape model manipulation on two behaviourally different woodland small mammals. *Oikos* **81**, 168–86.

Higgins, S. I. and M. L. Cain. 2002. Spatially realistic plant metapopulation models and the colonization–competition trade-off. *Journal of Ecology* **90**, 616–26.

Klinkhamer, P. G., T. J. de Jong, J. A. J. Metz, and J. Val. 1987. Life history tactics of annual organisms: the joint effects of dispersal and delayed germination. *Theoretical Population Biology* **32**, 127–56.

Léon-Cortés, J. L., J. J. Lennon, and C. D. Thomas. 2003. Ecological dynamics of extinct species in empty habitat networks. 1. The role of habitat pattern and quantity, stochasticity and dispersal. *Oikos* **102**, 449–64.

Levin, S. A., D. Cohen, and A. Hastings. 1984. Dispersal strategies in patchy environments. *Theoretical Population Biology* **26**, 165–91.

Levins, R. 1969. Some demographic and genetic consequences of environmental heterogeneity for biological control. *Bulletin of the Entomological Society of America* **15**, 237–40.

Levins, R. 1970. Extinction. Pages 77–107 in M. Gerstenhaber (ed.). *Lecture Notes on Mathematics in the Life Sciences 2*. Providence, RI: American Mathematics Society.

Lindenmayer, D. B., M. A. Burgman, H. R. Akçakaya, R. C. Lacy, and H. P. Possingham. 1995. A review of the generic computer programs ALEX, RAMAS-space and VORTEX for modelling the viability of wildlife metapopulations. *Ecological Modelling* **82**, 161–74.

Ovaskainen, O. and I. Hanski. 2003. How much does an individual habitat fragment contribute to metapopulation dynamics and persistence? *Theoretical Population Biology* **64**, 481–95.

Ovaskainen, O., K. Sato, J. Bascompte, and I. Hanski. 2002. Metapopulation models for extinction threshold in spatially correlated landscapes. *Journal of Theoretical Biology* **215**, 95–108.

Risser, P. G., J. R. Karr, and R. T. T. Forman. 1984. *Landscape Ecology: Directions and Approaches*. Special Publication Number 2. Champaign, IL: Illinois Natural History Survey.

Tischendorf, L., D. J. Bender, and L. Fahrig. 2003. Evaluation of patch isolation metrics in mosaic landscapes for specialist vs. generalist dispersers. *Landscape Ecology* **18**, 41–50.

Topping, C. J. and K. D. Sunderland. 1994. A spatial population dynamics model for *Lepthyphantes tenuis* (Araneae: Linyphiidae) with some simulations of the spatial and temporal effects of farming operations and land-use. *Agriculture, Ecosystems and Environment* **48**, 203–17.

Vandermeer, J. and R. Carvajal. 2001. Metapopulation dynamics and the quality of the matrix. *American Naturalist* **158**, 212–20.

Venier, L. and L. Fahrig. 1996. Habitat availability causes the species abundance–distrubution relationship. *Oikos* **76**, 564–70.

Wahlberg, N., A. Moilanen, and I. Hanski. 1996. Predicting the occurrence of endangered species in fragmented landscapes. *Science* **273**, 1536–8.

Walker, R. S., A. J. Novaro, and L. C. Branch. 2003. Effects of patch attributes, barriers, and distance between patches on the distribution of a rock-dwelling rodent (*Lagidium viscacia*). *Landscape Ecology* **18**, 187–94.

ROBERT H. GARDNER, JAMES D. FORESTER, AND
ROY E. PLOTNICK

6

Determining pattern–process relationships in heterogeneous landscapes

6.1   Introduction

Landscapes are now being altered at unprecedented rates (Forman and Alexander 1998), resulting in the loss and fragmentation of critical habitats (Gardner *et al.* 1993), declines in species diversity (Quinn and Harrison 1988, Gu *et al.* 2002), shifts in disturbance regimes (He *et al.* 2002, Timoney 2003), and threats to the sustainability of many ecosystems (Grime 1998, Simberloff 1999). Because the ecological consequences of landscape change are difficult to measure, especially at broad spatial and temporal scales, the quantification of landscape pattern has often been used as an indicator of potential biotic effects (e.g., Iverson *et al.* 1997, Wickham *et al.* 2000). It is hardly surprising, therefore, that the development of methods to measure landscape pattern has become an important endeavor in landscape ecology (see O'Neill *et al.* 1999 for a recent review).

Numerous landscape metrics have been developed and applied over the last 15 years or so, but relatively few studies have been successful in using metrics to establish pattern–process relationships at landscape scales. The first landscape metrics paper (Krummel *et al.* 1987) attempted to do this by presenting the hypothesis that the shape of small forest patches should be affected by human activities while large patches should follow natural topographic boundaries. The analytical results showed that this was the case, but causal relationships were never experimentally confirmed. The prospect of this first study stimulated the rapid development of additional indices (see O'Neill *et al.* 1988, Haines-Young and Chopping 1996), with progress in this arena frequently reviewed (e.g., Gustafson and Parker 1992, Riitters *et al.* 1996, Hargis *et al.* 1998, Fauth *et al.* 2000, He *et al.* 2000, Tischendorf 2001) and computational methods codified (e.g., Gardner 1999, McGarigal *et al.* 2002). However,

Key Topics in Landscape Ecology, ed. J. Wu and R. Hobbs.
Published by Cambridge University Press. © Cambridge University Press 2007.

the confirmation that pattern metrics reflect the changes in ecosystem processes as a result of landscape change has remained an elusive goal (but see Tischendorf 2001, Ludwig *et al.* 2002).

The absence of rigorous guidelines for the application of landscape metrics has raised additional concerns about their usefulness and validity (e.g., Fortin *et al.* 2003, Li and Wu 2004, Wu 2004). These concerns include that the use of multiple metrics to assess pattern change increases the probability of obtaining false positives (type I errors); that metrics with nonmonotonic relationships with pattern change are of limited usefulness and generality; and that the confidence levels associated with many metrics are difficult or impossible to estimate. In spite of these important issues, the results of landscape analyses using questionable measures of pattern are now driving costly programs to mitigate the effects of landscape change. Perhaps the most notable examples are the widespread use of corridors to link critical habitat areas in an effort to reduce extinction risks within fragmented landscapes (Anderson and Danielson 1997, Tewksbury *et al.* 2002).

In spite of the magnitude of efforts to increase the degree of habitat connectivity, the effectiveness of corridors as a species conservation tool remains controversial (Rosenberg *et al.* 1997, Beier and Noss 1998). The success of corridors is directly dependent on their use by target species to disperse to and populate otherwise unavailable patches of suitable habitat. Obtaining sufficient information on dispersal is notoriously difficult, resulting in a long history of using model simulations to define those features of the landscape which most impact the dispersal success (e.g., Murray 1967, Gustafson and Gardner 1996, Tischendorf *et al.* 1998).

The following analysis builds on past simulation methods to identify critical relationships between the landscape structure (corridor pattern) and ecosystem process (reestablishment of resident or invasion of exotic plant species), illustrating how pitfalls in analysis may be avoided. The subsequent results are both simple and robust, allowing a series of issues to be considered. Among these are the conditions under which corridors interact with the native biota to promote the reestablishment of endemics following disturbances or, alternatively, allow the invasion of exotics. The results of a factorial set of simulations are also used to offer recommendations for the general use of metrics to relate pattern and process within heterogeneous landscapes.

6.2   Methods

6.2.1 Model overview

The model used for simulating pattern–process within heterogeneous landscapes is CAPS, an individual based, spatially explicit model of plant competition, establishment, and dispersal (Plotnick and Gardner 2002). Plants simulated by CAPS may differ in life history, relative fecundity, habitat preferences, properties of propagule spread, and ability to compete for space in which to germinate and become established. The landscape is described as a grid with each grid site defined by 1 of n different habitat types. Maps may be randomly generated within CAPS or imported from landscape data. Competitive success which results in establishment and reproduction is simulated via a seed lottery with success randomly determined from the abundance of propagules at that site and the suitability of the local habitat to support that species (see details below).

CAPS is written and compiled in Lahey Fortran 95 to be executed under the Linux operating system. Full details regarding model formulation may be found in Plotnick and Gardner (2002) and program source code and executables may be downloaded from http://www.al.umces.edu/CAPS.htm.

6.2.2 Corridor generation

Corridors were randomly generated in CAPS using a fractal algorithm (Gardner 1999) to produce landscapes with a central, narrow habitat corridor (Fig. 6.1). Two parameters control the character of these landscapes: H sets the spatial dependence (or "roughness") of adjacent points (see Plotnick and Prestegaard 1993 for a full description of the role of H in the generation of fractal maps), and p controls the amount of each habitat type. Resulting landscapes were composed of 512 rows and columns (262 144 lattice sites) with each site representing a 2 m × 2 m habitat site (map size is 1024 by 1024 m or 104.9 ha). Differences in habitat types are considered as an abstract representation of the numerous biotic and abiotic factors (e.g., differences in soil type, moisture, elevation, light availability, etc.) that may negatively or positively affect the germination, survivorship, and reproduction of individual species.

6.2.3 Dispersal

CAPS allows a variety of probability distributions of propagule dispersal to be used in the dispersal kernel. The probability density functions (p.d.f.) for dispersal are uniquely specified for each species, and may be selected from either the uniform, normal, exponential, or Cauchy distributions (Fig. 6.2). The selected p.d.f. produces sets of values representing the probability, $d(i,r)$, of a viable seed released from a parental site, i, reaching a map site that is a distance r away. The p.d.f.s for the uniform, normal and exponential are more fully discussed in Plotnick and Gardner (2002), while the Cauchy distribution (Johnson and Kotz 1970) is a recent addition to the CAPS program. The

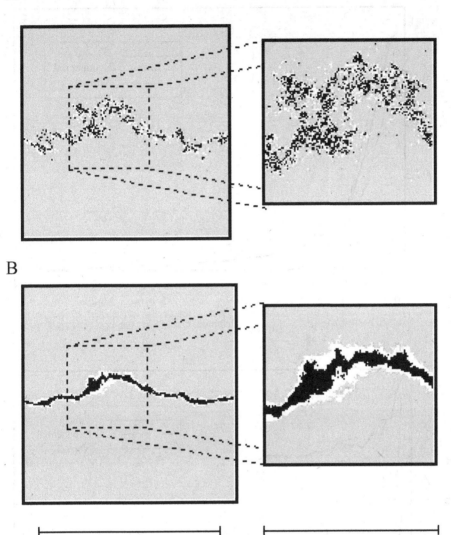

FIGURE 6.1
Example of two fractal maps (with expanded detail) used to generate random corridors. Both gridded maps have 256 rows and columns with p (the fraction of the map that is corridor) of 0.02. A: $H=0.3$; B: $H=0.7$ (see text for explanation of map generation procedure)

location parameter, θ, for the Cauchy was set to 0.0, allowing the Cauchy p.d.f. to be defined by the scale parameter, λ. Thus, the Cauchy p.d.f. $= 1/\pi\lambda/(\lambda^2 r^2)$, where r is the distance over which dispersal occurs.

Simulation efficiency was improved by setting finite limits on dispersal. For each distribution, a maximum dispersal distance, r', was defined (see the evaluation of this constraint discussed below) limiting dispersal to the area

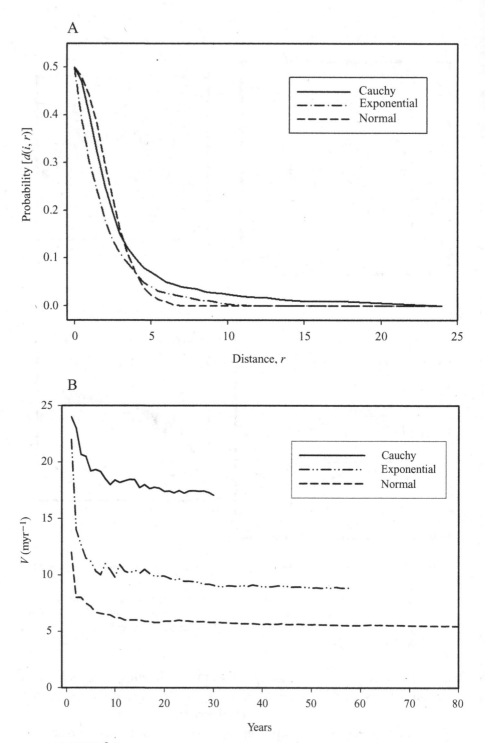

FIGURE 6.2
Comparison of the probability of movement (A) and the rate of movement (B) for the normal, exponential, and Cauchy probability distributions

around the parental site defined by the radius r'. The total number of sites, S, over which dispersal may occur is determined by r': $S = 5$, when $r' = 1$ (i.e., nearest neighbor dispersal), while $S = 441$ when $r' = 12$. All probabilities within the circle defined by the radius r' are summed and normalized so that the dispersal probabilities from a single site sums to 1.0. Our comparison of these discrete formulations with the equivalent continuous distributions used in a discrete time, spatially continuous integrodifference simulation of dispersal (Hart and Gardner 1997) has shown that the two methods are numerically equivalent provided that $d(i,r)$ is small when $r = r'$.

6.2.4 Competition

Competition for establishment, growth, and reproduction is simulated annually at each unoccupied site via a seed lottery (Lavorel *et al.* 1994, Plotnick and Gardner 2002). Sites are unoccupied if plant propagules have yet to reach that location, if the resident individual has died (this occurs yearly for annual plants), or if the habitat type is unsuitable for occupation (i.e., the optima, O_{ij}, for habitat j is 0.0 for all species, i). The seed lottery is performed as a two-step process:

1. The probability of viable seeds of species i landing on an unoccupied site of habitat type j is calculated and weighted by the suitability of that site for seed germination. This probability, T_{ij}, is estimated as:

$$T_{ij} = O_{ij} \left[\sum_{1}^{z} d(i, r) R_i \right]$$

where R_i is the relative fecundity of species i, z is the number of grid sites within the neighborhood defined by the radius r, and d is the dispersal kernel for species i. The suitability of each habitat type j for each species is described by the habitat optima matrix, O_{ij}: if $O_{ij} = 0.0$ then $T_{ij} = 0.0$ because species i cannot survive within habitat type j; if $0.0 < O_{ij} < 1.0$ then the probability of success, T_{ij}, is scaled according to the relative inhospitality of habitat j; if $O_{ij} \geq 1.0$ then the probability of successful establishment, T_{ij}, is proportionately increased for species i.
2. Finally, the values of T_{ij} are normalized by the sum across all species present so that ΣT_{ij} equals 1.0 for the site being considered. These distance and habitat weighted probabilities are then used in the seed lottery to randomly select the species that will establish at that site.

An average, nonspatial measure of competitive ability of each species within each habitat type may be estimated by: $\alpha_{ij} = O_{ij} R_j$, the product of the habitat optima and the relative fecundity. Calculation of α_{ij} ignores spatial effects considered by T_{ij} by assuming that seeds of all species will reach all unoccupied sites. For competition between two species, the ratio of the α_{ij}s provides a mean-field estimation of expected success in seed lottery competition.

6.2.5 Simulating invasion

Simulations of species invasion along corridors were performed either with or without a resident species present. All species were annual plants, with simulated differences in species achieved by varying the relative fecundity, R, the p.d.f used for dispersal, and the range of habitats (niche width or habitat optima, O_{ij}) that may be occupied. Maps were initialized by placing the invader along the east and west edges of the map, while residents, if simulated, were placed on all other map sites. The rate of movement of invaders, v, was measured as the maximum distance moved, c, per time interval, t. Thus, $v = c / t$, where t is number of years simulated. The use of the maximum distance moved has two advantages: (1) extensive calculations of mean squared distances (see Turchin 1998) are unnecessary because the direction of movement is known, (2) c can be estimated for all distributions – even for fat-tailed distributions which may lack finite moments (Clark *et al.* 2001), and (3) this statistic allows asymptotic rates of spread to be unambiguously estimated (see Fig. 6.3 of Clark *et al.* 2001). The initial conditions of all simulations produce a concentration of invaders along the east and west edges of the map, biasing initial estimates of v. Therefore the most reliable averages for v were obtained over the interval $t = 30$–35 years.

Four types of simulations were performed to evaluate model assumptions and determine the relative effects of landscape and biotic attributes on species invasions within habitat corridors. These simulations were: (1) an initial series to evaluate the performance of alternative dispersal kernels within lattice models, (2) a set of simulations with fixed species characteristics but variation in corridor width and degree of continuity, (3) a set of simulations of invasion with competition within homogeneous landscapes, and (4) a factorial set of simulations that systematically varied species characteristics, competition, and landscape structure to determine the relative importance of each set of factors in the invasion process. All simulations were run for 300 time steps (years), or until the invading species reached the center of the map. The initial landscape patterns and the final species distributions were analyzed with RULE (Gardner 1999).

6.2.5.1 *Truncation effects for different dispersal kernels*
The exponential distribution has frequently been used for modeling passive dispersal (Okubo and Levin 1989, Turchin 1998) and has been the foundation

upon which diffusive models are based (i.e., Skellam 1951, Okubo 1980). However, there are compelling arguments for using distributions whose extreme values do not decline exponentially with distance. These distributions, often termed "fat-tailed," are distinguished by the formation of new colonies at the extreme limits of dispersal (Clark *et al.* 1999, Wallinga *et al.* 2002) and result in higher rates of population expansion than exponential or normal distributions. Although seed dispersal data are rarely sufficient to unambiguously identify differences in the tails of the distribution (Wallinga *et al.* 2002), it is important to evaluate the effect of different p.d.f.s (including truncation effects) on the simulation of dispersal in CAPS.

Alternative forms of the dispersal kernel were simulated within a landscape composed of a single habitat type ($p=1.0$), or with maps containing a linear corridor 4 m wide (2 grid sites) that connected the east and west edges of the map. The habitat optima and relative fecundity of the invading species were held constant at $O_{ij} = 2.0, R=2.0$ and resident species were not simulated. The dispersal kernels were either the Cauchy, normal, or exponential distributions with the controlling parameter of each set to 1.0 and the p.d.f.s truncated at either 12 or 24 m (total of six sets of simulations). Invaders were initialized on the edge of the map and the rate of movement, v, measured until the invader reached the center of the map.

6.2.5.2 Structured landscapes

A second set of simulations was performed to evaluate the effect of corridor structure (i.e., variation in corridor width or gaps) on the rate of dispersal of an invading species. Two types of structured maps were created: (1) maps with a single line of habitat from the east to the west edge of the map (i.e., parallel to the directions of invasion) with the width of the lines set at 1, 2 or 4 map sites (2–8 m), and (2) maps with multiple vertical lines of habitat from north to south (i.e., perpendicular to the direction of invasion) with distances between lines of 1, 2, or 4 map sites (2–8 m). Resident species were not simulated, the p.d.f. of the invader was set to the exponential distribution ($r' = 12$ m, $2 = 1$), and O_{ij} constant at 2.0. For each map type a value of R was set at 1.0, 2.0 or 4.0 for a total of 18 sets of simulations. The invading species was initialized on the map edges and the rate of invasion, v, was measured until the center of the map was colonized. An average rate of invasion was estimated from the values of v recorded at $t=30$ to 35.

6.2.5.3 Effect of competition

The effect of competition on invasion was evaluated with a series of simulations within a homogeneous landscape. The relative fecundity of residents and invaders was set at $R=2.0$, maps were of a single habitat type ($p=1.0$), and invaders were initialized on the east and west edges of the maps. The rate of dispersal, v, was measured over a 300-year period, or until the invader reached the center of the map. The competitive abilities, α_{ij}, of each species was varied by altering only the habitat optima, O_{ij}, of the resident species to produce a series of simulations with the ratio of competitive abilities (resident/invader) ranging from 0.25 to 10.0.

6.2.5.4 *Landscape factorial*

The final set of simulations involved the examination of a wide range of landscape structures, life-history characteristics of the invading species, and competitive effects on the invasion process. The fractal map algorithm of RULE (Gardner 1999) was used to generate ten replicates of nine different landscapes types (a total of 90 maps), with landscape types differing in the value of H (either 0.3, 0.5, or 0.7; Fig. 6.1) and the fraction of the map occupied by the corridor, p (either 0.005, 0.01, or 0.02). A buffer habitat surrounding the corridor was held constant at $p = 0.02$ for all simulations. The remaining portion of the map was a third habitat type that could be occupied by a resident species, but would not support the invader (i.e., $O_{13} = 0.0$). These maps were designed to represent a broad range of corridor types from highly diffuse to highly concentrated (Fig. 6.1).

The invading species was an annual plant with fixed dispersal characteristics (p.d.f. = exponential, $2 = 1.0$, $r' = 6.0$), but variable levels of relative fecundity ($R = 1.0, 2.0,$ or 4.0). Because dispersal success is a function of the dispersal kernel and fecundity (Clark and Ji 1995, Higgins *et al.* 1996), varying just fecundity was sufficient for our purposes. But in reality, both the dispersal kernel and fecundity may be expected to vary. The niche width of invaders, defined by the values of O_{ij} for each of the three habitat types, was either narrow ($O_{ij} = 2.0$ for the corridor, but 0.0 elsewhere) or broad ($O_{ij} = 2.0$ for the corridor, 1.0 for the buffer, and 0.0 elsewhere). Invasion with competition was also simulated by initializing the map with a resident species with $R = 2$ and $O_{ij} = 0.0$ for the corridor, but $O_{ij} = 1.0$ elsewhere. The full factorial set of conditions resulted in a total of 810 simulations being performed.

6.3   Results

6.3.1 Truncation effects for different dispersal kernels

The comparison of truncation effects for three p.d.f.s is shown in Table 6.1. Other species characteristics were held constant ($O_{ij} = 2.0$, $R = 2.0$) and dispersal was observed within maps composed of either a single habitat type (A) or a corridor of suitable habitat constrained to a 4 m-wide region (C4).

TABLE 6.1. *Comparison of truncation effects for three probability distribution functions (p.d.f.). The probability, d(i,r), and rate of dispersal, v, are shown when the p.d.f. tails were truncated at the maximum range, r′, of 12 or 24 m*

p.d.f	Map type[a]	r′ = 12		r′ = 24	
		d(i,r)	v	d(i,r)	v
Normal	A	0.22e-6	5.42	0.13e-27	5.47
	C4	0.22e-6	4.08	0.13e-27	4.08
Exponential	A	0.39e-3	7.25	0.94e-6	8.87
	C4	0.39e-3	4.59	0.94e-6	5.07
Cauchy	A	0.23e-2	8.94	0.44e-3	17.39
	C4	0.23e-2	6.10	0.44e-3	8.77

[a] All maps were 512 rows and columns with a single habitat type (A) or a 4 m corridor (C4) connecting the east and west edges of the map.

The effect of truncation of the p.d.f.s was most noticeable for the Cauchy distribution, causing a 50 percent reduction in the rate of invasion when r′ was reduced from 24 to 12 m (Table 6.1). Even though the values of d(i,r) at r′ were always small, the fatter-tails of the Cauchy distribution (Fig. 6.2a) resulted in dispersal rates that were considerably larger than either the normal or exponential distributions (Table 6.1).

Truncation effects are barely noticeable for the normal distribution, but measurable for the exponential distributions in the solid (A) maps (Table 6.1). Reductions in the rate of dispersal due to the confines of the corridors were evident for all p.d.f.s, and considerably larger than those due to truncation of the p.d.f.s. The fatter-tails of the Cauchy distribution resulted in a larger number of propagules lost from the corridors, producing a 50 percent reduction in v when comparing results in the unconstrained maps (A) to those within the 4 m corridor when r′ = 24 m (Table 6.1). The effect of the corridor on dispersal was small but also evident for the normal distribution. The lower probabilities in the tails of the normal distribution (Fig. 6.2a) resulted in no appreciable changes in v due to truncation effects (Table 6.1). The exponential case showed an intermediate corridor effect with v reduced from 8.9 myr^{-1} in the solid map to 5.1 myr^{-1} within corridors when r′ = 24 m (Table 6.1). Because the following simulations considered the variation in a large number of species and landscape parameters, the dispersal kernel of the invading species was fixed to the exponential distribution ($\lambda = 1.0$, r′ = 12 m) for all subsequent simulations.

TABLE 6.2. *The effect of corridor width and dispersal barriers*
on the rate of dispersal, v, for species differing in relative fecundity, R

Dispersal case	$R = 1.0$	$R = 2.0$	$R = 4.0$
Solid maps (A)	7.21	7.31	8.25
Corridor width:			
2 m	1.77	3.28	4.97
4 m	3.36	5.17	6.24
8 m	4.65	5.95	7.21
Dispersal barriers:			
2 m	4.33	5.81	7.01
4 m	3.24	4.69	6.02
8 m	1.91	3.41	5.22

6.3.2 Effect of corridor width and gaps

Calculation of v within the solid map (A) provided an estimate of the maximum possible dispersal rate of 8.25 myr^{-1} when $R=4.0$ (Table 6.2). Differences due to variation in R were small, with an ~13 percent increase in v for a four-fold increase in R. When dispersal was constrained by narrow, 2 m corridors, with the highest level of fecundity ($R=4.0$), v was 40 percent slower then the solid (A) maps. Increasing corridor width resulted in proportionally fewer propagules dispersed into the nonhabitat areas surrounding the corridor and, therefore, an increase in v. The rate of invasion in an 8 m-wide corridor was only ~13 percent less than that of the A map when $R=4.0$ (Table 6.2).

Low fecundity and narrow corridors ($R = 1.0$, width$=2$ m, Table 6.2) had a nonadditive effect on dispersal, with v increased ~62 percent from 2 to 4 m, and ~64 percent when R increased from 1.0 to 4.0. If these effects were independent and additive, then v would be greater than 9.0 myr^{-1}; a rate greater than that observed for A maps (Table 6.2). This lack of additivity is probably due to the changing proportion of seeds falling into adjacent habitat as corridor width increases.

The results of the corridor gap simulations show that even the narrowest (2 m) gaps cause a decrease in dispersal rates (e.g., v reduced from 7.21 to 4.33 myr^{-1} when $R=1.0$, Table 6.2). The rate of invasion was very slow (1.91 myr^{-1}) when $R=1.0$ and the width of the barrier$=8$ m. However, barrier effects were diminished as R increased (Table 6.2). The truncation of the dispersal kernel at 12 m determined the maximum barrier that could be crossed.

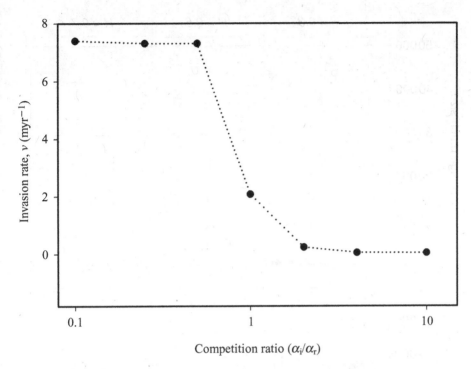

FIGURE 6.3
Changes in the invasion rate, v (myr^{-1}), as a result of increasing levels of competition from a resident species

6.3.3 Effect of competition on invasion

Systematic variation in the habitat optima of the resident species (O_r) allowed the competitive ability of the resident to be varied (e.g., α_i/α_r, Fig. 6.3). The rate of invasion, v, showed a threshold response to competition with a resident: v approached the maximum observed (7.31 myr^{-1}, Table 6.2) when the resident species was a relatively poor competitor ($\alpha_i/\alpha_r = 10.0$); but declined rapidly as the relative competitive ability of the resident increased (Fig. 6.3). When the two species were equal competitors ($\alpha_i/\alpha_r = 1.0$), v was reduced to ~ 2.0 myr^{-1}; and when the resident species was the superior competitor ($\alpha_r/\alpha_i = 0.1$), v was barely measurable (<0.05 myr^{-1}, Fig. 6.3).

6.3.4 Fractal maps factorial

The corridors produced by the fractal map generator vary as a function of p (the fraction of the map occupied by the corridor) and H (the parameter controlling the spatial dependence or "roughness" of corridor habitat, Fig. 6.1).

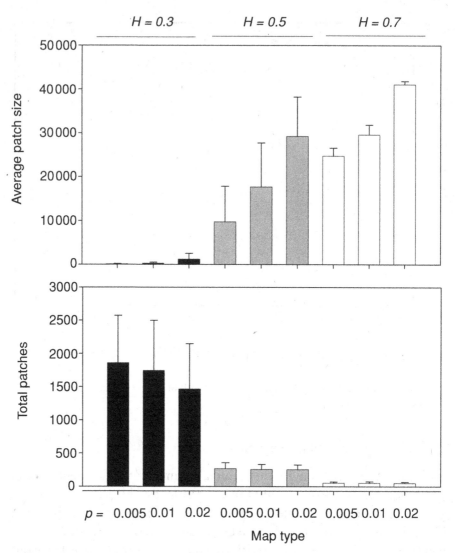

FIGURE 6.4
Changes in the mean number of habitat patches, $M(c)$, and average patch size, $S(c)$, within the randomly generated corridors as a function of p, the fraction of the map occupied by the corridor, and H, the spatial dependence of map habitat. Error bars represent one standard deviation above the mean.

The number and size of corridor patches were analyzed by RULE using the next-nearest neighbor criterion for patch identification (Gardner 1999). The results showed that p and H directly affected the average patch size within the corridors, $S(c)$, with the largest patches occurring when $p=0.02$ and $H=0.7$ (Fig. 6.4). The total number of patches, $M(c)$, was inversely related to patch size, with the greatest degree of fragmentation of corridor habitat occurring when $p=0.005$ and $H=0.3$ (Fig. 6.4).

TABLE 6.3. *The adjusted mean values estimated by SAS general linear model (Table 6.4)*

Independent Variable[a]	Level	v	S
H	0.3	0.73	~0.0
	0.5	2.52	0.52
	0.7	4.08	1.14
C	Without	1.53	0.66
	With	0.92	0.44
N	Wide	3.32	0.89
	Narrow	1.67	0.21
R	1.0	1.42	0.46
	2.0	2.48	0.56
	4.0	3.44	0.63
p	0.005	1.44	0.32
	0.01	2.37	0.47
	0.02	3.53	0.87

[a] H and p control the degree of map fragmentation and amount of habitat within the corridor, respectively; N defines the niche width (narrow or broad) of the invader; C indicates the presence or absence of competition; and R indicates the level of fecundity of the invader.

The adjusted mean rate of invasion, v, and the average patch size occupied by the invading species (analyzed at year 300) are presented in Table 6.3. The greatest effect on v was due to H, the parameter controlling map roughness (Fig. 6.1): v ranged from a low of 0.73 myr^{-1} when $H=0.3$ to 4.08 myr^{-1} at $H=0.7$. Relative fecundity of the invader, R, and the proportion of the map that was corridor habitat, p, also had a positive effect on v (Table 6.3); while the presence of a competitor reduced v by ~40 percent and increasing niche width changed the average invasion rate by ~50 percent.

The average patch size, S, occupied by the invading species (analyzed in year 300) was correlated with the invasion rate ($r=0.765$). Patch sizes of the invader were largest when H, p, and R were highest, the species niche was widest, and competitors were absent (Table 6.3). Simple correlation coefficients – which measure linear relationships uncorrected for other covariates – showed weak overall trends with R and p, but a strong effect due to H ($r=0.65$).

The analysis of variance of v and S (Table 6.4) showed that landscape pattern, species characteristics, and the degree of competition accounted for a high degree of the total variation in these response variables ($R^2=84.3$ and 73.7, respectively, Table 6.4). The variance terms for the ANOVA were estimated by

TABLE 6.4. *The relative sum of squares[a] for the rate of movement (v) and the mean cluster size (S) of the invading plant at year 300 of the simulation*

Source[b]	df	v I	v III	S I	S III
H	2	42.3	6.8	41.6	1.6
p	1	16.3	16.3	10.1	10.1
H p	2	0.9	0.1	6.1	6.1
N	1	4.3	8.9	13.5	14.6
C	1	5.7	5.7	1.6	1.6
R	1	14.8	14.8	0.7	0.7
R^2	–	84.3		73.7	

[a] The relative sum of squares are either the type I (uncorrected) or type III (the partial sum of squares) divided by the corrected total sum of squares and expressed as a percent. Values estimated by SAS (2001) generalized linear model procedure.

[b] H and p control the degree of map fragmentation and amount of habitat within the corridor, respectively; N defines the niche width (narrow or broad) of the invader; C indicates the presence or absence of competition; and R indicates the level of fecundity (1.0 or 2.0) of the invader. R^2 is the total relative sum of squares for the general linear model. All effects are significant at $P < 0.0001$. See text for additional details regarding these simulations.

the SAS generalized linear model (SAS 2001), with the relative uncorrected (type I) sum of squares and relative partial sum of squares (type III) reported in Table 6.4. The difference between these two sums of squares is an indication of the colinearity found when analyzing complex phonemena. H accounts for 42.3 and 41.6 percent of the variance for v and S, respectively, when it is the first variable in the model (type I, Table 6.4), but drops to 6.8 and 1.6 percent (type III, Table 6.4) when corrected for all other effects. The relative importance of the independent variables on v, as measured by the partial sum of squares, was greatest for p and least for C (ranking: $p > R > N > H > C$, with all effects significant). Interactions between H and p were small but significant, while interactions among other variables were not statistically significant.

Similar results were obtained for S, except that the most important parameter affecting the mean cluster size was N, the niche width of the invader. A wider niche (i.e., the ability to germinate and establish in multiple habitat types) allows more habitat to be occupied and, therefore, larger final cluster sizes. The parameter affecting competition, C, map roughness, H, and fecundity of the invader, R, had the least important effect on S. Examination of the correlations among predicted variables showed that the average patch size and the total

TABLE 6.5. *This ANOVA table shows the relationship between four landscape descriptors[a] and the rate of invasion (v) and mean cluster size (S) of invading species after 300 years*

Source[b]	df	v		S	
		I	III	I	III
H	2	42.3	0.5	41.6	0.1
p	1	16.3	2.7	10.1	0.7
H p	2	0.9	<0.1	6.1	1.4
S(c)	1	2.8	2.2	3.7	3.7
M(c)	1	1.7	1.7	<0.1	<0.1
R^2	–		63.9		61.6

[a] The relative sum of squares are either the type I (uncorrected) or type III (the partial sum of squares) divided by the corrected total sum of squares and expressed as a percent. Values estimated by SAS (2001) generalized linear model procedure. All effects > 0.1 are significant at $p < 0.01$.

[b] H and p control the degree of map fragmentation and amount of habitat within the corridor, respectively; $S(c)$ is the average patch size of available corridor habitat; $M(c)$ is the total number of habitat patches within the corridor estimated by RULE (Gardner 1999). R^2 is the total relative sum of squares for the general linear model. All effects are significant at $P < 0.0001$. See text for additional details regarding these simulations.

number of patches occupied by the invader were inversely related ($r = -0.732$) while the v and S were positively correlated ($r = 0.765$).

The role of landscape pattern on invasion was further examined by including other landscape metrics in the analysis of v and S (Table 6.5). Simple correlations showed that $S(c)$ (average patch size) was related to v ($r = 0.765$) and S ($r = 0.767$); while $M(c)$ (total number of patches) were inversely related to v ($r = -0.620$) but directly related to S ($r = 0.441$). The partial sum of squares accounts for the colinearity among landscape metrics, showing a drop among all responses to <4 percent. Although colinearity may always be expected among landscape metrics (Riitters *et al.* 1995), their presence makes the determination of cause–effect relationships problematic. For instance, if we drop H and p from the ANOVA then $S(c)$ and $M(c)$ alone explain 57.2 and 2.2 percent of the variance in v, respectively.

6.4 Conclusions and recommendations

Dispersal is a critical factor that may ensure population persistence at landscape scales even though local extinction events periodically occur (Hanski and Gilpin 1991, Hanski and Simberloff 1997). The extensive changes in landscape pattern that are being experienced in many areas (Malhi *et al.* 2002, Lambin *et al.* 2003, Parmenter *et al.* 2003) often create a matrix that will not support the successful dispersal of many organisms (Goodwin and Fahrig 2002). Under these circumstances corridors of suitable habitat may provide a link through which dispersal may continue to occur (Gonzalez *et al.* 1998). Although there is general agreement that maintaining connectance between populations via dispersal is critical, the effectiveness of corridors must be evaluated for each species and each landscape configuration. The lack of empirical evidence verifying the effectiveness of corridors within many landscapes has raised doubts about their general effectiveness (van Dorp *et al.* 1997, Tikka *et al.* 2001, Tewksbury *et al.* 2002).

Theoretical studies are considerably easier to perform, and support the general conclusion that corridors may effectively link spatially distinct populations (Merriam *et al.* 1990, Danielson and Hubbard 2000). However, in many of these studies, corridors have been regarded as simple links between isolated areas of natural habitat. Reality is much more complex. Corridors are rarely continuous or uniform in size. In addition, the effectiveness of corridors will vary by the specific requirements of the dispersing organism. Consequently the natural history (here represented as fecundity and dispersal kernel) and the ecology (competition and niche width) of the organism need to be explicitly considered within the context of varying landscape pattern. The goal of these simulations is to provide a sufficient understanding of these processes, and their relationships to landscape heterogeneity, to allow landscape metrics to be used to predict the usefulness of corridors for species that may differ greatly in their dispersal ability and life-history attributes.

The simulations reported here illustrate the potential range of effectiveness of corridors for plant dispersal. The simulations are useful because they allow a broad spectrum of factors to be considered, because dispersal is difficult to experimentally manipulate, because the effects of landscape pattern are poorly understood, and because community structure (i.e., the presence of competing species) can dramatically affect invasion success. Exploring these multiple factors and their interactions provided an opportunity to understand corridors and, perhaps of greater importance, to illustrate the linkage of pattern and process within heterogeneous landscapes. Although the particulars of each species, community, and corridor may result in unique outcomes, the broad scope of these simulations provide unusual insight into the relationships and interactions among annual plants and the pattern of corridors through which they may disperse. Four results are of particular note:

1. Pattern and process are scale dependent. The distribution of propagules around the parent plant differs greatly between species, being dependent on the morphology of the seed, its mode of transportation, and the environmental conditions which favor seed establishment (Harper *et al.* 1970). These species-specific characteristics interact with the local pattern of the landscape to produce scale-dependent patterns of dispersal and establishment (Table 6.2, Fig. 6.4). The scale-dependent nature of pattern and process within heterogeneous landscapes is a well-recognized phenomenon (Gardner *et al.* 1992, Keitt *et al.* 1997) and a key issue in landscape ecology (Levin 1992, Wu and Hobbs 2002). As the widths of corridors decline, the edge-to-area ratios increase. The loss of seeds across habitat boundaries can dominate seed dispersal events (e.g., van Dorp *et al.* 1997), with losses increasing as the widths of corridors decline and dispersal distances increase. Breaks in corridors do not prevent dispersal when fecundity is high (Table 6.2) and the tails of the dispersal kernel are long (Fig. 6.3). It is possible that this scale-dependency may allow island stepping stones to be sufficient to link widely separated populations (Keitt *et al.* 1997, Hewitt and Kellman 2002).

2. Effect of fecundity. Higher fecundity results in greater dispersal with a consequent increase in competitive advantage (Clark and Ji 1995, Lavorel and Chesson 1995). These simulation confirm the importance of fecundity, R (Table 6.2), on the velocity of dispersal, v, but also show that R is less important than landscape pattern (H and p) for determining the final pattern of species distribution, S (Table 6.4). Therefore, the consideration of the interaction between pattern and process is necessary to predict invasion and distribution of species within heterogeneous landscapes (Plotnick and Gardner 2002).

3. Competition dramatically reduces dispersal. The rate of dispersal dramatically declines as competition increases (Fig. 6.3). A critical threshold exists near the point where the competing species have an equal probability of establishment (i.e., a competitive ratio of 1.0), with the rate of invasion approaching zero when the resident species is a better competitor. If the form of this function is constant (and that remains to be determined), then it may be possible to reconcile differences in empirical measurements made within recently disturbed regions (i.e., in the absence of resident species) with those made in an established community (i.e., the equilibrium case). The former represents a maximum realized invasion rate, while the latter reflects reductions due to competition. Differences between these two cases are also measures of the importance of disturbance in the invasion process.

4. The importance of pattern. It is hardly surprising that variation in
landscape pattern is an important determinant of the speed and establishment of an invading plant. However, characterizing the attributes
of pattern that affect a particular process remains a challenging problem (Bartlett 1978). Because we generated landscapes with two parameters, H and p, we have the luxury of being able to assess the impact of
these parameters on the process of dispersal and establishment, v and S
($R^2 \sim 0.62$, Table 6.4). Patterns within actual landscapes are generated by
complex environmental and historical events that are difficult to express
with only a few parameters. Although the colinearity among landscape
metrics (see the type III sum of squares, Table 6.5) makes it difficult to
establish cause–effect relationships, only a few parameters are needed to
adequately characterize pattern–process dependencies.

The analysis of dispersal of annual plants within corridors illustrates the
larger issue of using metrics to identify pattern–process relationships in landscape ecology. The history of the development and use of landscape metrics
is a recent one, evolving rapidly since O'Neill *et al.* (1988) presented a catalog
of metrics. Simple metrics are certainly useful as a succinct description of spatial patterns, but pitfalls exist that hinder their use for determining pattern–
process relationships. The well-known issue of colinearity (Riitters *et al.* 1995)
noted above makes it difficult to define the most useful and robust set of pattern
metrics. Without a clearly specified hypothesis the danger of a false positive
(i.e., type II error) may be very large. A second pitfall is the existence of nonlinear relationships, including critical thresholds, where small changes in pattern induce disproportionately large changes in the process being studied. For
instance, the amount of edge within random landscapes is maximized when
$p=0.5$ and is minimized when $p=0.0$ and 1.0. Because actual landscapes show
similar patterns (Gardner *et al.* 1992, Turner *et al.* 2001) metrics based on patch
size and/or edge are descriptively useful but prescriptively dangerous. Many of
these problems stem from the widespread use of gridded integer maps which
make the use of classical spatial statistics for hypothesis testing much more difficult (Turner *et al.* 2001).

More rigor is needed in landscape ecology to avoid the propagation of error
when pattern–process relationships are studied within heterogeneous landscapes. Foremost is the need for identification of algebraic relationships among
similar landscape metrics. The existence of colinearity among metrics is a statistical indictor that metrics may be directly or indirectly developed from more
fundamental variables, such as p. Secondly, indices must be extensively tested
within random and real landscapes to assess their efficiency and usefulness for
description and prediction. Metrics that are not monotonic, or show critical

thresholds of change, should be separated from those whose sensitivity to change is constant across the range of patterns to be investigated.

An equal burden should be placed on the description of the landscapes used within each study. The source of data from which landscape maps were developed, the grain and extent of the study area, and the land-cover classification rules (or reclassification) rules should be thoroughly explained. In addition, the software-dependent neighborhood rules (Gardner 1999) used to identify pattern should be documented. Although these suggestions are familiar, a complete specification of analysis methods is required before results among landscape studies may be compared. Only when these conditions are met can we expect to see significant progress in determining pattern–process relationships within heterogeneous landscapes.

Acknowledgments

This research was supported by NSF grants EAR-9506639 to Plotnick and NSF EAR-9506606 to R.H. Gardner. James Forester was supported by an NSF grant DEB-0078138 to M.G. Turner. Developments to RULE required to analyze corridor patterns were supported by funds provided by the National Park Service through Cooperative Agreement T-3097-02-001.

References

Anderson, G.S. and B.J. Danielson. 1997. The effects of landscape composition and physiognomy on metapopulation size: the role of corridors. *Landscape Ecology* **12**, 261–71.

Bartlett, M.S. 1978. An introduction to the analysis of spatial patterns. *Advances in Applied Probability* **S10**, 1–13.

Beier, P. and R.F. Noss. 1998. Do habitat corridors provide connectivity? *Conservation Biology* **12**, 1241–52.

Clark, J.S. and Y. Ji. 1995. Fecundity and dispersal in plant-populations: implications for structure and diversity. *American Naturalist* **146**, 72–111.

Clark, J.S., M. Lewis and L. Horvath. 2001. Invasion by extremes: population spread with variation in dispersal and reproduction. *American Naturalist* **157**, 537–54.

Clark, J.S., M. Silman, R. Kern, E. Macklin, and J. HilleRisLambers. 1999. Seed dispersal near and far: patterns across temperate and tropical forests. *Ecology* **80**, 1475–94.

Danielson, B.J. and M.W. Hubbard. 2000. The influence of corridors on the movement behavior of individual *Peromyscus polionotus* in experimental landscapes. *Landscape Ecology* **15**, 323–31.

Fauth, P.T., E.J. Gustafson and K.N. Rabenold. 2000. Using landscape metrics to model source habitat for Neotropical migrants in the midwestern US. *Landscape Ecology* **15**, 621–31.

Forman, R.T.T. and L.E. Alexander. 1998. Roads and their major ecological effects. *Annual Review of Ecology and Systematics* **29**, 207–31.

Fortin, M.J., B. Boots, F. Csillag, and T.K. Remmel. 2003. On the role of spatial stochastic models in understanding landscape indices in ecology. *Oikos* **102**, 203–12.

Gardner, R.H. (ed.). 1999. *RULE: Map Generation and a Spatial Analysis Program*. New York: Springer-Verlag.

Gardner, R.H., R.V. O'Neill, and M.G. Turner. 1993. Ecological implications of landscape fragmentation. Pages 208–226 in S.T.A. Pickett and M.J. McDonnell (eds.). *Humans as Components of Ecosystems: The Ecology of Subtle Human Effects and Populated Areas*. New York: Springer-Verlag.

Gardner, R.H., M.G. Turner, R.V. O'Neill, and S. Lavorel. 1992. Simulation of the scale-dependent effects of landscape boundaries on species persistence and dispersal. Pages 76–89 in M.M. Holland, P.G. Risser, and R.J. Naiman (eds.). *The Role of Landscape Boundaries in the Management and Restoration of Changing Environments*. New York: Chapman and Hall.

Gonzalez, A., J.H. Lawton, F.S. Gilbert, T.M. Blackburn, and I. Evans-Freke. 1998. Metapopulation dynamics, abundance and distribution in a microecosystem. *Science* **281**, 2045–7.

Goodwin, B.J. and L. Fahrig. 2002. How does landscape structure influence landscape connectivity? *Oikos* **99**, 552–70.

Grime, J.P. 1998. Benefits of plant diversity to ecosystems: immediate, filter and founder effects. *Journal of Ecology* **86**, 902–10.

Gu, W.D., R. Heikkila, and I. Hanski. 2002. Estimating the consequences of habitat fragmentation on extinction risk in dynamic landscapes. *Landscape Ecology* **17**, 699–710.

Gustafson, E.J. and R.H. Gardner. 1996. The effect of landscape heterogeneity on the probability of patch colonization. *Ecology* **77**, 94–107.

Gustafson, E.J. and G.R. Parker. 1992. Relationships between landcover proportion and indexes of landscape spatial pattern. *Landscape Ecology* **7**, 101–10.

Haines-Young, R. and M. Chopping. 1996. Quantifying landscape structure: a review of landscape indices and their application to forested landscapes. *Progress in Physical Geography* **20**, 418–45.

Hanski, I. and M.E. Gilpin. 1991. Metapopulation dynamics: brief history and conceptual domain. *Biological Journal of the Linnean Society* **42**, 3–16.

Hanski, I. and D. Simberloff. 1997. The metapopulation approach, its history, conceptual domain and application to conservation. Pages 5–26 in I. Hanski and M.E. Gilpin (eds.). *Metapopulation Biology*. New York: Academic Press.

Hargis, C.D., J.A. Bissonette, and J.L. David. 1998. The behavior of landscape metrics commonly used in the study of habitat fragmentation. *Landscape Ecology* **13**, 167–86.

Harper, J.L., P.H. Lovell, and K.G. More. 1970. The shapes and sizes of seeds. *Annual Review of Ecology and Systematics* **1**, 327–56.

Hart, D.R. and R.H. Gardner. 1997. A spatial model for the spread of invading organisms subject to competition. *Journal of Mathematical Biology* **35**, 935–48.

He, H.S., B.E. DeZonia, and D.J. Mladenoff. 2000. An aggregation index (AI) to quantify spatial patterns of landscapes. *Landscape Ecology* **15**, 591–601.

He, H.S., D.J. Mladenoff, and E.J. Gustafson. 2002. Study of landscape change under forest harvesting and climate warming-induced fire disturbance. *Forest Ecology and Management* **155**, 257–70.

Hewitt, N. and M. Kellman. 2002. Tree seed dispersal among forest fragments: II. Dispersal abilities and biogeographical controls. *Journal of Biogeography* **29**, 351–63.

Higgins, S.I., D.M. Richardson, and R.M. Cowling. 1996. Modeling invasive plant spread: the role of plant–environment interactions and model structure. *Ecology* **77**, 2043–54.

Iverson, L.R., M.E. Dale, C.T. Scott, and A. Prasad. 1997. A GIS-derived integrated moisture index to predict forest composition and productivity of Ohio forests (USA). *Landscape Ecology* **12**, 331–48.

Johnson, N.L. and S. Kotz. 1970. *Continuous Univariate Distributions*. New York: Houghton Mifflin Company.

Keitt, T.H., D.L. Urban, and B.T. Milne. 1997. Detecting critical scales in fragmented landscapes. *Conservation Ecology* **1**, 4.

Krummel, J. R., R. H. Gardner, G. Sugihara, R. V. O'Neill, and P. R. Coleman. 1987. Landscape patterns in a disturbed environment. *Oikos* 48, 321–4.

Lambin, E. F., H. J. Geist, and E. Lepers. 2003. Dynamics of land-use and land-cover change in tropical regions. *Annual Review of Environment and Resources* 28, 205–41.

Lavorel, S. and P. Chesson. 1995. How species with different regeneration niches coexist in patchy habitats with local disturbances. *Oikos* 74, 103–14.

Lavorel, S., R. V. O'Neill, and R. H. Gardner. 1994. Spatio-temporal dispersal strategies and annual plant species coexistence in a structured landscape. *Oikos* 71, 75–88.

Levin, S. A. 1992. The problem of pattern and scale in ecology. *Ecology* 73, 1943–67.

Li, H. B. and J. G. Wu. 2004. Use and misuse of landscape indices. *Landscape Ecology* 19, 389–99.

Ludwig, J. A., R. W. Eager, G. N. Bastin, V. H. Chewings, and A. C. Liedloff. 2002. A leakiness index for assessing landscape function using remote sensing. *Landscape Ecology* 17, 157–71.

Malhi, Y., P. Meir, and S. Brown. 2002. Forests, carbon and global climate. *Philosophical Transactions of the Royal Society of London Series A: Mathematical Physical and Engineering Sciences* 360, 1567–91.

McGarigal, K., S. A. Cushman, M. C. Neel, and E. Ene. 2002. *FRAGSTATS: Spatial Pattern Analysis Program for Categorical Maps*. Amherst, MA: University of Massachusetts.

Merriam, G., K. Henein, and K. Stuart-Smith. 1990. Landscape dynamic models. Pages 399–416 in M. G. Turner and R. H. Gardner (eds.). *Quantitative Methods in Landscape Ecology*. New York: Springer-Verlag.

Murray, B. G. 1967. Dispersal in vertebrates. *Ecology* 48, 975–8.

O'Neill, R. V., J. Krummel, R. H. Gardner, *et al.* 1988. Indices of landscape pattern. *Landscape Ecology* 1, 153–62.

O'Neill, R. V., K. H. Riitters, J. D. Wickham, and K. B. Jones. 1999. Landscape pattern metrics and regional assessment. *Ecosystem Health* 5, 225–33.

Okubo, A. 1980. *Diffusion and Ecological Problems: Mathematical Models*. Berlin: Springer-Verlag.

Okubo, A. and S. A. Levin. 1989. A theoretical framework for data analysis of wind dispersal of seeds and pollen. *Ecology* 70, 329–38.

Parmenter, A. W., A. Hansen, R. E. Kennedy, *et al.* 2003. Land use and land cover change in the Greater Yellowstone Ecosystem: 1975–1995. *Ecological Applications* 13, 687–703.

Plotnick, R. E. and R. H. Gardner. 2002. A general model for simulating the effects of landscape heterogeneity and disturbance on community patterns. *Ecological Modelling* 147, 171–97.

Plotnick, R. E. and K. L. Prestegaard (eds.). 1993. *Fractal Analysis of Geologic Time Series*. Englewood Cliffs, NJ: Prentice Hall.

Quinn, J. F. and S. P. Harrison. 1988. Effects of habitat fragmentation and isolation on species richness: evidence from biogeographic patterns. *Oecologia* 75; 132–40.

Riitters, K. H., R. V. O'Neill, C. T. Hunsaker, *et al.* 1995. A factor analysis of landscape pattern and structure metrics. *Landscape Ecology* 10, 23–39.

Riitters, K. H., R. V. O'Neill, J. D. Wickham, and K. B. Jones. 1996. A note on contagion indices for landscape analysis. *Landscape Ecology* 11, 197–202.

Rosenberg, D. K., B. R. Noon, and E. C. Meslow. 1997. Biological corridors: form, function, and efficacy. *Bioscience* 47, 677–87.

SAS. 2001. *The SAS System for Windows*, Version 8.01. Cary, NC: SAS Institute Inc.

Simberloff, D. 1999. The role of science in the preservation of forest biodiversity. *Forest Ecology and Management* 115, 101–11.

Skellam, J. G. 1951. Random dispersal in theoretical populations. *Biometrika* 38, 196–218.

Tewksbury, J. J., D. J. Levey, N. M. Haddad, *et al.* 2002. Corridors affect plants, animals, and their interactions in fragmented landscapes. *Proceedings of the National Academy of Sciences of the United States of America* 99, 12923–6.

Tikka, P. M., H. Hogmander and P. S. Koski. 2001. Road and railway verges serve as dispersal corridors for grassland plants. *Landscape Ecology* 16, 659–66.

Timoney, K. P. 2003. The changing disturbance regime of the boreal forest of the Canadian Prairie Provinces. *Forestry Chronicle* 79, 502–16.

Tischendorf, L. 2001. Can landscape indices predict ecological processes consistently? *Landscape Ecology* 16, 235–54.

Tischendorf, L., U. Irmler, and R. Hingst. 1998. A simulation experiment on the potential of hedgerows as movement corridors for forest carabids. *Ecological Modelling* 106, 107–18.

Turchin, P. 1998. *Quantitative Analysis of Movement: Measuring and Modeling Population Redistribution in Animals and Plants*. Sunderland: Sinauer.

Turner, M. G., R. H. Gardner, and R. V. O'Neill. 2001. *Landscape Ecology in Theory and Practice: Pattern and Process*. New York: Springer-Verlag.

van Dorp, D., P. Schippers, and J. M. van Groenendael. 1997. Migration rates of grassland plants along corridors in fragmented landscapes assessed with a cellular automation model. *Landscape Ecology* 12, 39–50.

Wallinga, J., M. J. Kropff, and L. J. Rew. 2002. Patterns of spread of annual weeds. *Basic and Applied Ecology* 3, 31–8.

Wickham, J. D., R. V. O'Neill, and K. B. Jones. 2000. A geography of ecosystem vulnerability. *Landscape Ecology* 15, 495–504.

Wu, J. 2004. Effects of changing scale on landscape pattern analysis: scaling relations. *Landscape Ecology* 19, 125–38.

Wu, J. and R. Hobbs. 2002. Key issues and research priorities in landscape ecology: an idiosyncratic synthesis. *Landscape Ecology* 17, 355–65.

7

Scale and scaling: a cross-disciplinary perspective

7.1   Introduction

Scale and heterogeneity are two key concepts in landscape ecology which are inherently related. Scale would matter little in a world where entities and relationships remain invariant across space or time, or in a landscape that is spatially or temporally homogeneous (i.e., uniform or random). However, real landscapes are heterogeneous biophysically and socioeconomically, and they must be treated as such for most questions and problems that interest us as scientists or citizens. Spatial heterogeneity – the diversity of entities and their spatial arrangement – is one of the most essential and unifying features of all natural and anthropogenic systems. Landscape heterogeneity is the manifestation of patchiness (discrete patterns) and gradients (continuous variations) that are intertwined across multiple spatial scales. Thus, scale is indispensable for describing and understanding landscape pattern.

It is not surprising, therefore, that scale has become one of the most fundamental concepts in landscape ecology, a field that focuses prominently on spatial heterogeneity and its ecological consequences (Risser *et al.* 1984, Forman and Godron 1986, Forman 1995, Turner *et al.* 2001). In fact, landscape ecology has been widely recognized by biologists, geographers, and even social scientists for its leading role in studying scale issues (McBratney 1998, Marceau 1999, Withers and Meentemeyer 1999, Meadowcroft 2002, Sayre 2005). However, it was not until the 1980s that the notion of scale began to gain its prominence in landscape ecology (and in ecology in general). Also, landscape ecology is not the only discipline that deals with scale and spatial pattern. The goal of geographical research is to describe and explain the spatial patterns of natural and anthropogenic features on the Earth's surface (Harvey 1968), and scale as a

Key Topics in Landscape Ecology, ed. J. Wu and R. Hobbs.
Published by Cambridge University Press. © Cambridge University Press 2007.

geographic variable is "almost as sacred as distance" (Watson 1978). However, geographers have long opted for single-scale studies without adequate justification (Watson 1978, Meentemeyer 1989).

Nevertheless, studies explicitly dealing with spatial scale in both ecology and geography date back to several decades ago. For example, plant community ecologists have used various block-variance methods to investigate multiple-scale patterns of vegetation since the 1950s (Greig-Smith 1952, Dale 1999). On the other hand, insightful discussions on the relationships among pattern, process, and scale were provided by several prominent geographers in the 1960s and the 1970s (e.g., Haggett 1965, Harvey 1968, Miller 1978), when the field of landscape ecology was still unknown to most ecologists around the world. The most notable research on scale issues in the geographic literature, however, is the study of the so-called "modifiable areal unit problem" or the MAUP (Openshaw 1984, Jelinski and Wu 1996). The MAUP is quite relevant to scale issues in landscape ecology and will be further discussed later.

Even those disciplines that do not focus explicitly on spatial patterns have not been able to completely ignore the role of scale. For example, economists have long made the distinction between microeconomics and macroeconomics that correspond to fine-scale and coarse-scale economic patterns and processes, whereas different levels of institutions or organizational hierarchies (e.g., household, community, regional, national, and international) often define the scope and objectives of sub-disciplines and research topics in social and political sciences. In these cases, however, scale has often been treated implicitly or rather coarsely. Although scale is as important in social sciences as in natural sciences, greater progress has been made in ecological and physical sciences in recent decades. To date, efforts to compare and integrate scale issues across disciplines are lacking, but urgently needed (Wu and Hobbs 2002, Sayre 2005).

The main goal of this chapter is to provide an overview of the key concepts, methods, and state-of-the-science of scale and scaling issues that are relevant to landscape ecology. Obviously, this is an extremely ambitious goal because of the enormous scope and complexity of this topic. I shall discuss both the conceptual and technical issues of scale and scaling, and identify major research questions and challenges in scaling across heterogeneous landscapes. Although the principal emphasis is placed upon spatial scale, most of the concepts and methods also apply to temporal scale.

7.2 Concepts of scale and scaling

The terms scale and scaling have acquired a number of connotations from various disciplines. No matter how it is defined, scale generally "implies a certain level of perceived detail" (Miller 1978), which most commonly pertains to time, space, or levels of organization. Scale definitions may be grouped into three classes: dimensions, kinds, and components of scale (Table 7.1). Space, time, and organizational hierarchies represent three primary dimensions of scale, of which space and time are most fundamental. Organizational hierarchies, when nested, generally follow the space–time correspondence principle: higher levels correspond to broader spatial and longer temporal scales, whereas lower levels are associated with finer spatial and shorter temporal scales (Simon 1962, Urban *et al.* 1987, Wu 1999). Within each scale dimension, one can distinguish between different kinds of scale: intrinsic scale, observation scale, experimental scale, analysis/modeling scale, and policy scale (see Table 7.1 for definitions). Except for intrinsic scale, all other types of scale are defined or imposed by the investigator. To quantify variations of a pattern or process across scale, however, one must specify scale components that are operational. Common components of scale include cartographic (map) scale, grain (resolution, support), extent, coverage (sampling density, intensity), and spacing (interval, lag). While cartographic scale remains a fundamentally important concept in mapping science, grain and extent have firmly established themselves as the most frequently used, operational concepts of scale in ecology. Specifically, grain refers to the finest level of spatial or temporal resolution of a pattern or a data set, and extent is the spatial or temporal span of a phenomenon or a study (Allen *et al.* 1984, Turner *et al.* 1989a, Wiens 1989).

The term scaling is sometimes also known as scale transfer or scale transformation (Blöschl and Sivapalan 1995, Bierkens *et al.* 2000). In physical sciences, scaling has traditionally referred to the derivation of power laws, and this narrow definition has been adopted in biology and ecology for decades. In particular, biological allometry involves deriving power-law relationships between the size of organisms and biological processes (Schmidt-Nielsen 1984, Niklas 1994). Some researchers treat ecological scaling simply as the search for power laws in the biological world (e.g., Calder 1983, Brown and West 2000). However, a broader definition of scaling, i.e., the translation of information across scales or organizational levels, has widely been used in ecology, geography, and environmental sciences (Turner *et al.* 1989a, Wiens 1989, King 1991, Rastetter *et al.* 1992, Blöschl and Sivapalan 1995, Marceau 1999, Wu 1999, Bierkens *et al.* 2000). Accordingly, the process of transferring information from finer to broader scales is called scaling up or upscaling, whereas translating information from broader to finer scales is known as scaling down or downscaling. In general, scaling involves changing grain size, extent, or both (Allen *et al.* 1984, Turner *et al.* 1989a, King 1991, Wu 1999). Note that hierarchical levels and scales in time and space are different but closely related concepts. All levels can be characterized in terms of specific spatiotemporal scales, but not all scales represent organizational levels of hierarchical systems. Nevertheless,

TABLE 7.1. *A three-tiered conceptual framework for scale definitions. While all the definitions are useful for different purposes, only scale components are operational in the practice of scaling*

Dimensions of scale	
Time	A fundamental dimension that allows for fast or frequent events to be distinguished from those that are slow or infrequent
Space	A fundamental dimension whereby large and small entities can be distinguished and their configurations can be discerned
Organizational hierarchy	A directional ordering of interacting entities that have distinctive process rates, thus forming different levels. As ecological organizations exist in space and time, levels always correspond to certain spatial and temporal scales
Kinds of scale	
Intrinsic scale	Scale at which a pattern or process actually operates
Observation scale	Scale at which measurements are made or sampling is conducted
Experimental scale	Scale at which an experiment is performed
Analysis/modeling scale	Scale at which an analysis is conducted or a model is constructed
Policy scale	Scale at which policies are intended to be implemented
Components of scale	
Grain	Finest level of spatial or temporal resolution of a pattern or a data set; equivalent or similar to resolution, support, or minimum mapping unit (MMU)
Extent	Spatial or temporal span of a phenomenon or a study; equivalent to the study area or study duration
Coverage	Proportion of the study area or duration actually sampled; also called sampling density or intensity
Spacing	Distance between two neighboring sampling units; also called sampling interval or lag
Cartographic scale	Ratio of map distance to actual distance on the Earth's surface; also called map scale

"a change in scale often necessitates consideration of new levels of organization" (O'Neill and King 1998).

7.3   Scale effects, MAUP, and "ecological fallacy"

Scale-related studies in landscape ecology during the past two decades have focused on three distinctive but intrinsically linked issues: characteristic scales, scale effects, and scaling. In this section I shall discuss the first two, with an emphasis on scale effects. Scaling approaches and methods will be the subject of the next section. In particular, this section makes a deliberate effort to compare and contrast scale effects in ecology with the MAUP and the so-called "ecological fallacy" in geography and the social sciences.

7.3.1   Characteristic scales and scale effects

The characteristic scale of an ecological phenomenon is the spatial and temporal scale on which the phenomenon principally operates and thus can be most appropriately studied. The background assumption of characteristic scales is that many, if not most, patterns and processes each take place on a finite range of scales (or scale domains), and thus different phenomena can be characterized by their distinctive scale domains. A number of empirically constructed space–time diagrams, in which phenomena are plotted against the space and time scales of their occurrences, corroborate this assumption (e.g., Stommel 1963, Clark 1985, Urban *et al.* 1987, Delcourt and Delcourt 1988, Blöschl and Sivapalan 1995). On the other hand, different phenomena may overlap in their scale domains to varying degrees, and this scale overlap can tell us the nature of the relationship between the different processes of interest. For example, processes operating on commensurate scales may interact frequently, whereas processes with disparate rates (e.g., a few orders of magnitude apart) may have no direct effect on each other. From this perspective, identifying the characteristic scales of relevant patterns and processes is a critical first step in designing a successful research project. Hierarchy theory has provided a conceptual framework as well as practical guidelines for the search of characteristic scales, whose detection is often associated with scale breaks (e.g., O'Neill *et al.* 1991, Cullinan *et al.* 1997, Wu 1999, Hay *et al.* 2001, Hall *et al.* 2004).

A phenomenon may not be observed or gauged properly if the scale of observation is not commensurate with the characteristic scale of the phenomenon. While the scale of observation is a choice by the observer, characteristic scales are intrinsic to that being observed. Across a landscape, changing the "lenses" of observation may lead to a series of different patterns, and the same phenomenon may be manifested differently on different scales. These are scale effects, reflective of both the scale multiplicity of landscape structure and artifacts in pattern analyses (Wu 2004, Li and Wu, Chapter 2, this volume). More specifically, scale effects may occur in any statistical analyses or dynamic models that use area-based data when grain size or extent is changed. Although the effects of quadrat size and position on observed vegetation pattern were explicitly investigated in the 1950s by plant ecologists, it was not until the late 1980s that landscape ecologists began to investigate the various effects of changing grain size and extent on landscape pattern analysis and, to a lesser extent, on landscape modeling. Turner *et al.* (1989b) were among the first to systematically study how changing grain size and extent could affect three landscape indices (diversity, dominance, and contagion). Since then, numerous studies have examined scale effects in landscape pattern analysis (Benson and Mackenzie 1995, Wickham and Riitters 1995, Jelinski and Wu 1996, O'Neill *et al.* 1996, Saura 2004, Wu 2004) and spatial modeling (King *et al.* 1991, Wu and Levin 1994, Ciret and Henderson-Sellers 1998, Kersebaum and Wenkel 1998, Jenerette and Wu 2001).

7.3.2 The MAUP

When landscape ecologists were busy "discovering" scale effects with new pattern metrics and remote sensing data in the 1990s, studies of the MAUP-related issues had existed for several decades in geography and the social sciences. The root of the MAUP, as the name suggests, is the use of areal units that are "modifiable" or arbitrary. Area-based data include census data, remote sensing data, and raster-based maps of soil, vegetation, land use, and other themes. The MAUP has two components: the scale effect and the zoning effect (Openshaw 1984, Jelinski and Wu 1996). The scale effect here refers to the variation in the results of statistical analysis caused by spatially aggregating data into fewer and larger areal units (i.e., reducing the spatial resolution or coarse-graining). This is equivalent to the effect of changing grain size in landscape ecology (Turner *et al.* 1989b, Wu 2004). The zoning effect is the variation in the results of statistical analysis due solely to different ways of aggregating areal units to a given scale of analysis (i.e., changing the boundaries and configurations of areal units at a given spatial resolution). In landscape ecology, variability in statistical results due to the aggregation of pixels along different directions is an example of the zoning effect (Jelinski and Wu 1996, Wu 2004).

The phenomenon of arbitrarily defined areal units affecting statistical results was first noticed in electoral geography more than 100 years ago when politicians purposefully manipulated the local boundaries of electoral districts to alter the outcome of an election without changing the individual votes themselves (gerrymandering). As the earliest study of the MAUP, Gehlke and Biehl

(1934) conducted a correlation analysis between male juvenile delinquency and median monthly income from 252 census tracts in Cleveland, USA, and found that the correlation coefficient increased considerably as the areal units were aggregated contiguously. While early studies were sporadic, the resurgence of interest in the MAUP in the 1980s was evident from the flurry of stimulating studies by Openshaw and his associates (e.g., Openshaw and Taylor 1979, Openshaw 1984). Numerous MAUP studies have been published ever since, most of which were concerned with correlation and regression analyses (Arbia 1989, Goodchild and Gopal 1989, Fotheringham and Wong 1991, Wrigley 1994, Amrhein 1995).

After decades of research, however, geographers still have different views on the nature and scope of the MAUP (Goodchild and Gopal 1989, Wrigley 1994, Jelinski and Wu 1996, Marceau 1999). One extreme view regards the MAUP as simply a consequence of using "bad" or improper methods, and thus the solution is to find "scale-independent" or "frame-independent" methods. But most other views recognize that the MAUP is a result of the interactions between the methods and the data used. That is, spatial effects are not just artifacts, and the MAUP can provide useful information on the multiple-scaled patterns embedded in the data (Jelinski and Wu 1996, Marceau 1999, Hay *et al.* 2001).

7.3.3 The "ecological fallacy"

The existence of the MAUP implies that statistical relationships from area-based data may change with the scale of analysis, and thus cross-scale inferences are unwarranted. This point was made clearly and loudly by Robinson (1950) when he introduced the distinction between "an individual correlation" and "an ecological correlation." In individual correlations the variables are descriptive properties of indivisible individuals, whereas in ecological correlations the variables are descriptive properties of groups of individuals. A striking example in Robinson (1950) was the correlation between nativity and illiteracy for the USA in 1930. The analysis using individual-level data produced a positive correlation between foreign birth and illiteracy (i.e., the individual correlation $= 0.118$), supporting the common observation that the native-born generally had a better command of American English. However, the same analysis using the state-level aggregated data indicated that the percent illiterate was negatively correlated with percent foreign-born (the ecological correlation $= -0.619$). Apparently, this aggregate-level result could lead to a wrong inference at the individual level that the foreign-born were more likely to be literate of American English than the native-born. In reality, this aggregate-level correlation was due largely to the fact that most foreign-born lived in states where the native-born were relatively literate (Freedman 2001).

Robinson (1950) concluded that individual and ecological correlations were almost always different in practice because the ecological correlations were usually stronger than the individual correlations. Since then, the phenomenon of improper inferences of individual behavior from an analysis of groups has been known as the "ecological fallacy" (Wrigley *et al.* 1996, King 1997). Note that the word, "ecological," in this case means "of groups" or "of aggregates," not really related to the interrelationship between organisms and their environment. Unfortunately, this connotation of "ecology," a dangerously misleading distortion of the original meaning of the word, has long been used in the social and behavioral sciences, such as "ecological correlations," "ecological regressions," "ecological inferences," and "ecological fallacies" (e.g., Dogan and Rokkan 1969, Poole 1994, Wrigley *et al.* 1996, King 1997, Freedman 2001). Alker (1969) attempted to develop "a typology of ecological fallacies" to include several types of inappropriate inferences from aggregated areal data. In particular, the individualistic fallacy referred to the improper generalization of aggregate-level relationships from individual-level results, a somewhat converse problem of the "ecological fallacy." The rest of the "ecological fallacies" identified by Alker (1969) were related to different kinds of sampling and conceptual errors in statistical inferences. The "ecological" and individualistic fallacies are both cross-level inference fallacies, and really should have been called as such.

Robinson's (1950) study has attracted a great deal of attention particularly because quantitative social and political studies (and thus policies and actions based on such studies) at the time were based primarily on aggregate areal data. It "startled, dismayed, and even infuriated many users" of areal data (Alker 1969), and "sent two shock waves through the social sciences that are still being felt, causing some scholarly pursuits to end and another to begin" (King 1997). Unfortunately, a number of unintended yet misleading consequences have resulted from Robinson's (1950) study. In particular, the notion of "ecological fallacy" has led to several misguided conceptions: individual-level models are always better specified and more accurate than aggregate-level models, aggregate-level relationships are always intended as substitutes of individual-level relationships, and aggregate-level variables have no relevance to causal relationships and mechanistic explanations of individual-level activities (Allardt 1969, Schwartz 1994). In fact, aggregate-level relationships can be quite useful for defining the context, generating potential hypotheses, and identifying the relevance for studying individual-level phenomena. Frequently, aggregate-level variables may not only be constraints on, but also direct causes of, individual-level processes. For example, population-level studies are crucial for identifying important public health problems, and certain risk factors for diseases genuinely operate at the population level (Pearce 2000).

7.3.4 Towards a more comprehensive understanding of scale effects

The numerous studies of the MAUP and cross-level fallacies are evidently relevant to understanding scale issues faced by landscape ecologists as well as other scientists. The literature of the social sciences on these issues is a rich source of information for learning how scale can help elucidate complex processes, identify hierarchical linkages, or create spurious patterns in human landscapes where social, economic, and political forces are dominant drivers. Findings of the effects of MAUP on correlation analysis, regression analysis, and geospatial models (Openshaw 1984, Arbia 1989, Goodchild and Gopal 1989, Fotheringham and Wong 1991, Amrhein 1995) should be relevant for similar types of landscape ecological analyses.

The "ecological fallacy" is a problem of disaggregation (or downscaling) in which inferences about a lower level are made from knowledge of an upper level. Ecologists are frequently faced with such challenges to predict the properties of "trees" using information on the "forest." Developed in the social sciences over the past several decades, the various methods for solving the problem of cross-level fallacies may prove to be useful for solving genuinely ecological problems as well. These methods are collectively known as the "ecological inference" methods, including "ecological regression" (Goodman 1953, Freedman 2001), the neighborhood method (Freedman *et al.* 1991), and the EI method (King 1997). However, analogous to deciphering land cover composition within a pixel of a remote sensing image, inferring the behavior of lower levels from higher-level data is inherently difficult because: (1) aggregate data usually do not contain explicit information on subgroup behavior, and (2) the characteristics of aggregates may be outcomes of nonlinear interactions among subgroups or emergent properties, so that they cannot be simply "decomposed" using reductionist methods. Like other downscaling methods (more in the next section), none of the ecological inference methods can work well in all circumstances. Both impressive progress and thorny problems in cross-level inference research are evident in a series of exchanges between some leading scholars in this area (Freedman *et al.* 1998, Freedman *et al.* 1999, King 1999).

In spite of its relevance to ecology, the term MAUP seemed completely absent in the ecological literature until the mid-1990s when Jelinski and Wu (1996) discussed the implications of the MAUP for landscape ecology. Even today, the enormous literature on the MAUP and cross-level fallacies continues to

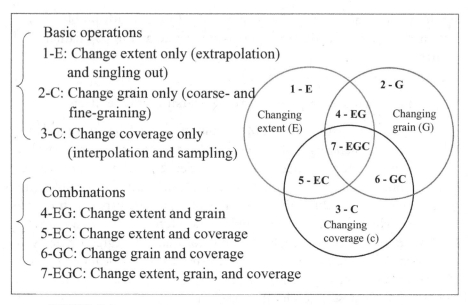

Basic operations
  1-E: Change extent only (extrapolation)
      and singling out)
  2-C: Change grain only (coarse- and
      fine-graining)
  3-C: Change coverage only
      (interpolation and sampling)

Combinations
  4-EG: Change extent and grain
  5-EC: Change extent and coverage
  6-GC: Change grain and coverage
  7-EGC: Change extent, grain, and coverage

FIGURE 7.1
Seven different kinds of scaling operations. In practice, a scaling project may often
involve two or more operations in combination

be ignored by biological and physical scientists, including the most scale-cognizant landscape ecologists. This situation is puzzling because ecological analyses frequently use area-based data and because landscape ecology is actually known for being highly interdisciplinary. In geography and the social sciences, on the other hand, even the recent literature on the MAUP and cross-level fallacies seldom cites any of the scale-related studies in ecology. This is equally disappointing given that geography and landscape ecology both emphasize spatial views and approaches.

7.4   Theory and methods of scaling

Spatial scaling is about translating information across heterogeneous landscapes. The significance and challenges for spatial scaling both reside in the fact that landscape patterns and processes are spatially heterogeneous, non-linearly interactive, and replete with feedbacks and threshold dynamics. Thus, to move from one scale to another in such complex landscapes, one has to either assume away heterogeneity, nonlinearity, and feedbacks, or deal with them explicitly and effectively. In practice, spatial scaling is done through seven basic operations (Bierkens *et al.* 2000). Changing extent, grain size, and coverage are the three basic operations, whereas the other four are different combinations of the three (Fig. 7.1). Strictly speaking, extrapolation is to increase the extent of an observation set, while interpolation is to increase the coverage of a study area. In landscape ecology and geography, scaling frequently involves the change of extent (extrapolation) and grain size or resolution (fine-graining and coarse-graining). The key to spatial scaling is to figure out ways to implement these scaling operations, i.e., scaling approaches and methods.

Scaling methods may be grouped into two general approaches (Blöschl and Sivapalan 1995, Bierkens *et al.* 2000): the similarity-based scaling approach (SBS) and the dynamic model-based scaling approach (MBS). SBS is based on the principles of similarity, and often characterized by power-law scaling functions derived either analytically or empirically. In contrast, MBS transfers information between different scales through changing the input, parameters, and formulation of dynamic models. MBS tends to be more comprehensive, and usually does not lead to simple scaling functions as does SBS.

7.4.1 The SBS approach

Similarity has long been used as the background assumption in a number of scaling methods. Two systems are said to be similar if they share some properties that can be related across the systems by a simple conversion factor (Blöschl and Sivapalan 1995). These similarities can be of different kinds, including geometric, dynamic, and functional similarities. An important and relatively new concept in the SBS approach is self-similarity, which is the key idea in fractal geometry (Mandelbrot 1982, Hastings and Sugihara 1993). Self-similarity refers to the phenomenon that the whole is composed of smaller parts resembling the whole itself and that patterns remain similar at different scales. In the following, I discuss two commonly used SBS methods: similarity analysis and allometric scaling.

7.4.1.1 *Similarity analysis*

Similarity analysis aims to reduce dimensional quantities required for describing a phenomenon based on the known governing equations (Blöschl and Sivapalan 1995). Barenblatt (1996) provided a "general recipe" for similarity analysis that included seven steps: (1) to specify a system of governing variables that are necessary to describe the phenomenon of interest, such that a mathematical relation of the form, $a = f(a_1, \ldots, a_k, b_1, \ldots, b_m)$, can be assumed to hold; (2) to determine the dimensions of variables and select those variables whose dimensions are independent of each other; (3) to represent or transform the relations under study as products of powers (or dimensionless ratios) of variables with independent dimensions; (4) to estimate the numerical values of the similarity parameters (dimensionless variables) using empirical data; (5) to formulate scaling laws (i.e., relationships between nondimensional groups) under the assumption of complete similarity, and test them against empirical data; (6) if the test in step 5 fails, then formulate scaling laws under the assumption of incomplete similarity (or self-similarity) and test them against empirical data (in this case, scaling analysis cannot be completed using dimensional analysis because the power laws are fractal); and (7) to formulate similarity laws with as few similarity parameters as possible.

There have been a number of successful applications of similarity analysis in geophysical sciences. The Monin–Obukhov theory assumes that atmospheric boundary-layer flows can be viewed as being dynamically similar across scales and relates turbulent fluxes to a mean vertical gradient of wind, temperature, and specific humidity (Brutsaert 1982, Wu 1990). Thus, the gradient-diffusion theory (or K-theory), originally developed for molecular-level diffusion processes, has been used to estimate broader-scale turbulent transfer of heat and mass based on the small-eddy concept that treats turbulent transport as a result of local mixing by small eddies. In other words, even though the turbulent diffusivity (about $1\,m^2\,s^{-1}$) is as much as 10^5 times greater than the molecular diffusivity (about 10 to $20\,mm^2\,s^{-1}$), turbulent transfer may still be treated as a dynamically similar process to molecular diffusion. Of course, this is not always a valid treatment. While the K-theory has been successful in modeling turbulent transfer for the boundary layer above vegetation, its success is limited within plant canopies where the small-eddy concept is less appropriate (Brutsaert 1982, Wu 1990).

Similarity analysis has widely been used in soil and hydrological sciences (Blöschl and Sivapalan 1995, Sposito 1998, Bierkens et al. 2000). A well-known example is the derivation of scaling equations for soil-water transport on the basis of the fine-scale similar-media concept known as the Miller–Miller similitude (Miller and Miller 1956, Sposito 1998). Similarity analysis has not been widely used in ecology maybe because the required governing equations for most ecological processes are either nonexistent or analytically intractable.

7.4.1.2 Allometric scaling

Allometry usually refers to the study of the relationship of biological form and process to the size of organisms (LaBarbera 1989, Niklas 1994, Brown and West 2000). The allometric scaling relations are usually based on assumptions of similarity (e.g., geometric similarity and self-similarity), and take the form of a power law: $Y = Y_0 M^b$, where Y is some variable representing a pattern or process of interest, Y_0 is a normalization (or scaling) constant, M is some size-related variable (e.g., body mass), and b is the scaling exponent. There are two ways of obtaining allometric scaling relations: the analytical and empirical approaches. This dichotomy may be generalized for all SBS methods in physical and biological sciences.

The analytical approach derives scaling relations from the existing theory of similarity using techniques such as dimensional analysis, and thus has the ability to explain and predict cross-scale relationships. However, these analytically derived scaling relations must be tested against empirical observations for their validity. The empirical approach is descriptive and inductive, and usually employs two kinds of regression analysis. Ordinary least squares regression (OLS) can be used when the purpose of a study is only to predict one variable based on the other, or to find out if the relationship is statistically significant (in this case simple correlation analysis can also be used). However, if the purpose is to determine the exact value of the scaling exponent (i.e., the slope of the regression line in a log–log plot), OLS regression is generally inadequate especially when the coefficient of correlation is small (Niklas 1994). In this case, reduced major axis (RMA) regression is more appropriate (LaBarbera 1989, Niklas 1994) because it treats the two variables in the allometric equation in the same way (i.e., no "independent" variables in the regression equation and both variables have an error term).

Brown *et al.* (2002) discussed three classes of power laws that describe a variety of biological and ecological phenomena. Power laws of the first class have a rather limited range of variation in the scaling constant (Y_0) and the scaling exponent (b), and are mostly quarter-power laws (e.g., animal metabolic rates, developmental time, life span, maximum rate of population growth, and other organism-level allometric relations). The second class has a wide range of values of Y_0 and b (e.g., population densities of different species). For the third class of power laws (e.g., species-area relationship, species-time relationship, and species-abundance distribution), not only are the scaling parameters not well constrained, but also the power laws themselves do not hold up over many orders of magnitude. Brown *et al.* (2002) asserted that the first class "apparently reflects the fractal-like designs of resources distribution networks," whereas the third class "may not represent examples of self-similar behavior over a wide range of scales." In the past decade, there has been a resurgence of interest in biological allometry which has generated much excitement and controversy (e.g., Dodds *et al.* 2001, Bokma 2004, Brown *et al.* 2004, Cyr and Walker 2004, Kozlowski and Konarzewsk 2004). More than 15 years ago, LaBarbera (1989) commented, "Whether a power law reflects a basic biological truth, the underlying structure of the universe we are embedded in, or whether it is simply fairly robust at approximating a variety of data relations is yet to be determined." Alas, this statement still seems to hold true today.

Biological allometry is not always relevant to spatial scaling unless spatial scale is incorporated into the allometric equation. Schneider (2001, 2002) provided a number of examples of spatial allometry for lake ecosystems and aquatic mesocosms in terms of the geometric attributes of the systems (e.g., the volume, area, perimeter, and depth of lakes or mesocosms) and biological properties (e.g., fish catch and primary production). Landscape-scale studies using this approach have been increasing in recent years. For example, Hood (2002) identified several allometric scaling relations between slough attributes (e.g., area, outlet width, perimeter, and length) for rivers, and showed that detrital insect flotsam density was also allometrically related to slough perimeter. Similarly, Belyea and Lancaster (2002) found that the area, depth, width, and length of peatland bog pools were allometrically related. It is tempting to jump from empirically derived power-law relations to ecological explanations of underlying mechanisms by invoking the theory of self-similarity and self-organization. But this is unwarranted, be it in vogue. Nonetheless, spatial allometry provides a general method to summarize and extrapolate observed patterns over a range of scales, and to suggest underlying processes (Wu 2004).

7.4.2 The MBS approach

Unlike the SBS approach in which similarity goes both ways, MBS methods for upscaling versus downscaling differ in terms of both general perspectives and detailed procedural steps (Fig. 7.2). Thus, they are discussed separately here although both may be used interactively in a given scaling project.

7.4.2.1 Upscaling methods

Upscaling with dynamic models typically consists of two major steps: characterizing heterogeneity, and aggregating information by scaling up local (or patch-level) models (Fig. 7.2). Characterizing heterogeneity usually involves the classification and quantification of spatial patterns, which is a way of simplifying the complexity of scaling by partitioning the heterogeneous landscape into a limited number of relatively homogeneous patches. The second step is to aggregate information from the finer to the broader (target) scale through manipulating the input and parameters or altering the formulation of the local-scale model. Depending on the scaling context, this process may correspond to one of two basic scaling operations: coarse-graining (increasing grain size) or extrapolation (increasing extent). A number of upscaling methods have been developed in geophysical and biological sciences during the past decades. King (1991) presented four methods: extrapolation by lumping (EL), direct extrapolation (DE), extrapolation by expected value (EEV), and explicit integration (EI). This list can be expanded to include additional methods, such as extrapolation by effective parameters (EEP), spatially interactive modeling (SIM), and the scaling ladder method (SL). Each method is described below.

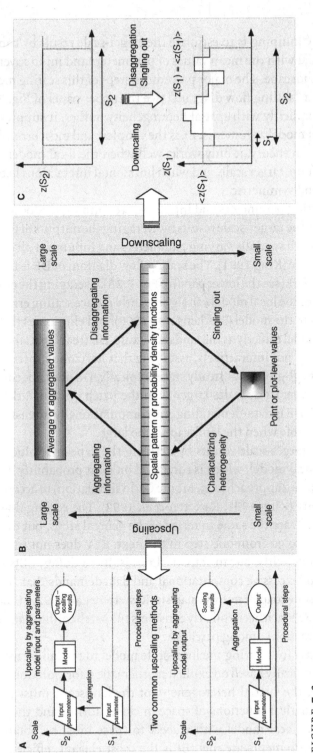

FIGURE 7.2

Illustration of general procedures in spatial scaling using dynamic models. (A) Flow diagrams for upscaling by aggregating model input and parameters (upper panel), and by aggregating model output (lower panel; adapted from Bierkens *et al.* 2000); (B) A two-step scheme for both upscaling and downscaling (modified from Blöschl and Sivapalan 1995); and (C) A schematic representation of the essence of downscaling involving disaggregation and singling out (adapted from Bierkens *et al.* 2000)

Extrapolation by lumping is to estimate the target-scale result by running the local-scale model with the mean values of parameters and inputs averaged across the entire landscape. The major procedural steps of this scaling method can be depicted in a "scaling flow diagram" (see the upper panel of Fig. 7.2A). EL does not deal explicitly with spatial heterogeneity; rather, it suppresses it in average values of model arguments. EL is the simplest and most error-prone upscaling method. In theory, it only works well when the local model is linear and still valid at the target scale, and when horizontal interactions between patches are weak and symmetric.

Instead of averaging parameters and inputs before running the local model as in EL, DE obtains the target-scale results by averaging the outputs of the local model that is run, with spatially varying parameters and inputs, for all patches of the entire landscape (King 1991). The scaling flow diagram of DE is in sharp contrast with that of EL (see the lower panel of Fig. 7.2A). Averaging the outputs rather than inputs of the local model can significantly reduce scaling errors due to the nonlinearity in the model (Bierkens *et al*. 2000), and eliminates the need to apply the local model directly at the landscape scale. DE treats spatial heterogeneity explicitly but not interactively, assuming that horizontal interactions and feedbacks are negligible or at steady state. Typically, DE does not consider any processes that operate at scales larger than the patch on which the local model is developed. DE is data-demanding and computationally intensive, and thus may not be feasible when the landscape is too large.

EEV obtains the target-scale results by deriving the expected value of the outputs from the local model, which is run based on joint probability density functions or a sampling approach (e.g., Monte Carlo simulation) to account for spatial heterogeneity (King 1991, Rastetter *et al*. 1992). The scaling flow diagrams of EEV and DE are the same in terms of the general steps, but differ in the specifics of how to go from one step to the next. EEV does not treat spatial heterogeneity explicitly, but in statistical terms. By so doing, EEV alleviates the problems of excessive computational and data demands that DE may suffer, and is amenable to uncertainty analysis (Rastetter et al. 1992, Li and Wu 2006). As with DE, EEV neither explicitly considers the patch configuration nor feedbacks and interactions among patches.

EI refers to directly integrating the local-scale model to the landscape scale analytically or numerically based on explicit mathematical formulations (King 1991). In this case, the spatial heterogeneity of the landscape must be represented as mathematical functions of space in closed forms, and the indefinite integral of the local model with respect to space must be obtainable. When all of its requirements are met, EI is the most elegant, efficient, and accurate upscaling method. Unfortunately, this is rarely the case with real landscapes.

Similar to EL, EEP also assumes that the local model applies to the target scale, but uses "effective" or "representative" parameters, instead of spatial averages, to produce the target-scale estimates (Blöschl and Sivapalan 1995, Bierkens *et al.* 2000). Because both methods run the local-scale model with landscape-scale input and parameters, EEP and EL share the same scaling flow diagram (Fig. 7.2A). EEP has been widely used in soil physics, hydrology, and micrometeorology, and finding effective parameters can be quite difficult for nonlinear models (Blöschl and Sivapalan 1995, Bierkens *et al.* 2000).

When horizontal or lateral interactions must be considered explicitly (e.g., metapopulation processes, disturbance spread, and land–water interactions), spatially interactive modeling seems to be the only option (Judson 1994, Wu and Levin 1997, Tenhunen and Kabat 1999, Rastetter *et al.* 2003, Peters *et al.* 2004). SIM is able to incorporate feedbacks, time delays, and new features on larger scales. Spatially interactive models include variables, parameters, and input at multiple scales. Thus, the scaling flow diagram of SIM would be different from those in Fig. 7.2A; rather it needs to reflect the multi-scaled nature of the models themselves (e.g., Fig. 2 in Wu and Levin 1997). Such models can easily become ecologically too complex and computationally overwhelming (Levin *et al.* 1997, Levin and Pacala 1997). This is particularly true when the number of scales becomes more than just a few. In this case, a hierarchical scaling scheme is useful to simplify complexity and reduce aggregation errors.

All the upscaling methods discussed above typically are of "short-range" because the assumptions behind them are less likely to be satisfied over a broad range of scales and because they become technically less feasible when multiple scale breaks (or thresholds) are encountered. In these cases, the scaling ladder method may be used (Wu 1999). SL is based on the hierarchical patch dynamics (HPD) paradigm, which integrates hierarchy theory and patch dynamics (Wu and Loucks 1995). The basic idea is to establish a spatial patch hierarchy consisting of a series of nested scale domains, and then use it as a scaling ladder to move information between two adjacent scales one step a time (Wu 1999, Wu and David 2002). Thus, the short-range scaling methods discussed above can all be used in a hierarchical scaling framework. Examples of patch hierarchies for upscaling purposes include levels of biological organization (e.g., leaf–plant–stand as in Reynolds *et al.* 1993) and different types of nested landscape units (e.g., Reynolds and Wu 1999, Wu and David 2002, Hall *et al.* 2004).

7.4.2.2 Downscaling methods

Downscaling also consists of two general steps: disaggregating information and singling out (Fig. 7.2). The goal of disaggregating coarse-grained information is to derive the fine-scale pattern within a given areal unit (e.g., pixel or patch), a process also known as fine-graining (Blöschl and Sivapalan 1995, Bierkens *et al.* 2000). Downscaling often uses stochastic or probabilistic methods with auxiliary information on the finer scale. Singling out is simply to locate the site of interest in the disaggregated pattern. Much of the research on downscaling in the past few decades has been done in the context of global climate change, and the primary goal is to translate general circulation model (GCM) output into regional-scale predictions for scientific research as well as decision-making purposes. These methods are usually classified into two general categories: empirically based statistical and process model-based downscaling approaches (Hewitson and Crane 1996, Wilby and Wigley 1997, Kidson and Thompson 1998, Wilby *et al.* 1998, Murphy 1999, 2000).

The empirically based statistical downscaling approach aims to derive regional climate conditions (e.g., temperature, precipitation, and wind velocity) from large-scale synoptic circulation features (e.g., upper-level winds, geopotential heights, and sea-level pressure) predicted by GCMs. This is usually done through "transfer functions" which are obtained through multiple linear regression, artificial neural networks, classification and regression trees, or other statistical methods (Hewitson and Crane 1996, Wilby *et al.* 1998, Li and Sailor 2000, Crane *et al.* 2002). The empirical downscaling approach works well for temporally continuous variables such as temperature, but much less effectively for temporally discontinuous and highly intermittent variables such as precipitation (Li and Sailor 2000).

The process model-based downscaling approach, on the other hand, uses nested dynamic models of different scales to disaggregate information downward. For example, a higher-resolution regional climate model may be embedded within a global GCM, so that the GCM output drives the regional model which in turn produces downscaled results. The models are coupled either through one-way or two-way nesting schemes. In a two-way nesting scheme, GCM and the embedded regional climate model are run simultaneously and interact with each other across scales (Hewitson and Crane 1996, Kidson and Thompson 1998, Murphy 1999).

While the current literature on downscaling is dominated by meteorological and climatologic studies, other methods exist in hydrological and soil research that focus on regional down to local scales. These methods may also be grouped into the two general downscaling approaches discussed above. Examples of disaggregating information on soil properties, hydrological time series, and precipitation patterns are abundant (Blöschl and Sivapalan 1995, Bierkens *et al.* 2000). Also, in the social sciences, as mentioned earlier in this chapter, the methods of "ecological inferences," including "ecological regression," the neighborhood model, and the EI model (Freedman *et al.* 1998, Freedman *et al.* 1999, King 1999), may also be used for ecological downscaling, particularly, when the research goal is to decipher the behavior of lower-level elements from higher-level aggregate relationships.

In addition, the common problem of "pixel mixing" in remote sensing arises from the fact that a single pixel is often a mixture of multiple spectrally unique land cover types (i.e., the "endmembers"), which leads to errors in image classification. Remote-sensing scientists have developed a series of subpixel analysis methods to "unmix" individual pixels to estimate the relative areal proportions of different land-cover types within a pixel. The most widely used has been the linear spectral unmixing model, which assumes that the reflectance spectrum of any pixel is the result of linear combinations of the spectra of all constituent land-cover types within that pixel (Rosin 2001, Song 2005). The relative abundance of each land-cover type within a pixel is obtained by solving a closed system of n linear equations where n is the number of bands in an image. In recent years, a number of other methods have been developed for pixel unmixing, including fuzzy membership functions (Foody 2000), the least median of squares method (Rosin 2001), and wavelet and neural network-based methods (e.g., Mertens *et al.* 2004). The potential of these pixel unmixing methods for ecological downscaling studies is yet to be explored.

7.4.3 Uncertainty analysis

Scaling practices always come with uncertainties because of spatial heterogeneity, nonlinearity, data inadequacy, and problems with scaling techniques. The main purposes of uncertainty analysis (or error propagation analysis) are to identify the various sources of uncertainties and quantify their effects on scaling results (Rastetter *et al.* 1992, Heuvelink 1998a, 1998b). Different scaling methods are amenable to different uncertainty analysis techniques. For example, many empirically based statistical scaling methods produce scaling results with some relevant information on uncertainty (e.g., variance, confidence intervals, and regression or correlation coefficients). Monte Carlo techniques may be used with dynamic modeling methods, such as extrapolation by expected value and other stochastic models, to estimate confidence intervals. While uncertainty analysis can be quite challenging, a number of methods have been developed in recent years (Rastetter *et al.* 1992, Heuvelink 1998a).

Li and Wu (2006) reviewed different aspects of uncertainty analysis, including sources of uncertainty in scaling, evaluation of scaling algorithms, error propagation from parameters and input data to scaling results, and presentation of prediction accuracy and error partitioning. Several techniques for uncertainty analysis have been used in ecology and environmental sciences, including probability theory, Taylor series expansion, Monte Carlo simulation, generalized likelihood uncertainty estimation, Bayesian statistics, and sequential partitioning. They recommended the following desirable outputs of uncertainty analysis: (1) measures of model adequacy, (2) full probability distributions of model outputs (e.g., density function and probability-weighted values), (3) reliability of model results (e.g., accuracy, confidence level, and error), (4) relative contribution or importance of each factor as an error source to total uncertainty, (5) the likelihood of different scenarios (probability or ranking), and (6) identification of the least understood or critical components of the model.

From the above discussion it is clear that uncertainty analysis should be regarded as an essential part of the scaling process. But this has not been the case in ecological studies. Given the increasing importance of cross-scale studies in today's scientific research and environmental decision-making, it is crucial to properly quantify and report uncertainties with scaling results.

7.5 Discussion and conclusions

The increasing prominence of scale issues in ecology and other sciences since the 1980s seems inevitable for several reasons. First, ecology as a science has become progressively more explanatory, and mechanistic explanations inevitably invoke multiple scales in space and time as well as multiple levels of organization. Second, for the increasing need to understand and solve broad-scale environmental problems, scientists have to translate information across spatial and temporal scales or organizational hierarchies. Third, the past two decades have witnessed significant advances in theory and methodology for tackling the complexity of spatially extended, heterogeneous systems such as landscapes. Important theories and methods for scaling include hierarchy theory (Allen and Starr 1982, O'Neill *et al.* 1986), fractal geometry (Mandelbrot 1982), phase transition and percolation theory (Gardner *et al.* 1987, Milne 1992), cellular automata (Wolfram 1984), self-organized criticality (Bak 1996), and complex adaptive systems (Cowan *et al.* 1994, Levin 1999). Fourth, recent advances in remote sensing, geographic information systems (GIS), and computing technologies have equipped scientists with powerful tools for dealing with issues of heterogeneity and scale. In addition, the rapid development of landscape ecology since the 1980s has certainly contributed to the widespread recognition of the importance of scale within ecology and beyond.

Today, landscape ecologists are generally aware that scale may directly influence the results of a study whenever spatial heterogeneity cannot, or should not, be assumed away. Heterogeneity makes no sense without the explicit consideration of scale, and scale matters little without heterogeneity. There seems to be a consensus among landscape ecologists today that, whenever possible, a multiple-scale or hierarchical approach is preferable to a single-scale approach. Scaling, as the process of translating information across spatiotemporal scales and organizational levels, has been increasingly emphasized in ecological studies. Indeed, scaling is the essence of understanding and prediction, and has a central role in ecological theory and application (Levin 1992, Levin and Pacala 1997).

There are still numerous problems and challenges in dealing with scale and scaling issues across disciplines. I conclude this chapter by highlighting several of them as follows:

- First, scale and scaling are unifying concepts that cut across all disciplines in both natural and social sciences, and the diversity of connotations presents both problems and opportunities. To avoid unnecessary confusion, these terms should always be specified when they are used. Beyond that, to take a leading role in developing a science of scale, landscape ecologists must familiarize themselves with the scale-related terminology and methods developed in other fields, such as geography, soil science, hydrology, and the social sciences.

- Second, while scale effects are pervasive in the study of heterogeneous landscapes, we must move beyond simply reporting the occurrences of scale effects, which would be an endless effort. Instead, the emphasis should be placed on the search for scaling relations that can be used to identify underlying processes and translate information across scales. A straightforward and powerful approach is to construct empirical scalograms in which variations of patterns, processes, and their relationships are plotted directly against scale (Turner *et al.* 1989b, Ludwig *et al.* 2000, Wu 2004). Such scalograms provide not only direct evidence to test scaling theories, but also a simple yet reliable way of scaling up and down information across landscapes.

- Third, the two general scaling approaches, similarity-based and dynamic modeling, need to be better understood and integrated in ecological studies. No matter how authentic it may sound, the classic definition of scaling that hinges on power laws is not adequate; it only covers part of what has actually taken place in ecological scaling. The two scaling approaches are not contradictory, but complementary to each other. Scale-invariance theory and hierarchy theory may seem at odds, but they are simply different perspectives on the same multiscaled world. A hierarchical system is composed of a number of scale domains within which scale-invariance may well exist. Similarly, a hierarchical scaling scheme may include similarity-based methods. Future scaling studies in landscape ecology should clearly recognize the pros and cons of both approaches, and emphasize the integration between the two whenever necessary. Neither brutal forces with overwhelmingly complex models nor scale-free power laws with elegantly simplistic equations alone would be adequate for understanding and predicting the dynamics of landscapes.

- Fourth, for both approaches it is important to properly identify scaling thresholds at which scaling relations change abruptly. These thresholds suggest fundamental shifts in underlying processes or controlling factors (Gardner *et al.* 1989, Turner *et al.* 1989a, King *et al.* 1991, Wu and Loucks 1995), and define the domains of applicability of the various scaling methods.
- Fifth, one of the greatest challenges for scaling in real landscapes is to integrate biophysical with socioeconomic processes. This is especially true for human-dominated landscapes (e.g., agricultural and urban landscapes) where natural and anthropogenic processes are intertwined and often operate on different scales. The mechanics and rules of scaling for different processes may also vary dramatically. When it comes to the practice of scaling, universality is elegant, but more of utopia; idiosyncrasy is torturous, but more of reality. Complex interdisciplinary issues call for a hierarchical, pluralistic scaling strategy that integrates both empirical statistical and dynamic modeling methods.
- Finally, scaling without known accuracy is unreliable, and uncertainty analysis needs to be an integral part of the scaling process. Ecological scaling, especially with dynamic models, has rarely been done with rigorous accuracy assessment. While it is challenging, uncertainty analysis should be emphasized in future scaling studies because it provides critical information about the accuracy of scaling results. This uncertainty issue of scaling becomes particularly important when scaling results are expected to be used for management and policy-making purposes.

Acknowledgments

I sincerely thank Helene Wagner, Darrel Jenerette, Richard Hobbs, and Alex Buyantuyev for their comments on an earlier version of the chapter. My research on scaling has been supported by US EPA, US NSF, USDA, and NSF of China.

References

Alker, H.R. 1969. A typology of ecological fallacies. Pages 69–86 in M. Dogan and S. Rokkan (eds.). *Quantitative Ecological Analysis in the Social Sciences*. Cambridge, MA: MIT Press.

Allardt, E. 1969. Aggregate analysis: the problem of its informative value. Pages 41–51 in M. Dogan and S. Rokkan (eds.). *Quantitative Ecological Analysis in the Social Sciences*. Cambridge, MA: MIT Press.

Allen, T. F. H., R. V. O'Neill, and T. W. Hoekstra. 1984. *Interlevel Relations in Ecological Research and Management: Some Working Principles from Hierarchy Theory*. USDA Forest Service General Technical Report RM-110. Fort Collins, Co: Rocky Mountain Forest and Range Experiment Station.

Allen, T. F. H. and T. B. Starr. 1982. *Hierarchy: Perspectives for Ecological Complexity*. Chicago: University of Chicago Press.

Amrhein, C. G. 1995. Searching for the elusive aggregation effect: evidence from statistical simulations. *Environment and Planning A* **27**, 105–19.

Arbia, G. 1989. *Spatial Data Configuration in the Statistical Analysis of Regional Economic and Related Problems*. Dordrecht: Kluwer.

Bak, P. 1996. *How Nature Works: The Science of Self-Organized Criticality*. New York: Copernicus.

Barenblatt, G. I. 1996. *Scaling, Self-Similarity, and Intermediate Asymptotics*. Cambridge: Cambridge University Press.

Belyea, L. R. and J. Lancaster. 2002. Inferring landscape dynamics of bog pools from scaling relationships and spatial patterns. *Journal of Ecology* **90**, 223–34.

Benson, B. J. and M. D. Mackenzie. 1995. Effects of sensor spatial resolution on landscape structure parameters. *Landscape Ecology* **10**, 113–20.

Bierkens, M. F. P., P. A. Finke, and P. de Willigen. 2000. *Upscaling and Downscaling Methods for Environmental Research*. Dordrecht: Kluwer Academic Publishers.

Blöschl, G. and M. Sivapalan. 1995. Scale issues in hydrological modelling: a review. *Hydrological Processes* **9**, 251–90.

Bokma, F. 2004. Evidence against universal metabolic allometry. *Functional Ecology* **18**, 184–7.

Brown, J. H., J. F. Gillooly, A. P. Allen, V. M. Savage, and G. B. West. 2004. Toward a metabolic theory of ecology. *Ecology* **85**, 1771–89.

Brown, J. H., V. K. Gupta, B.-L. Li, *et al.* 2002. The fractal nature of nature: Power laws, ecological complexity and biodiversity. *Philosophical Transactions of the Royal Society (London B)* **357**, 619–26.

Brown, J. H. and G. B. West (eds.). 2000. *Scaling in Biology*. New York: Oxford University Press.

Brutsaert, W. 1982. *Evaporation into Atmosphere: Theory, History, and Applications*. Dordrecht: D. Reidel Publishing Company.

Calder, W. A. 1983. Ecological scaling: mammals and birds. *Annual Reviews Ecology and Systematics* **14**, 213–30.

Ciret, C. and A. Henderson-Sellers. 1998. Sensitivity of ecosystem models to the spatial resolution of the NCAR community climate model CCM2. *Climate Dynamics* **14**, 409–29.

Clark, W. C. 1985. Scales of climate impacts. *Climatic Change* **7**, 5–27.

Cowan, G. A., D. Pines, and D. Meltzer (eds.). 1994. *Complexity: Metaphors, Models, and Reality*. Reading, MA: Perseus Books.

Crane, R. G., B. Yarnal, E. J. Barron, and B. Hewitson. 2002. Scale interactions and regional climate: examples from the Susquehanna River Basin. *Human and Ecological Risk Assessment* **8**, 147–58.

Cullinan, V. I., M. A. Simmons, and J. M. Thomas. 1997. A Bayesian test of hierarchy theory: scaling up variability in plant cover from field to remotely sensed data. *Landscape Ecology* **12**, 273–85.

Cyr, H. and S. C. Walker. 2004. An illusion of mechanistic understanding. *Ecology* **85**, 1802–4.

Dale, M. R. T. 1999. *Spatial Pattern Analysis in Plant Ecology*. Cambridge: Cambridge University Press.

Delcourt, H. R. and P. A. Delcourt. 1988. Quaternary landscape ecology: relevant scales in space and time. *Landscape Ecology* **2**, 23–44.

Dodds, P. S., D. H. Rothman, and J. S. Weitz. 2001. Re-examination of the "3/4-law" of metabolism. *Journal of Theoretical Biology* **209**, 9–27.

Dogan, M. and S. Rokkan (eds.). 1969. *Quantitative Ecological Analysis in the Social Sciences*. Cambridge, MA: MIT Press.

Foody, G. M. 2000. Estimation of sub-pixel land cover composition in the presence of untrained classes. *Computers and Geosciences* **26**, 469–78.

Forman, R. T. T. 1995. *Land Mosaics: The Ecology of Landscapes and Regions*. Cambridge: Cambridge University Press.

Forman, R. T. T. and M. Godron. 1986. *Landscape Ecology*. New York: John Wiley & Sons, Inc.

Fotheringham, A. S. and D. Wong. 1991. The modifiable areal unit problem in multivariate statistical analysis. *Environment and Planning A* **23**, 1026–44.

Freedman, D. A. 2001. Ecological inference and the ecological fallacy. *International Encyclopedia for the Social and Behavioral Sciences* **6**, 4027–30.

Freedman, D. A., S. P. Klein, M. Ostland, and M. R. Roberts. 1998. A solution to the ecological inference problem (book review). *Journal of the American Statistical Association* **93**, 1518–22.

Freedman, D. A., S. P. Klein, J. Sacks, C. A. Smyth, and C. G. Everett. 1991. Ecological regression and voting rights. *Evaluation Review* **15**, 673–711.

Freedman, D. A., M. Ostland, M. R. Roberts, and S. P. Klein. 1999. Response to King's comment. *Journal of the American Statistical Association* **94**, 355–7.

Gardner, R. H., B. T. Milne, M. G. Turner, and R. V. O'Neill. 1987. Neutral models for the analysis of broad-scale landscape pattern. *Landscape Ecology* **1**, 19–28.

Gardner, R. H., R. V. O'Neill, M. G. Turner, and V. H. Dale. 1989. Quantifying scale-dependent effects of animal movement with simple percolation models. *Landscape Ecology* **3**, 217–27.

Gehlke, C. E. and K. Biehl. 1934. Certain effects of grouping upon the size of the correlation coefficient in census tract material. *Journal of the American Statistical Association* **29**, 169–70.

Goodchild, M. F. and S. Gopal (eds.). 1989. *Accuracy of Spatial Databases*. London: Taylor and Francis.

Goodman, L. 1953. Ecological regression and the behavior of individuals. *American Sociological Review* **18**, 663–4.

Greig-Smith, P. 1952. The use of random and contiguous quadrats in the study of the structure of plant communities. *Annals of Botany* **16**, 293–316.

Haggett, P. 1965. Scale components in geographical problems. Pages 164–185 in R. J. Chorley and P. Hagget (eds.). *Frontiers in Geographical Teaching*. London: Methuen.

Hall, O., G. J. Hay, A. Bouchard, and D. J. Marceau. 2004. Detecting dominant landscape objects through multiple scales: an integration of object-specific methods and watershed segmentation. *Landscape Ecology* **19**, 59–76.

Harvey, D. W. 1968. Pattern, process and the scale problem in geographical research. *Transactions of the Institute of British Geographers* **45**, 71–8.

Hastings, H. M. and G. Sugihara. 1993. *Fractals: A User's Guide for the Natural Sciences*. Oxford: Oxford University Press.

Hay, G., D. J. Marceau, P. Dubé, and A. Bouchard. 2001. A multiscale framework for landscape analysis: object-specific analysis and upscaling. *Landscape Ecology* **16**, 471–90.

Heuvelink, G. B. M. 1998a. *Error Propagation in Environmental Modelling with GIS*. Bristol: Taylor & Francis Inc.

Heuvelink, G. B. M. 1998b. Uncertainty analysis in environmental modelling under a change of spatial scale. *Nutrient Cycling in Agroecosystems* **50**, 255–64.

Hewitson, B. C. and R. G. Crane. 1996. Climate downscaling: techniques and application. *Climate Research* **7**, 85–95.

Hood, W. G. 2002. Application of landscape allometry to restoration of tidal channels. *Restoration Ecology* **10**, 213–22.

Jelinski, D. E. and J. Wu. 1996. The modifiable areal unit problem and implications for landscape ecology. *Landscape Ecology* **11**, 129–40.

Jenerette, G. D. and J. Wu. 2001. Analysis and simulation of land-use change in the central Arizona–Phoenix region, USA. *Landscape Ecology* **16**, 611–26.

Judson, O. P. 1994. The rise of the individual-based model in ecology. *TREE* **9**, 9–14.

Kersebaum, K.C. and K.-O. Wenkel. 1998. Modelling water and nitrogen dynamics at three different spatial scales: influence of different data aggregation levels on simulation results. *Nutrient Cycling in Agroecosystems* **50**, 313–19.

Kidson, J.W. and C.S. Thompson. 1998. A comparison of statistical and model-based downscaling techniques for estimating local climate variations. *Journal of Climate* **11**, 735–53.

King, A.W. 1991. Translating models across scales in the landscape. Pages 479–517 in M.G. Turner and R.H. Gardner (eds.). *Quantitative Methods in Landscape Ecology*. New York: Springer-Verlag.

King, A.W., A.R. Johnson, and R.V. O'Neill. 1991. Transmutation and functional representation of heterogeneous landscapes. *Landscape Ecology* **5**, 239–53.

King, G. 1997. *A Solution to the Ecological Inference Problem: Reconstructing Individual Behavior from Aggregate Data*. Princeton: Princeton University Press.

King, G. 1999. The future of ecological inference research: a comment on Freedman *et al. Journal of the American Statistical Association* **94**, 352–5.

Kozlowski, J. and M. Konarzewsk. 2004. Is West, Brown and Enquist's model of allometric scaling mathematically correct and biologically relevant? *Functional Ecology* **18**, 283–9.

LaBarbera, M. 1989. Analyzing body size as a factor in ecology and evolution. *Annual Reviews Ecology and Systematics* **20**, 97–117.

Levin, S.A. 1992. The problem of pattern and scale in ecology. *Ecology* **73**, 1943–67.

Levin, S.A. 1999. *Fragile Dominions: Complexity and the Commons*. Reading, MA: Perseus Books.

Levin, S.A., B. Grenfell, A. Hastings, and A.S. Perelson. 1997. Mathematical and computational challenges in population biology and ecosystems science. *Science* **275**, 334–43.

Levin, S.A. and S.W. Pacala. 1997. Theories of simplification and caling of spatially distributed processes. Pages 271–95 in D. Tilman and P. Kareiva (eds.). *Spatial Ecology*. Princeton: Princeton University Press.

Li, H. and J. Wu. 2006. Uncertainty analysis in ecological studies: an overview. Pages 45–66 in J. Wu, K.B. Jones, H. Li, and O.L. Loucks (eds.). *Scaling and Uncertainty Analysis in Ecology: Methods and Applications*. New York: Columbia University Press.

Li, X. and D. Sailor. 2000. Application of tree-structured regression for regional precipitation prediction using general circulation model output. *Climate Research* **16**, 17–30.

Ludwig, J.A., J.A. Wiens, and D.J. Tongway. 2000. A scaling rule for landscape patches and how it applies to conserving soil resources in Savannas. *Ecosystems* **3**, 84–97.

Mandelbrot, B.B. 1982. *The Fractal Geometry of Nature*. New York: W.H. Freeman and Company.

Marceau, D.J. 1999. The scale issue in social and natural sciences. *Canadian Journal of Remote Sensing* **25**, 347–56.

McBratney, A.B. 1998. Some considerations on methods for spatially aggregating and disaggregating soil information. *Nutrient Cycling in Agroecosystems* **50**, 51–62.

Meadowcroft, J. 2002. Politics and scale: some implications for environmental governance. *Landscape and Urban Planning* **61**, 169–79.

Meentemeyer, V. 1989. Geographical perspectives of space, time, and scale. *Landscape Ecology* **3**, 163–73.

Mertens, K.C., L.P.C. Verbeke, T. Westra, and R.R. De Wulf. 2004. Sub-pixel mapping and sub-pixel sharpening using neural network predicted wavelet coefficients. *Remote Sensing of Environment* **91**, 225–36.

Miller, D.H. 1978. The factor of scale: ecosystem, landscape mosaic, and region. Pages 63–88 in K.A. Hammond, G. Macinio, and W.B. Fairchild (eds.). *Sourcebook on the Environment: A Guide to the Literature*. Chicago: University of Chicago Press.

Miller, E.E. and R.D. Miller. 1956. Physical theory for capillary flow phenomena. *Journal of Applied Physics* **27**, 324–32.

Milne, B. T. 1992. Spatial aggregation and neutral models in fractal landscapes. *American Naturalist* **139**, 32–57.

Murphy, J. 1999. An evaluation of statistical and dynamical techniques for downscaling local climate. *Journal of Climate* **12**, 2256–84.

Murphy, J. 2000. Predictions of climate change over Europe using statistical and dynamical downscaling techniques. *International Journal of Climatology* **20**, 489–501.

Niklas, K. J. 1994. *Plant Allometry: The Scaling of Form and Process*. Chicago: University of Chicago Press.

O'Neill, R. V., D. L. DeAngelis, J. B. Waide, and T. F. H. Allen. 1986. *A Hierarchical Concept of Ecosystems*. Princeton: Princeton University Press.

O'Neill, R. V., R. H. Gardner, B. T. Milne, M. G. Turner, and B. Jackson. 1991. Heterogeneity and spatial hierarchies. Pages 85–96 in J. Kolasa and S. T. A. Pickett (eds.). *Ecological Heterogeneity*. New York: Springer-Verlag.

O'Neill, R. V., C. T. Hunsaker, S. P. Timmins, *et al.* 1996. Scale problems in reporting landscape pattern at the regional scale. *Landscape Ecology* **11**, 169–80.

O'Neill, R. V. and A. W. King. 1998. Homage to St. Michael; or, why are there so many books on scale? Pages 3–15 in D. L. Peterson and V. T. Parker (eds.). *Ecological Scale: Theory and Applications*. New York: Columbia University Press.

Openshaw, S. 1984. *The Modifiable Areal Unit Problem*. Norwich: Geo Books.

Openshaw, S. and P. Taylor. 1979. A million or so correlation coefficients: three experiments on the modifiable areal unit problem. Pages 127–144 in N. Wrigley (ed.). *Statistical Application in the Spatial Sciences*. London: Qion.

Pearce, N. 2000. Editorial: the ecological fallacy strikes back. *Journal of Epidemiology and Community Health* **54**, 326–7.

Peters, D. P. C., J. E. Herrick, D. L. Urban, R. H. Gardner, and D. D. Breshears. 2004. Strategies for ecological extrapolation. *Oikos* **106**, 627–36.

Poole, C. 1994. Editorial: ecologic analysis as outlook and method. *American Journal of Public Health* **84**, 715–16.

Rastetter, E. B., J. D. Aber, D. P. C. Peters, D. S. Ojima, and I. C. Burke. 2003. Using mechanistic models to scale ecological processes across space and time. *BioScience* **53**, 68–76.

Rastetter, E. B., A. W. King, B. J. Cosby, *et al.* 1992. Aggregating fine-scale ecological knowledge to model coarser-scale attributes of ecosystems. *Ecological Applications* **2**, 55–70.

Reynolds, J. F., D. W. Hilbert, and P. R. Kemp. 1993. Scaling ecophysiology from the plant to the ecosystem: a conceptual framework. Pages 127–40 in J. R. Ehleringer and C. B. Field (eds.). *Scaling Physiological Processes: Leaf to Globe*. San Diego: Academic Press.

Reynolds, J. F. and J. Wu. 1999. Do landscape structural and functional units exist? Pages 273–96 in J. D. Tenhunen and P. Kabat (eds.). *Integrating Hydrology, Ecosystem Dynamics, and Biogeochemistry in Complex Landscapes*. Chichester: John Wiley & Sons, Ltd.

Risser, P. G., J. R. Karr, and R. T. T. Forman. 1984. *Landscape Ecology: Directions and Approaches*. Special Publication 2, Champaign, IL: Illinois Natural History Survey.

Robinson, W. S. 1950. Ecological correlations and the behavior of individuals. *American Sociological Review* **15**, 351–7.

Rosin, P. L. 2001. Robust pixel unmixing. *IEEE Transactions on Geoscience and Remote Sensing* **39**, 1978–83.

Saura, S. 2004. Effects of remote sensor spatial resolution and data aggregation on selected fragmentation indices. *Landscape Ecology* **19**, 197–209.

Sayre, N. F. 2005. Ecological and geographical scale: parallels and potential for integration. *Progress in Physical Geography* **29**, 276–90.

Schmidt-Nielsen, K. 1984. *Scaling: Why is Animal Size so Important?* Cambridge: Cambridge University Press.

Schneider, D.C. 2001. Spatial allometry. Pages 113–53 in R.H. Gardner, W.M. Kemp, V.S. Kennedy, and J.E. Petersen (eds.). *Scaling Relations in Experimental Ecology*. New York: Columbia University Press.

Schneider, D.C. 2002. Scaling theory: application to marine ornithology. *Ecosystems* **5**, 736–48.

Schwartz, S. 1994. The fallacy of the ecological fallacy: the potential misuse of a concept and the consequences. *American Journal of Public Health* **84**, 819–24.

Simon, H.A. 1962. The architecture of complexity. *Proceedings of the American Philosophical Society* **106**, 467–82.

Song, C. 2005. Spectral mixture analysis for subpixel vegetation fractions in the urban environment: how to incorporate endmember variability? *Remote Sensing of Environment* **95**, 248–63.

Sposito, G. (ed.). 1998. *Scale Dependence and Scale Invariance in Hydrology*. Cambridge: Cambridge University Press.

Stommel, H. 1963. Varieties of oceanographic experience. *Science* **139**, 572–6.

Tenhunen, J.D. and P. Kabat (eds.). 1999. *Integrating Hydrology, Ecosystem Dynamics, and Biogeochemistry in Complex Landscapes*. Chichester: John Wiley & Sons, Ltd.

Turner, M.G., V.H. Dale, and R.H. Gardner. 1989a. Predicting across scales: theory development and testing. *Landscape Ecology* **3**, 245–52.

Turner, M.G., R.H. Gardner, and R.V. O'Neill. 2001. *Landscape Ecology in Theory and Practice: Pattern and Process*. New York: Springer-Verlag.

Turner, M.G., R.V. O'Neill, R.H. Gardner, and B.T. Milne. 1989b. Effects of changing spatial scale on the analysis of landscape pattern. *Landscape Ecology* **3**, 153–62.

Urban, D.L., R.V. O'Neill, and H.H. Shugart. 1987. Landscape ecology: a hierarchical perspective can help scientists understand spatial patterns. *BioScience* **37**, 119–27.

Watson, M.K. 1978. The scale problem in human geography. *Geografiska Annaler* **60B**, 36–47.

Wickham, J.D. and K.H. Riitters. 1995. Sensitivity of landscape metrics to pixel size. *International Journal of Remote Sensing* **16**, 3585–95.

Wiens, J.A. 1989. Spatial scaling in ecology. *Functional Ecology* **3**, 385–97.

Wilby, R.L. and T.M.L. Wigley. 1997. Downscaling general circulation model output: a review of methods and limitations. *Progress in Physical Geography* **21**, 530–48.

Wilby, R.L., T.M.L. Wigley, D. Conway, *et al.* 1998. Statistical downscaling of general circulation model output: a comparison of methods. *Water Resources Research* **34**, 2995–3008.

Withers, M. and V. Meentemeyer. 1999. Concepts of scale in landscape ecology. Pages 205–52 in J.M. Klopatek and R.H. Gardner (eds.). *Landscape Ecological Analysis: Issues and Applications*. New York: Springer.

Wolfram, S. 1984. Cellular automata as models of complexity. *Nature* **311**, 419–24.

Wrigley, N. 1994. Revisiting the modifiable areal unit problem and the ecological fallacy. Pages 1–35 in A.D. Cliff, P.R. Goulc, and A.G. Hoare (eds.). *Festschrift for Peter Haggett*. Oxford: Blackwell.

Wrigley, N., T. Holt, D. Steel, and M. Tranmer. 1996. Analyzing, modelling, and resolving the ecological fallacy. Pages 23–40 in P. Longley and M. Batty (eds.). *Spatial Analysis: Modelling in a GIS Environment*. Cambridge: GeoInformation International.

Wu, J. 1990. Modelling the energy exchange processes between plant communities and environment. *Ecological Modelling* **51**, 233–50.

Wu, J. 1999. Hierarchy and scaling: extrapolating information along a scaling ladder. *Canadian Journal of Remote Sensing* **25**, 367–80.

Wu, J. 2004. Effects of changing scale on landscape pattern analysis: Scaling relations. *Landscape Ecology* **19**, 125–38.

Wu, J. and J.L. David. 2002. A spatially explicit hierarchical approach to modeling complex ecological systems: theory and applications. *Ecological Modelling* **153**, 7–26.

Wu, J. and R. Hobbs. 2002. Key issues and research priorities in landscape ecology: an idiosyncratic synthesis. *Landscape Ecology* 17, 355–65.

Wu, J. and S. A. Levin. 1994. A spatial patch dynamic modeling approach to pattern and process in an annual grassland. *Ecological Monographs* 64(4), 447–64.

Wu, J. and S. A. Levin. 1997. A patch-based spatial modeling approach: conceptual framework and simulation scheme. *Ecological Modelling* 101, 325–46.

Wu, J. and O. L. Loucks. 1995. From balance-of-nature to hierarchical patch dynamics: a paradigm shift in ecology. *Quarterly Review of Biology* 70, 439–66.

8

Optimization of landscape pattern

8.1 Introduction

Wu and Hobbs (2002) state that:

A fundamental assumption in landscape ecology is that spatial patterns have significant influences on the flows of materials, energy, and information while processes create, modify, and maintain spatial patterns. Thus, it is of paramount importance in both theory and practice to address the questions of landscape pattern optimization ... For example, can landscape patterns be optimized in terms of both the composition and configuration of patches and matrix characteristics for purposes of biodiversity conservation, ecosystem management, and landscape sustainability?

Physical restructuring of landscapes by humans is a prominent stress on ecological systems (Rapport *et al.* 1985). Landscape restructuring occurs primarily from land-use conversions or alteration of native habitats through natural resource management. A common faunal response to such land-use intensification is an increased dominance of opportunistic species leading to an overall erosion of biological diversity (Urban *et al.* 1987). Slowing the loss of biodiversity in managed systems will require interdisciplinary planning efforts that meld analysis approaches from several fields including landscape ecology, conservation biology, and management science. Again from Wu and Hobbs (2002), "Such studies are likely to require theories and methods more than those in traditional operations research (e.g., different types of mathematical programming), as well as the participation of scientists and practitioners in different arenas."

Key Topics in Landscape Ecology, ed. J. Wu and R. Hobbs.
Published by Cambridge University Press. © Cambridge University Press 2007.

The objective of this chapter is to review emerging methods from this set of disciplines that allow analysts to make explicit recommendations (prescriptions) concerning the placement of different features in managed landscapes. We will refer to this general set of methods as "spatial optimization." As used here, spatial optimization refers to methods that capture spatial relationships between different land areas in the process of maximizing or minimizing an objective function subject to resource constraints. We will begin with a terse review of the state-of-the-science in spatial optimization and then proceed to a discussion of prominent research questions in need of attention.

8.2   State-of-the-science in spatial optimization

In our view, four basic approaches have been taken to address optimization of landscape pattern: adjacency constraints (with augmentation), spatial enhancement of the reserve-selection problem, direct spatial-optimization approaches, and automated manipulation of simulation models. We will review each of these in turn.

8.2.1   Adjacency constraints

In the forestry literature, the predominant perception of spatial considerations has been that of adjacency relationships. Concerns regarding adjacent timber harvests emerged from legal and regulatory restrictions on the effective size of harvests – if two seemingly legal-size clearcuts, for example, occur within a short time of each other in adjacent areas, the effective size of the clearcut may not be legal. The basic formulation for avoiding adjacent harvests is:

$$X_1 + X_2 \leq 1, X_1, X_2 \in (0, 1)$$

where the X_1 and X_2 are binary (integer) management prescriptions that have harvests within the prohibited time limit of each other, for areas that are adjacent. If "n-tuples" (triplets, quadruplets, etc.) of mutually adjacent areas can be defined for $i = 1, \ldots, I$, then:

$$\sum_i X_i \leq 1, X_i \in (0, 1) \; \forall i$$

can cover the set. Many other sophistications are possible. These approaches require that the management variables be binary (integer), and considerable effort has been expended towards solution of these problems – typically with heuristic methods such as tabu search or simulated annealing (see, for example, Martell *et al.* 1988, Murray and Church 1995, Boston and Bettinger 1999).

Because the original motivations for the clearcut size limits included concerns for aesthetics, wildlife habitat, water quality, and so forth, addressing adjacency constraints has often been interpreted (or even advertised) as addressing those underlying concerns. Adjacency constraints are most clearly applicable when the problem is avoidance of adjacent management actions, per se. They do not necessarily address any other spatial relationship. However, there have been a few recent attempts at augmenting the basic adjacency formulation that show more promise in addressing the underlying ecological problems (see, for example, Bettinger *et al.* 1997, Barrett *et al.* 1998, Falcao and Borges 2001). We will not discuss this approach further here, because we believe that as spatial relationships in addition to adjacency are captured, this approach becomes a special (simple) case of direct spatial optimization or heuristic manipulation of simulation models, as described below.

8.2.2 Spatial enhancement of the natural reserve-selection models

It has been said that the next great challenge for conservation science is the design and implementation of comprehensive and ecologically adequate reserve networks (Lubchenco *et al.* 2003, Williams *et al.* 2004). Because conservation reserve areas are rare (at least in areal extent) and human impacts on natural ecosystems are increasing, there is a growing realization that when there are alternative locations for reserve lands the choice should be, in some sense, optimal (Pressey *et al.* 1993, Reid 1998). Quite a large body of literature has developed based on the set-covering formulation (Toregas and ReVelle 1973) applied to the natural reserve-selection problem (see, for example, Williams and ReVelle 1997, Polasky *et al.* 2001, and Rodrigues and Gaston 2002). The basic formulation is as follows:

$$\text{Minimize:} \quad \sum_i c_i X_i$$

$$\text{Subject to:} \quad \sum_{i \in \Omega_j} X_i \geq k_j \ \forall j$$

$$X_i \in \{0, 1\} \ \forall i$$

where $X_i = 1$ if site i is selected and 0 if site i is not selected for the reserve system, c_i = the cost of selecting site i for the reserve network, Ω_j = the set of sites that contain species j, and k_j = the number of sites required for adequate representation of species j ($k = 1$ for the most basic set-covering problem; $k > 1$ if some degree of redundancy is desired).

Because the choice variables are often defined in a fairly spatially explicit manner, this formulation is sometimes referred to as a form of spatial optimization. This provides a good opportunity for us to draw a distinction between "spatially explicit optimization" (of which this is an example) and what we refer to as "spatial optimization," which (as defined above), refers to methods that capture spatial *relationships* between different land areas in the process of maximizing or minimizing an objective function subject to resource constraints.

Just as attempts have been made to augment the adjacency constraint formulation, recent contributions in the literature (for a review see Williams *et al.* 2004) have endeavored to add spatial relationships to the set-covering reserve selection model (some examples are Possingham *et al.* 2000, Nalle *et al.* 2002, Onal and Briers 2002, 2003, Fischer and Church 2003). And, similar to our point about the augmented adjacency constraint formulation, we will not discuss this approach further here, because we believe that as spatial relationships are captured, this approach will be best described as being one of the next two approaches.

8.2.3 Direct approaches to spatial optimization

The basic approach that we favor is to directly include the spatial relationships of concern in a formulation that is focused on the optimization of landscape pattern, per se. We would characterize the approach as having a closed-form formulation with a formal solution method. Hof and Bevers (1998, 2002) explore a number of formulations along these lines, including static models, dynamic models, models of spatial autocorrelation, and models of sustainability. Other authors that have taken relatively direct approaches include Nevo and Garcia (1996), Farmer and Wiens (1999), and Loehle (1999).

As an example, wildlife habitat fragmentation (spatial division into disaggregated patches) is a common concern with regard to placement of timber harvests. The approach is to directly model the wildlife population growth and dispersal patterns that make habitat connectivity (nonfragmentation) important. This is a dynamic problem where management activities must be scheduled over time, wildlife habitat (determined by forest age) must be tracked as forest stands age and grow, and different wildlife species respond differently to those habitats. The method is related to the classic "reaction-diffusion models" (Skellum 1951, Kierstead and Slobodkin 1953, Allen 1983).

First, the land is divided into cells, where the cell could be scaled to the ecology of the species (e.g., average home range or territory size) or could be scaled simply to provide adequate spatial resolution for the optimization problem. Then, a set of choice variables is defined for each cell, each of which represents a complete scheduled management prescription. For example, each prescription could define the time periods for harvesting the given cell, including a no-harvest option. Any harvest would reset the forest age class to zero, and would change the habitat for each wildlife species accordingly. Initial forest age classes are assigned to each cell, as well as initial population numbers for each wildlife species included.

The following definitions will be used in our discussion of these approaches: i indexes species, k indexes the management prescription, h indexes the cells, as does n, t indexes the time period, q_h = the number of potential management prescriptions for cell h, T = the number of time periods, X_{kh} = the area in cell h that is allocated to management prescription k, K_h = the total area in cell h, S_{iht} = the expected population of species i in cell h at time period t, a_{ihtk} = a coefficient set that gives the expected carrying capacity of species i in cell h at time period t, if management prescription k is implemented (based on forest age class), N_{ih} = the initial population numbers for species i in cell h, and g_{inh} = the probability that an individual of species i will disperse from cell n in any time period to cell h in the subsequent time period. This includes a probability for $n = h$, so that the g_{inh} sum to one for each combination of h and i. In addition, r_i is the "r value" of population growth rate (net of mortality not related to dispersal) for species i, and F_{it} is the total population for species i in time period t.

The simplest objective function would be to maximize a given species' expected total population:

$$\text{Maximize } \sum_t F_{it} \text{ for a given } i$$

The minimum population over all time periods (m), for a given species could be maximized as follows:

$$\text{Maximize } m$$
$$\text{Subject to } m \le F_{it} \quad t = 1, \ldots, T \text{ for a given } i$$

Or, a weighted (V_i) sum of multiple species' populations could be maximized:

$$\text{Maximize } \sum_i V_i \left(\sum_t F_{it} \right).$$

Many other objective functions are also possible. The basic constraint set for such a model is:

$$\sum_{k=1}^{q_k} X_{kh} = K_h \quad \forall h \tag{8.1}$$

$$S_{ih0} = N_{ih} \quad \substack{\forall i \\ \forall h} \tag{8.2}$$

$$S_{iht} \leq \sum_k a_{ihtk} X_{kh} \quad \begin{matrix} \forall i \\ \forall h \end{matrix} \tag{8.3}$$

$$t = 1, \ldots, T$$

$$S_{iht} \leq \sum_n g_{inh} \left[(1+r) S_{in(t-1)} \right] \quad \begin{matrix} \forall i \\ \forall h \end{matrix} \tag{8.4}$$

$$t = 1, \ldots, T$$

$$F_{it} = \sum_h S_{iht} \quad \begin{matrix} \forall i \\ \forall t \end{matrix} \tag{8.5}$$

$$0 \leq X_{kh} \leq K_h \quad \begin{matrix} \forall k \\ \forall h \end{matrix} \tag{8.6}$$

This model is linear, with continuous variables, and can be solved with standard solution software. Equation (8.1) limits the total management prescription allocation to the area in each cell. The management prescriptions are defined with no action in the first time period ($t = 0$), which is used simply to set initial conditions. Equation (8.2) sets the initial population numbers for each species, by cell. The S_{iht} (expected population by species by cell, for each time period) is determined by whichever of Equation (8.3) or Equation (8.4) is binding. The constraint set of Equation (8.3) limits each cell's population to the carrying capacity of the habitat in that cell, determined by forest age classes. The constraint set of Equation (8.4) limits each cell's population according to the growth and dispersal from other cells and itself in the previous time period. The growth and dispersal characteristics of each species are reflected by the parameters in constraint set of Equation (8.4). The constraint set of Equation (8.4) adds up the expected value of the population dispersing from all cells in the previous time period to the given cell in the given time period. It is important to note that whenever Equation (8.3) is binding for a cell, some of the animals assumed to disperse into that cell are lost because of limited carrying capacity. Reaction–diffusion models assume that organisms dispersing into unsuitable regions will perish. This mechanism provides a probabilistic basis for the expectation that mortality (beyond the nondispersal-related mortality accounted for in the r value) occurs in proportion to the usage of inhospitable surroundings. Thus, actual population growth is determined by a combination of potential growth, dispersal, and spatially located carrying capacities determined by the management prescription allocations. The constraint set of Equation (8.5) defines the total population of each species, in each time period. And finally, the constraint set of Equation (8.6) limits the choice variables to be between zero and the area in each cell.

Hof and Bevers (1998, 2002) applied this basic type of model structure to problems of habitat placement for the black-footed ferret, which also accounted for release schedules of captive-born animals. As a follow up, it was applied to the black-tailed prairie dog, accounting for population-dependent dispersal behavior. Hof and Bevers also modified this structure to account for water-borne seed dispersal and for habitat edge effects, converted the model to optimize the location of control measures in managing exotic pests, and applied these basic ideas to problems other than organism management (especially stormflow management and fire management).

8.2.4 Heuristic manipulation of simulation models

The primary criticism that can be leveled at the approaches in the previous section is that there are limits to the complexity of ecological relationships that can be captured in a closed-form optimization formulation. An alternative that has been suggested by several authors (see, for example, Haight and Travis 1997, Boston 1999, Jager and Gross 2000, Calkin *et al.* 2002) is to start with a more complex ecological model and use heuristic procedures to direct repeated predictions with different management regimes, hopefully converging on a near-optimum.

A number of ecological modeling approaches are available to serve this purpose, including empirical models derived from statistical estimation (e.g., Morrison *et al.* 1987), analytical models that have exact solutions derived from a few fundamental mathematical relationships (e.g., Lande 1987), and simulation models derived from mechanistic mathematical relationships whose number and complexity requires that solutions be explored numerically with digital computers (e.g., Fahrig 1997). Because empirical models tend to be correlative in nature and have not performed well in prediction (Block *et al.* 1994), and because analytical models tend to become intractable in all but the simplest cases of spatial heterogeneity (Fahrig 1991), simulation modeling has become the method of choice for management problems (Simberloff 1988). It is hard to provide a general formulation for this approach because it depends on the simulation model chosen.

The primary shortcoming of this approach is, of course, that the outcome is only "the best" alternative from among the landscape layouts investigated. Even with a large number of layouts, near-optimality is not assured. To demonstrate the point, suppose we have 100 land units (for example, in a 10×10 grid). Even if we must consider only one action (versus none), with no scheduling component, there are still 2^{100} or 1.2676×10^{30} possible spatial layouts. Even if 99.9999999 percent (all but a trillionth) of the layouts can be eliminated as undesirable, we still have 1.2676×10^{21} options. Even if there are a trillion layouts that are acceptable, we only have a 7.886×10^{-13} ($1 \times 10^{9} \div 1.2676 \times 10^{21}$) chance of hitting an acceptable

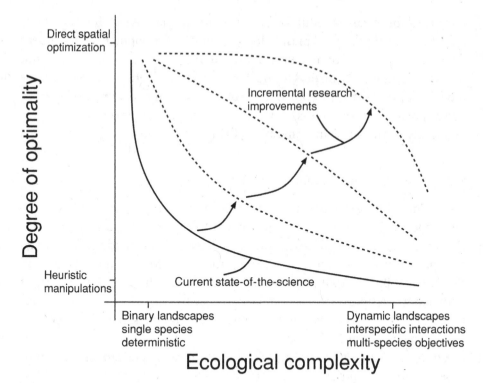

FIGURE 8.1
The trade-off between degree of optimality assured and the ecological complexity captured

solution if we randomly arrange our management actions. This suggests the need for optimization procedures in all but the simplest spatial problems. In addition, the implicit response surface may or may not be convex, such that a solution that appears to be near-optimal may actually not be at all.

Thus, the choice is between a precise optimum to a simplified model and an approximate optimum to a more precise model. Figure 8.1 depicts this trade-off and how it might change as progress is made in this research area. We have previously drawn the analogy between these two alternatives and the two alternatives faced early in the US space program between the X-15 plane and the Mercury program rocket as the direction for future manned space exploration (von Braun *et al.* 1985). The simulation approach, like the Mercury rocket, can do more immediately (especially in terms of ecological complexity and uncertainty). The closed-form optimization, however, is more like the X-15 plane in that it is of the basic structure that we would eventually like to get to (it finds global optima and is of the mathematical form that most closely captures the concept of landscape pattern optimization in the first place). Like the choice faced by the US space program, it is really hard to say which approach might be the most fruitful in the long run, and both are probably worth researching until proven otherwise. In a given planning application, the answer may depend on the questions being asked and the circumstances surrounding the planning problem. For example, habitat placement choices may be limited because the pattern of land development has reduced the configuration possibilities (Saunders *et al.* 1991). When habitat-placement options are restricted to a small set, simulation modeling may offer a very useful approach for ranking alternative configurations. If, however, placement choices are numerous, then formal spatial optimization may be more useful in determining a layout that really is "the best" given the objectives and constraints of the planning problem. Joint use of both strategies as in Hof and Raphael (1997) or Haight *et al.* (2002) might offer planners the opportunity to take advantage of both the ecological detail captured by simulation models and the analytical power of formal spatial optimization to select the best solution.

A very pragmatic version of the heuristic approach that is commonly applied is to use simulation modeling to evaluate population response to alternative landscapes a posteriori. Management choices are thus made by ranking a small number of landscape alternatives according to some criterion (e.g., organism abundance, persistence) and selecting the strategy that ranks highest. This use of simulated predictions in conservation planning has been termed "relative ranking" by Turner *et al.* (1995) and is a form of prescriptive planning because the result specifies a habitat arrangement that will "best" address some set of management objectives. This approach is severely limited (as just discussed) by the small number of landscape layouts investigated (Conroy 1993).

8.3   Critical research questions

It should be fairly clear from the preceding section that the state-of-the-science in optimization of landscape pattern borrows heavily from the operations-research or management-sciences fields. Thus, ecology in general and landscape ecology in particular could probably benefit from more utilization of management-science methodology. It should be pointed out that the flow of knowledge also goes the other way. The study of natural processes has inspired several heuristic procedures in optimization. In particular, a class of heuristic solution algorithms called "genetic algorithms" (Reeves 1993) solves mathematical programming problems by emulating evolutionary processes. In the heuristic search, new trial solutions are created by "mating" previous solutions so as to emphasize positive traits much like natural selection promotes evolution in natural systems. Another example is "simulated annealing" (Reeves 1993) which was originally developed to simulate the annealing process of cooling metals, but which is now commonly used as an optimization search routine. There is sufficient interest in this line of thought that an advertisement appeared in OR/MS Today (April 1999), the online version of the INFORMS magazine of the Institute for Operations Research and the Management Sciences, for an endowed chair in industrial and systems engineering at a major university that focused on "the design of sustainable manmade systems drawing upon understanding of efficiencies of natural systems" and "the introduction of efficiencies from natural systems to industrial operations." At any rate, our focus here is on the application of optimization to manage natural systems (and not the other way around).

8.3.1 Randomness

The most glaring weakness in the state-of-the-science in spatial optimization is in the treatment of stochastic variables. In Hof and Bevers (1998), we treat randomness with "chance constraint" formulations that assume normal distributions, and knowledge of the means and variances for all random variables (as well as the spatial covariances between them). With knowledge of random variable distributions, stochastic programming as well as Monte Carlo and other numerically intensive approaches may show promise in both evaluating the impact of risk on optimization model solutions and in modifying those solutions to better account for the randomness in the systems being modeled. This knowledge allows rigorous analysis of random variables, but we must admit that it is rarely available in real-world resource management and planning problems.

In order to discuss the less rigorous possibilities, let us make sure that a few definitions are clear. Under conditions of "risk," we face randomness in the systems that we must manage, but we have knowledge of the probabilities of the possible outcomes, and we know the effects that our alternative courses of action have on those probabilities. Under "uncertainty" we do not know the probabilities or the effects of our potential actions on those probabilities. And, there is a third condition, usually referred to (rather unsympathetically) as "ignorance," where we do not even know what outcomes are possible with our alternative courses of action. Obviously, uncertainty is more difficult to deal with than risk, and ignorance is more difficult than uncertainty. Analytical approaches are quite potent in managing risk, but are less so with uncertainty and ignorance. Stochastic programming, Monte Carlo simulations, expert systems, etc. all need information on probabilities of outcomes in order to manage randomness rigorously. The fundamental research problem for landscape (and other) ecologists is to reduce ignorance and uncertainty, but we often must manage systems that are inherently stochastic so that the best we could ever hope to do is convert ignorance and uncertainty into risk. At least, if the problem can be so-reduced, it is conducive to rigorous analysis. Under conditions of significant uncertainty or ignorance, there is really not very much that can be done analytically to spatially optimize landscape pattern.

The best that can probably be done from a management viewpoint is to diversify the land management portfolio in the short term, and use that as a start to an adaptive management process (Walters and Holling 1990) in the long term. This approach is quite analogous to how a portfolio manager diversifies a client's investments in order to reduce the chance of a catastrophic loss. The two basic principles of diversifying a land management portfolio in the short term are: (1) within areas that are subject to the same random events, we should do different things that will respond differently to those random events, and (2) when we do the same things in different areas, we should do them far enough apart that they are subject to different random events.

In both cases, the objective is to reduce the random variability of system response by diversifying what we are doing across elements that are relatively noncovariant (that have covariances that are smaller than the variances of the individual elements). We can do this (imperfectly) without knowing the actual variances or covariances, but just knowing that landscapes are spatially auto-correlated, such that areas are more independent the farther they are away from each other. This diversification also presents a good beginning for an adaptive management process where we use management actions as experiments so as to learn more about the systems that we are managing. By diversifying our approach to management, we will learn more, sooner. It will be critical that monitoring systems be put into place if we are going to learn anything from these "management experiments" (monitoring will be discussed as a separate research need below). With this approach, we will adapt our management strategy as we go along, learning what we can in a continuous process. Thus, research is integrated with management in this approach to dealing with uncertainty and ignorance.

8.3.2 Organism movement

The second most obvious set of research questions emanates from the most fundamental reason that spatial pattern matters in ecological systems – the fact that organisms move around on the nonhomogeneous landscape. The recognition that environmental heterogeneity and organism movement (in particular organism dispersal) interact to affect population dynamics has been key to what Turchin (1998) called a "... paradigmatic shift from the aspatial equilibrium view ... to a spatially explicit view." Unfortunately, organism movement is poorly understood and is an extremely difficult area of empirical research (Clobert *et al.* 2001), particularly at the landscape scale (Harrison and Bruna 1999). It may seem trivial to highlight species movement as a knowledge gap given its oft-cited importance, but we would be remiss not to emphasize the need for more empirical data regarding how (e.g., random or biased diffusion), when (e.g., what ecological conditions promote dispersal), and to what degree (e.g., how often and how far) species move in landscapes.

Techniques for studying animal movement are less well developed than are methods for estimating demographic parameters (Turchin 1998). Choice of methods will depend to a large degree on whether one needs to quantify the movement paths or simply the population-level pattern of redistribution. The kind of movement data required will be dictated by the information needed to address the planning problem. For example, quantifying species-specific search rules used to locate vacant territories would require information on the actual movement paths taken by individual organisms. Alternatively, estimating the distribution of dispersal distances or the resistance of the interpatch environment to movement may only require information on the rate and directionality of population spread. Turchin (1998) and Clobert *et al.* (2001) provide comprehensive reviews of data collection and analysis methods that are appropriate for both types of questions.

Although the use of new approaches to study species movement will extend our abilities to prescribe landscape configurations, the spatial and temporal extent of conservation plans often make it very difficult to detect population response, replicate conservation treatments, and identify mechanisms underlying population change. These problems pervade empirical testing of conservation plans and they have been addressed with several atypical research protocols. Carpenter (1990) reviews pre- and post-treatment time-series analyses to infer nonrandom change in system response. Similarly, Hargrove and Pickering (1992) outline the use of "quasi-experiments" to take advantage of natural disturbances that alter habitat configuration. An approach directed specifically at increasing replication is to repeat unreplicated studies in different systems (Carpenter 1990). Such an approach can be approximated by accumulating the efforts of others and analyzing independent research efforts with meta-analysis procedures (Arnqvist and Wooster 1995). And again, where replication and experimental controls within the planning problem are feasible, active adaptive planning (Walters and Holling 1990) could provide opportunities for stronger inferences from management experiments.

8.3.3 Monitoring of spatially explicit plans

Even if landscape planners had access to complete information on species movement, a more fundamental empirical gap is the paucity of tests of spatially explicit conservation plans. An especially critical need is for well-designed and active monitoring programs that are linked to plan implementation. Monitoring information is needed to assess the success or failure of a given plan, to test basic ecological concepts upon which the plan is based, and to identify the mechanisms underlying population response to habitat prescriptions (Hansson and Angelstam 1991). For example, critical thresholds have been demonstrated theoretically, but there have been very few attempts to establish the existence of critical thresholds in field studies of species distribution and abundance (but see With and Crist 1995 and Trzcinski *et al.* 1999).

Regardless of the protocol used, it is critical that a more concerted effort be directed at designing and implementing monitoring strategies that allow conservation plans to be tested (Havens and Aumen 2000). Because long time periods are required to quantify species response to modified landscapes, researching the relative merits of alternative monitoring designs will be difficult. Wennergren *et al.* (1995) offer an intriguing suggestion for researchers – use spatially explicit population models to test the ability of alternative monitoring designs to predict the consequences of habitat layouts prior to implementing them in the field. Failure to devote more effort toward monitoring will perpetuate our current reliance on untested concepts and will heighten the contention and uncertainty that surrounds the use of planning models to derive spatially explicit habitat prescriptions.

8.3.4 Multiple species/community level models

Another important theoretical gap concerns how one prescribes habitat layouts that are relevant to the broader species assemblage occupying the landscape. The discussion above was limited to the response of individual populations to varying habitat arrangements, avoiding the issue of conflicting habitat requirements among multiple species. Even if planning models could be developed for some suite of species inhabiting the planning area, the feasibility of deriving habitat prescriptions that are relevant to the group is questionable due to species interactions. The extension to biodiversity conservation as a whole is even less clear (Turner *et al.* 1995).

Moving from single-species to multi-species conservation planning is certainly difficult and we know of no definitive strategy. One approach that has emerged repeatedly is the use of indicator species (related incarnations include keystone, umbrella, or focal species) to reflect the status of the overall species assemblage. Unfortunately, indicators are often chosen opportunistically (e.g., well-studied and well-surveyed taxa). Tests of this strategy using broad taxonomic groups (e.g., birds, mammals, butterflies) have not supported its general applicability (Flather *et al.* 1997). However, rejection at broad taxonomic levels should not preclude the search for indicators among finer sets of species. MacNally and Fleishman (2002) are developing statistical modeling approaches to defining efficient sets of indicator species within taxonomic groups (e.g., butterflies) that are showing promise, at least among butterflies. A more refined search to identify life-history attributes that could serve to group species based on critical vital rates (e.g., reproductive potential, dispersal ability) may offer an alternative approach to defining indicators because species that share similar life histories may respond to habitat geometry in a similar fashion (see Noon *et al.* 1997). In a slight variation of this approach, mathematical taxonomy and ordination techniques (Gauch 1982) could be used to define clusters of species based on measured life-history attributes such that the species pool for a given planning area is partitioned into sets that may be similarly sensitive to habitat arrangement effects.

8.3.5 Synthesis

Even though theoretical developments in spatial ecology are relatively recent, theoretical research on the concepts reviewed in this chapter has far outpaced empirical research. The disparity between theory and empirical research has resulted in a baffling array of models and results that can appear contradictory to conservation planners (de Roos and Sabelis 1995). Consequently, the lack of theoretical synthesis is an important knowledge gap hindering the application of these concepts in conservation planning. There is a need to summarize how different modeling approaches and parameterizations affect habitat prescriptions so that domains of applicability can be recognized by those who develop and implement conservation plans. In the absence of a more comprehensive synthesis, landscape planners will continue to be reluctant to implement the habitat prescriptions derived.

Although synthesis may lack the intellectual excitement associated with new theoretical developments, it is nonetheless important. We identify two broad approaches to theoretical synthesis. The first is through structured reviews of the literature. This will not be an easy task. There are numerous nuances to spatial planning models that can quickly overwhelm attempts to distill rules for application (de Roos and Sabelis 1995). It is not uncommon for models of similar ecological phenomena to reach divergent conclusions (e.g., compare the conclusions reached by Hill and Caswell 1999 and Fahrig 1998 regarding the importance of habitat arrangement in facilitating population persistence). Part of the problem is that both structure and parameterization differ among models, making it difficult to determine what is causing the variability in model results. It is this complexity that may render traditional narrative reviews for theoretical synthesis flawed, suggesting that the more structured analyses formalized in meta-analysis might be better suited to identifying common patterns among published models (Arnqvist and Wooster 1995).

A second approach to synthesis involves the systematic manipulation of specific models using sensitivity analysis (Conroy *et al.* 1995). This would involve the quantification of model response following purposeful and systematic alteration of model parameters (singly and in combination). Comprehensive exploration of model behavior under a wide range of ecologically relevant parameter settings is important for at least two reasons. First, it can help identify those aspects of species life history or habitat arrangement that are critically important for understanding population response to landscape changes. Second, it can determine whether there are fundamental shifts in habitat arrangements in response to relatively small changes in the parameter space.

8.4   Conclusion

The most fundamental research need in optimizing landscape pattern is the ability to better capture the relevant ecological relationships in an optimization analysis. The other chapters in this book provide summaries of the state-of-the-science in the most important research areas in landscape ecology. Thus, an appropriate concluding remark would be that research in spatial optimization of landscape pattern should strive to capture the state-of-the-science described elsewhere in this book, wherever appropriate. Also, the other chapters identify the most pressing research questions in landscape ecology, so obviously the future challenge will be to continue to capture this landscape ecology research in spatial optimization models.

A natural reaction to the idea of optimizing spatial pattern across a landscape is that we simply do not know enough about ecological systems to actually optimize them. Indeed, we will probably never know as much about ecology as we would like to. Our reaction is that it is important to apply spatial optimization in the context of an adaptive learning process (as we have noted previously). We will probably never have a level of knowledge that is adequate to find a permanent optimal strategy for a managed ecosystem in a one-time optimization analysis. On the other hand, an adaptive management process that does not take advantage of optimization methods is much less likely to make progress either in learning about the ecological system or in managing it.

Applied in a careful, learning process, spatial optimization of landscape pattern has the potential to illuminate new hypotheses for landscape ecology research as well as providing a mechanism to apply landscape ecology research in landscape management. We will close this chapter as we began it, with a quote from Wu and Hobbs (2002): "Research into the spatial optimization of landscape pattern . . ." as it relates to ecological structure (e.g., species distribution, community composition) and function (e.g., disturbance spread, species dispersal) across landscapes, ". . . presents a new and exciting direction for landscape ecology." We definitely hope so.

References

Allen, L. J. S. 1983. Persistence and extinction in single-species reaction–diffusion models. *Bulletin of Mathematical Biology* **45**, 209–27.

Arnqvist, G. and D. Wooster. 1995. Meta-analysis: synthesizing research findings in ecology and evolution. *Trends in Ecology and Evolution* **10**, 236–40.

Barrett, T. M., J. K. Gilless, and L. S. Davis. 1998. Economic and fragmentation effects of clearcut restrictions. *Forest Science* **44**, 569–77.

Bettinger, P., J. Sessions and K. Boston. 1997. Using tabu search to schedule timber harvests subject to spatial wildlife goals for big game. *Ecological Modelling* **42**, 111–23.

Block, W. M., M. L. Morrison, J. Verner, and P. N. Manley. 1994. Assessing wildlife–habitat-relationships models: a case study with California oak woodlands. *Wildlife Society Bulletin* **22**, 549–61.

Boston, K. 1999. Review: spatial optimization for managed ecosystems. *Forest Science* **45**, 595.

Boston, K. and P. Bettinger. 1999. An analysis of Monte Carlo integer programming, simulated annealing, and tabu search heuristics for solving spatial harvest scheduling problems. *Forest Science* **45**, 292–301.

Calkin, D. E., C. A. Montgomery, N. H. Schumaker, *et al.* 2002. Developing a production possibility set of wildlife species persistence and timber harvest value. *Canadian Journal of Forest Research* **32**, 1329–42.

Carpenter, S. R. 1990. Large-scale perturbations: opportunities for innovation. *Ecology* **71**, 2038–43.

Clobert, J., E. Danchin, A. A. Dhondt, and J. D. Nichols (eds.). 2001. *Dispersal*. Oxford: Oxford University Press.

Conroy, M. J. 1993. The use of models in natural resource management: prediction, not prescription. *Transactions of the North American Wildlife and Natural Resources Conference* **58**, 509–19.

Conroy, M. J., Y. Cohen, F. C. James, Y. G. Matsinos, and M. B. A. 1995. Parameter estimation, reliability, and model improvement for spatially explicit models of animal populations. *Ecological Applications* **5**, 17–19.

de Roos, A. M. and M. W. Sabelis. 1995. Why does space matter? In a spatial world it is hard to see the forest before the trees. *Oikos* **74**, 347–8.

Fahrig, L. 1991. Simulation methods for developing general landscape-level hypotheses of single-species dynamics. Pages 416–42 in M. G. Turner and R. H. Gardner (eds.). *Quantitative Methods in Landscape Ecology*. New York: Springer-Verlag.

Fahrig, L. 1997. Relative effects of habitat loss and fragmentation on population extinction. *Journal of Wildlife Management* **61**, 603–10.

Fahrig, L. 1998. When does fragmentation of breeding habitat affect population survival? *Ecological Modelling* **105**, 273–92.

Falcao, A. O. and J. Borges. 2001. Combining random and systematic search heuristic procedures for solving spatially constrained forest management scheduling models. *Forest Science* **48**, 608–21.

Farmer, A. H. and J. A. Wiens. 1999. Models and reality: time–energy trade-offs in pectoral sandpiper (*Calidris melanotos*) migration. *Ecology* **80**, 2566–80.

Fischer, D. T. and R. L. Church. 2003. Clustering and compactness in reserve site selection: an extension of the biodiversity management area selection model. *Forest Science* **49**, 555–65.

Flather, C. H., K. R. Wilson, D. J. Dean, and W. C. McComb. 1997. Identifying gaps in conservation networks: of indicators and uncertainty in geographic-based analyses. *Ecological Applications* **7**, 531–42.

Gauch, H. G. 1982. *Multivariate Analysis in Community Ecology*. Cambridge: Cambridge University Press.

Haight, R. G., B. Cypher, P. A. Kelly, *et al.* 2002. Optimizing habitat protection using demographic models of population viability. *Conservation Biology* **16**, 1386–97.

Haight, R. G. and L. E. Travis. 1997. Wildlife conservation planning using stochastic optimization and importance sampling. *Forest Science* **43**, 129–39.

Hansson, L. and P. Angelstam. 1991. Landscape ecology as a theoretical basis for nature conservation. *Landscape Ecology* **5**, 191–201.

Hargrove, W. W. and J. Pickering. 1992. Pseudoreplication: a *sine qua non* for regional ecology. *Landscape Ecology* **6**, 251–8.

Harrison, S. and E. Bruna. 1999. Habitat fragmentation and large-scale conservation: what do we know for sure? *Ecography* **22**, 225–32.

Havens, K. E. and N. G. Aumen. 2000. Hypothesis-driven experimental research is necessary for natural resource management. *Environmental Management* **25**, 1–7.

Hill, M. F. and H. Caswell. 1999. Habitat fragmentation and extinction thresholds on fractal landscapes. *Ecological Letters* **2**, 121–7.

Hof, J. and M. Bevers. 1998. *Spatial Optimization for Managed Ecosystems*. New York: Columbia University Press.

Hof, J. and M. Bevers. 2002. *Spatial Optimization in Ecological Applications*. New York: Columbia University Press.

Hof, J. and M. G. Raphael. 1997. Optimization of habitat placement: a case study of the Northern Spotted Owl in the Olympic Peninsula. *Ecological Applications* **7**, 1160–9.

Jager, H. I. and L. J. Gross. 2000. Spatial control: the final frontier in applied ecology (a book review of Spatial Optimization for Managed Ecosystems). *Ecology* **81**, 1473–4.

Kierstead, H. and L. B. Slobodkin. 1953. The size of water masses containing plankton blooms. *Journal of Marine Research* **12**, 141–7.

Lande, R. 1987. Extinction thresholds in demographic models of territorial populations. *American Naturalist* **130**, 624–35.

Loehle, C. 1999. Optimizing wildlife habitat mitigation with a habitat defragmentation algorithm. *Forest Ecology and Management* **120**, 245–51.

Lubchenco, J., S. R. Palumbi, S. D. Gaines, and S. Andelman. 2003. Plugging a hole in the ocean: the emerging science of marine reserves. *Ecological Applications* **13**, S3–7.

MacNally, R. and E. Fleishman. 2002. Using "indicator" species to model species richness: model development and predictions. *Ecological Applications* **12**, 79–92.

Martell, D., E. Gunn, and A. Weintraub. 1998. Forest management challenges for operational researchers. *European Journal of Operations Research* **104**, 1–17.

Morrison, M. L., I. C. Timossi, and K. A. With. 1987. Development and testing of linear regression models predicting bird-habitat relationships. *Journal of Wildlife Management* **51**, 247–53.

Murray, A. T. and R. L. Church. 1995. Heuristic approaches to operational forest planning problems. *OR Spektrum* **17**, 193–203.

Nalle, D. J., J. L. Arthur, and J. Sessions. 2002. Designing compact and contiguous reserve networks with a hybrid heuristic algorithm. *Forest Science* **48**, 59–68.

Nevo, A. and L. Garcia. 1996. Spatial optimization of wildlife habitat. *Ecological Modelling* **91**, 271–81.

Noon, B., K. McKelvey, and D. Murphy. 1997. Developing an analytical context for multispecies conservation planning. Pages 43–59 in S. T. A. Pickett, R. S. Ostfeld, M. Shachak, and G. E. Likens (eds.). *The Ecological Basis of Conservation: Heterogeneity, Ecosystems, and Biodiversity*. New York: Chapman and Hall.

Onal, H. and R. A. Briers. 2002. Incorporating spatial criteria in optimum reserve selection. *Proceedings of the Royal Society of London B* **269**, 2437–41.

Onal, H. and R. A. Briers. 2003. Selection of a minimum boundary reserve network using integer programming. *Proceedings of the Royal Society of London* **270**, 1487–91.

Polasky, S., J. D. Camm, and B. Garber-Yonts. 2001. Selecting biological reserves cost-effectively: an application to terrestrial vertebrate conservation in Oregon. *Land Economics* **77**, 68–78.

Possingham, H., I. Ball, and S. Andelman. 2000. Mathematical methods for identifying representative reserve networks. Pages 291–306 in S. Ferson and M. Burgman (eds.). *Quantitative Methods for Conservation Biology*. New York: Springer-Verlag.

Pressey, R. L., C. J. Humphries, C. R. Margules, R. I. Vane-Wright, and P. H. Williams. 1993. Beyond opportunism: key principles for systematic reserve selection. *Trends in Ecology and Evolution* **8**, 124–8.

Rapport, D. J., H. A. Regier, and T. C. Hutchinson. 1985. Ecosystem behavior under stress. *American Naturalist* **125**, 617–40.

Reeves, C. R. 1993. *Modern Heuristic Techniques for Combinatorial Problems*. New York: Halsted Press.

Reid, W. V. 1998. Biodiversity hotspots. *Trends in Ecology and Evolution* **13**, 275–80.

Rodrigues, A. S. L. and K. J. Gaston. 2002. Optimisation in reserve selection procedures: why not? *Biological Conservation* **107**, 123–9.

Saunders, D. A., R. J. Hobbs, and C. R. Margules. 1991. Biological consequences of ecosystem fragmentation: a review. *Conservation Biology* **5**, 18–32.

Simberloff, D. 1988. The contribution of population and community biology to conservation science. *Annual Review of Ecology and Systematics* **19**, 473–511.

Skellum, J. G. 1951. Random dispersal in theoretical populations. *Biometrika* **38**, 196–218.

Toregas, C. and C. ReVelle. 1973. Binary logic solutions to a class of location problems. *Geographical Analysis* **5**, 145–55.

Trzcinski, M. K., L. Fahrig, and G. Merriam. 1999. Independent effects of forest cover and fragmentation on the distribution of forest breeding birds. *Ecological Applications* **9**, 586–93.

Turchin, P. 1998. *Quantitative Analysis of Movement: Measuring and Modeling Population Redistribution in Animals and Plants*. Sunderland, MA: Sinauer Associates.

Turner, M. G., G. J. Arthaud, R. T. Engstrom, *et al.* 1995. Usefulness of spatially explicit population models in land management. *Ecological Applications* **5**, 12–16.

Urban, D. L., R. V. O'Neill, and H. H. Shugart. 1987. Landscape ecology. *BioScience* **37**, 119–27.

von Braun, W., F. I. Ordway III, and D. Dooling. 1985. *Space Travel: A History*, 4th edn. New York: Harper and Row.

Walters, C. J. and C. S. Holling. 1990. Large-scale management experiments and learning by doing. *Ecology* **71**, 2060–8.

Wennergren, U., M. Ruckelshaus, and P. Kareiva. 1995. The promise and limitations of spatial models in conservation biology. *Oikos* **74**, 349–56.

Williams, J. C. and C. S. ReVelle. 1997. Applying mathematical programming to reserve selection. *Environmental Modeling and Assessment* **2**, 167–75.

Williams, J. C., C. S. ReVelle, and S. A. Levin. 2004. Using mathematical optimization models to design nature reserves. *Frontiers in Ecology and the Environment* **2**, 98–105.

With, K. A. and T. O. Crist. 1995. Critical thresholds in species' responses to landscape structure. *Ecology* **76**, 2446–59.

Wu, J. and R. Hobbs. 2002. Key issues and research priorities in landscape ecology: an idiosyncratic synthesis. *Landscape Ecology* **17**, 355–65.

9

Advances in detecting landscape changes at multiple scales: examples from northern Australia

9.1 Introduction

As we move into a new century, changes in the cover and condition (or "state of health"; see Section 9.3) of different landscapes continue under human influences. Our responses to landscape changes are often slow because subtle problems often go undetected, resulting in costly long-term environmental and socio economic problems (Scheffer *et al.* 2003). Landscape changes are also spatially heterogeneous and occur at different scales, and it is difficult to select the appropriate scale for analysis (Gustafson 1998). As landscape ecologists, we are perhaps best at quantifying biophysical changes at fine scales (e.g., plots of 1–$100\,m^2$) using ground-based data and at coarse scales (e.g., regions and catchments of $>100\,km^2$) using satellite imagery (Ludwig *et al.* 2000). However, between fine plots and coarse regions, that is, at local watershed and pasture scales (e.g., 1–$100\,km^2$), where management actions are often most urgently needed, changes in land condition often go undetected. A key challenge for us is to develop methods and indicators that will readily detect changes in land cover and condition at these watershed and pasture scales, as well as at fine plot and coarse regional scales.

Thus, landscape change information is needed at multiple scales, from fine to coarse. In Australia, land cover and condition changes are used at the very coarse national scale (i.e., 7.7 million km^2) for periodic reporting on the "State of the Environment" (Hamblin 2001), which is used to guide national environmental policies. At state-territory, regional, and property scales in northern Australia, information on changes in land cover and condition are used to develop natural resource management plans (Landsberg *et al.* 1998, Karfs *et al.* 2000) and to set land-use regulations, for example, on clearing remnant vegetation on pastoral leases (Wilson *et al.* 2002). Qualitatively monitoring land

Key Topics in Landscape Ecology, ed. J. Wu and R. Hobbs.
Published by Cambridge University Press. © Cambridge University Press 2007.

condition at the paddock scale (e.g., 100–1000 km^2) has aided rangeland managers in developing strategies and tactics for improving the condition of their grazing lands (Landsberg *et al.* 1998). However, as landscape ecologists, our goal is to quantify changes in land cover and condition at multiple scales so that landscape changes can be better managed across all scales.

Land condition changes at one scale have flow-on or secondary effects at other scales. For example, in northern Australia, changes in land cover and condition in upper watersheds have impacts on streamside riparian zone vegetation and on in-stream water quality and aquatic organisms lower in the watershed (Burrows and Butler 2001, Douglas *et al.* 2003). Some land condition changes can have far-reaching environmental, social, and economic effects as, for example, soil erosion in catchments draining into the Great Barrier Reef estuary of Australia is causing environmental damage (Prosser *et al.* 2001), which reduces the aesthetic and economic value of this reef for tourism. Identifying such flow-on effects of land condition change across a range of scales, and predicting their consequences using tools such as models (Strayer *et al.* 2003), is another key challenge facing landscape ecologists.

Identifying changes in land cover and condition, and their flow-on effects, is important, but perhaps a greater challenge for landscape ecologists is to understand the ecological processes that drive landscape changes, and the interactions between patterns of change and processes across a range of scales (Turner *et al.* 2001). A good understanding of pattern–process relationships is needed, along with tools such as simulation models, if we want to counteract problems or restore landscapes by appropriate management. For example, simulation models have been used to explore how pastoral activities influence the interactions between fire and grazing, and how these factors change savanna landscapes in northern Australia (Liedloff *et al.* 2001).

In this paper, I provide my perspective on recent advances in developing methods for addressing three key challenges facing landscape ecologists: (1) detecting changes in land cover and condition, and its spatial heterogeneity, at multiple scales, (2) identifying flow-on effects from landscape changes at different scales, and (3) understanding the ecological processes that drive changes in landscape condition so that effective management actions can be taken. To illustrate these three challenges, I use examples from the savanna landscapes of northern Australia.

9.2   Examples of detecting landscape changes from northern Australia

9.2.1   Source of examples

For the sake of brevity and author familiarity, examples will be drawn from recent ecological research in the landscapes of northern Australia. These landscapes are mostly savannas, as comprehensively described on a tropical savannas website (http://savanna.cdu.edu.au/information/). Briefly, savannas vary from nearly treeless grasslands to open woodlands (Mott *et al.* 1985). *Corymbia* and *Eucalyptus* species dominate the tree layer and tall tropical grasses characterize the grass layer. The climate is tropical with distinctive wet and dry seasons. Near the north coast of Australia, rainfall often exceeds 1000 mm during the wet season (December to March), declining inland to less than 500 mm towards the center at Alice Springs. Soils vary from sands and calcareous loams to cracking clays, depending on geological parent materials and topography.

9.2.2 Defining landscape condition

Before proceeding with examples, some definitions are provided. Landscapes in good condition, or "healthy" landscapes, have been defined (Whitehead *et al.* 2000) as those landscapes in northern Australia that:

- maintain basic functions at all spatial scales, including functions such as nutrient cycling, water capture, and provision of food and shelter for fauna;
- maintain viable populations of all native species of plants and animals at appropriate spatial scales and time scales; and
- reliably meet the long-term material, aesthetic, and cultural needs of people who have an ongoing interest in the savannas.

At the time this definition was developed, through a series of workshops held in Darwin, Australia, in 1999, it was considered a working definition, that is, one subject to change and refinement. In 2003, we refined our view of what constitutes healthy savannas to include the notion that landscapes in good condition also provide a "home" for different groups of people, that is, a place where they can enjoy their own culture and freely share their culture with others.

The condition of landscapes falls along a continuum, which can be simply illustrated as varying from good to poor (Fig. 9.1a). To evaluate where a particular landscape falls on this continuum, one can assess specific attributes and indicators of the landscape, for example, the capacity of the landscape to retain resources (e.g., capture runoff). Good condition savannas, for example, have a dense ground-layer of perennial grasses under an open tree layer (Fig. 9.1b), which functions to retain soil sediments and organic matter flowing in runoff (Tongway and Ludwig 1997). In contrast, poor condition savannas have little vegetation cover and mostly bare soils, which do not retain, but leak resources. Other landscape attributes and indicators can also be assessed, such as natural soil fertility and the complexity of habitats, and these indicators can be used to position a landscape along this continuum of condition or health. The position of a landscape along this continuum based on one attribute or indicator can be

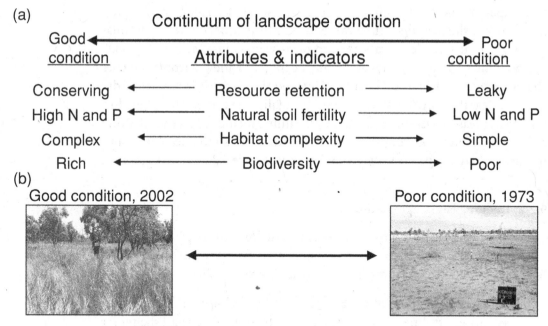

FIGURE 9.1

(a) Landscapes can be positioned along a continuum from good to poor condition using attributes and indicators related to their functionality (see text), which is illustrated by fixed-point photographs, (b) where a landscape was in good condition in 2002 after recovering from a poor condition due to an exclosure being established in 1973

different from that for another attribute or indicator. For example, a fertilized site (a field) with high N and P could have a simple habitat structure (a single crop) with poor biological diversity.

9.3   Key challenges

9.3.1 Key challenge 1: detecting changes in landscape condition at multiple scales

Ground-based methods are used to detect changes in land cover and condition. For example, changes in vegetation density and biomass have been monitored using fixed plots (1400 m^2) in the savannas of northern Australia (Bastin *et al.* 2003). Fixed photo points were also used to record and illustrate landscape changes. The two photos in Fig. 9.1 were taken at the same point in 2002 and 1973 (left to right), which illustrates vegetation recovery within an exclosure built in 1973 after the land was subject to 90 years of cattle, horse, and donkey grazing. Land condition data are also being collected on ground-based monitoring plots in the savannas of the Northern Territory (Karfs *et al.* 2000).

Although useful for detecting changes in landscape cover and condition at specific plot locations, this fine-scale, ground-based monitoring does not provide a broader view of vegetation change, which is needed for improving land management at coarser watershed and property scales (i.e., 100–1000 km^2) and at regional scales (i.e., >1000 km^2). At watershed, property and regional scales, remote-sensing data have proven useful for monitoring changes in rangeland cover and condition, and the spatial heterogeneity of these changes (e.g., Karfs *et al.* 2000). The usefulness of different ground- and remote-based methods for monitoring changes in Australia's rangelands at different scales has been comprehensively reviewed (TS-CRC 2000).

An example of a useful method, applicable at the property and regional scales, is the use of time sequences of Landsat TM imagery. Karfs *et al.* (2000) acquired this imagery from 1987 to 1998 to assess changes in land cover and condition across the Victoria River District of the Northern Territory of Australia. With this "cover-change" method, the region is spatially partitioned into different lithology types and, within each type, the ground cover reflectance data from Landsat TM time-series imagery are first corrected and then statistically evaluated to define spatial trends and patterns of heterogeneity (Wallace and Campbell 1998). Changes in cover over different time periods are color-coded and regional maps are produced (Karfs *et al.* 2000). These cover-change assessments are generated in a timely manner to provide land managers with early warnings of land condition problems so that they can take remedial action. Regional, color-coded maps also help land managers better understand the dynamics of their property relative to the general region, and relative to natural rainfall patterns and their own land management practices.

At finer watershed scales (i.e., a few km^2), Landsat TM imagery has proven useful for investigating landscape patterns of soil erosion and deposition in the Northern Territory of Australia (Pickup 1985). Imagery at this scale has also been used for assessing trends in the condition of pastoral lands in the Territory (Pickup *et al.* 1998). With this method, Landsat TM time-series data, reflecting changes in ground cover near cattle watering-points, are compared between wet and dry periods. If time-series data document that only ephemeral vegetation occurs near watering-points, this indicates that the pasture is likely to be in poor condition, whereas good condition is indicated if the imagery data suggests that perennial vegetation occurs near these watering-points. Such land condition assessments assist land managers set stocking levels for their pastures.

These Landsat TM land cover and condition indicators are based on well-established remote-sensing methods. New advances are aimed at developing indicators that reflect the functionality of landscapes in terms of how well resources are likely to be retained in, not leaked from, landscapes (Ludwig *et al.* 2000). Ideally, indicators of landscape functionality should be based on remotely sensed land-cover data, and on the spatial configuration and heterogeneity of this cover. A landscape leakiness index has been developed based on high-resolution, remotely sensed imagery (aerial videography) from fine-scale (i.e., 1000 m^2), relatively uniform hillslopes in northern Australia (Ludwig *et al.* 2002). This "directional" leakiness index strongly reflected the observed condition or health of these uniform hillslopes.

This landscape leakiness index also has a number of conceptual and computational advantages over related landscape metrics such as the Lacunarity index, which correctly ranked sites from poor to good condition, but did not strongly distinguish the poor condition sites (Bastin *et al.* 2002). Current work by these authors is to develop a new landscape leakiness index based on remotely sensed imagery of coarser-scale (i.e., 1–100 km^2), relatively rough-terrain landscapes (i.e., with variable topography). The aim is to derive a leakiness index that is more widely applicable and that can be rapidly assessed to provide early warning indicators of undesirable changes in landscape functionality so that effective action can be taken.

9.3.2 Key challenge 2: flow-on effects at multiple scales

Identifying how land-use, cover and condition changes affect various components of the landscape remains a challenging research topic in landscape ecology (Wu and Hobbs 2002). Ecological studies on how natural and anthropogenic disturbances affect organisms and ecosystems have been conducted for decades, and books (e.g., Pickett and White 1985) and journals (e.g., *Restoration Ecology*) are dedicated to this topic. However, placing such disturbance effects into a landscape context requires a broad perspective of patch patterns and ecological processes (Turner *et al.* 2001).

For example, in northern Australia's savannas, the killing of individual or clumps of trees (Fig. 9.2a,b) and the general clearing of trees for creating more open pastures (Fig. 9.2c) is a common practice (Ash *et al.* 1997). Tree clearing greatly alters the patterning of vegetation patches and the action of hydrological processes in savanna landscapes (Ludwig and Tongway 2002). Vegetation patchiness is reduced and the cleared landscape is typically sown with exotic grasses such as buffel grass (*Cenchrus ciliaris*). This loss of trees and the new open grassy habitat has a flow-on effect to favor birds such as the red-backed fairy-wren (*Malurus melanocephalus*).

A review of the literature by Ludwig and Tongway (2002) also documented other flow-on or secondary effects. Cleared savanna landscapes can have

FIGURE 9.2
Photos of savannas in northern Australia where (a) individual trees and (b) clumps of trees have been killed, and (c) where tree have been cleared and exotic grasses sown to create open pastures

significantly greater rates of runoff and soil erosion than uncleared lands on similar topographies and soils. Further, soil carried in runoff can pollute other systems such as the Great Barrier Reef estuary (Prosser *et al.* 2001).

In the rangelands of northern Australia, the disturbance of vegetation patterns and ecological processes in savannas and grasslands can have flow-on effects over a range of scales (Ludwig *et al.* 2004). For example, at fine plot-scales (i.e., 1–100 m^2), the composition and diversity of fauna such as grasshoppers and spiders is strongly reduced near cattle watering-points due to a loss of vegetation patches, that is, habitat (Ludwig *et al.* 1999, Churchill and Ludwig 2004). These findings suggest that the structure and spatial pattern of vegetation patches can be used as indicators of how well landscapes provide habitats for a diverse fauna. Advances are being made in identifying habitat indicators for monitoring biodiversity at these fine plot-scales, and such habitat indicators are now being incorporated into rangeland monitoring programs (Woinarski and Fisher 2003, Smyth and James 2004). However, further advances are needed on how these habitat indicators can be related to biodiversity at coarser property and regional scales (i.e., 100–1000 km^2) (TS-CRC 2000, Ludwig *et al.* 2004).

9.3.3 Key challenge 3: ecological processes driving landscape change

Another key research challenge is to advance our understanding of how ecological processes drive the changes and flow-on effects we are observing at different landscape scales. For example, as noted above, we have a basic understanding of how reducing the ground cover, and the patchiness of this cover, on a hillslope can increase runoff and erosion (Ludwig and Tongway 2002), and how this can have downstream or flow-on effects (Burrow and Butler 2001, Prosser *et al.* 2001). However, when it comes to accurately predicting the consequences of altering landscape processes by our management actions (e.g., Liedloff *et al.* 2001), we need an in-depth ecological and socioeconomic understanding.

An example of research that is advancing our understanding on how landscape processes control and drive changes at different scales comes from a hierarchical geo-ecological approach (Pringle and Tinley 2001, Pringle 2002). This approach aims to identify key geomorphic "nick-points" in the landscape where incisions, such as head-cutting gullies caused by cattle impacts, are altering surface-flow processes, which then drive changes in vegetation (Fig. 9.3). In this example from northern Australia, note how head-cutting gullies have formed to "nick" water flowing off higher slopes (foreground), starving savanna trees of vital water, causing their death. The grassy ground-layer has also been lost. Such drastic vegetation and hydrological changes will also

FIGURE 9.3
An oblique aerial photo of a landscape in northern Australia where head-cutting gullies have altered water flows causing tree death, hence, a change from a savanna to a barren, eroded plain

alter other components in this landscape, such as the composition of flora and fauna assemblages, as found for other savanna systems (e.g., Ludwig *et al.* 1999, Woinarski *et al.* 2002). This understanding of how such "nick-points" drive landscape changes can be used to predict how repairing landscape incisions could restore former hydrologic and geomorphic processes and hence return the vegetation to something similar to, but not necessarily identical to, what was there before.

9.4   Summary

In this chapter, I have presented my perspective on what I consider to be three key research topics in landscape ecology that need advancement. First, we need new indicators that are sensitive to subtle changes in landscape cover and condition, and that can be derived from remotely sensed imagery at different scales of resolution so that we can take a fine-to-coarse view of landscapes. Ideally, these new indicators should reflect the functionality of landscapes, that is, how they retain vital natural resources, provide habitats for our native flora and fauna, and meet the needs of people, and they should provide early warnings of negative changes so that appropriate management action can be taken. Second, we need to improve our understanding of how changes in a landscape at one scale can have flow-on or secondary effects at other scales. Third, we also need to better understand the ecological and socioeconomic processes that drive landscape changes so that we can improve our predictive capacity and, hence, our management of changing landscapes.

Acknowledgments

I am indebted to my colleagues in the Tropical Savannas Cooperative Research Centre and in the Commonwealth Scientific and Industrial Research Organisation for helping me develop the views expressed in this essay, although I take full responsibility for any errors in my interpreting and expressing their work.

References

Ash, A.J., J.G. McIvor, J.J. Mott, and M.H. Andrew. 1997. Building grass castles: integrating ecology and management of Australia's tropical tallgrass rangelands. *Rangeland Journal* **19**, 123–44.

Bastin, G.N., J.A. Ludwig, R.W. Eager, *et al.* 2003. Vegetation changes in a semiarid tropical savanna, northern Australia: 1973–2002. *Rangeland Journal* **25**, 3–19.

Bastin, G.N., J.A. Ludwig, R.W. Eager, V.H. Chewings, and A.C. Liedloff. 2002. Indicators of landscape function: comparing patchiness metrics using remotely sensed data from rangelands. *Ecological Indicators* **1**, 247–60.

Burrows, D.W. and B. Butler. 2001. Managing livestock to protect the aquatic resources of the Burdekin Catchment, North Queensland. Pages 95–101 in I. Rutherford, F. Sheldon, G. Brierley, and C. Kenyon (eds.). *The Value of Healthy Streams, Proceedings of the Third Australian Stream Management Conference.* Melbourne: Cooperative Research Centre for Catchment Hydrology.

Churchill, T.B. and J.A. Ludwig. 2004. Changes in spider assemblages along grassland and savanna grazing gradients in northern Australia. *Rangeland Journal* **26**, 3–16.

Douglas, M.M., S.A. Townsend, and P.S. Lake. 2003. Streams. Pages 59–78 in A.A. Andersen, G.D. Cook, and R.J. Williams (eds.). *Fire in Tropical Savannas: the Kapalga Experiment.* New York: Springer.

Gustafson, E.J. 1998. Quantifying landscape spatial patterns: what is the state of the art? *Ecosystems* **1**, 143–56.

Hamblin, A.P. 2001. *Australia State of the Environment Report 2001: Land Theme Report.* Department of the Environment and Heritage, Commonwealth of Australia. Melbourne: CSIRO Publishing.

Karfs, R., R. Applegate, R. Fisher, *et al.* 2000. Regional land condition and trend assessment in tropical savannas. In *Impacts on Biophysical Resources. National Land and Water Resources Audit.* Canberra: Australian Government, Natural Heritage Trust Ministerial Board (http://audit.ea.gov.au/ANRA/rangelands/docs/project.html).

Landsberg, R.G., A.J. Ash, R.K. Shepard, and G.M. McKeon. 1998. Learning from history to survive in the future: management evolution on Trafalgar Station, northeast Queensland. *Rangeland Journal* **20**, 104–18.

Liedloff, A.C., M.B. Coughenour, J.A. Ludwig, and R. Dyer. 2001. Modelling the trade-off between fire and grazing in a tropical savanna landscape, northern Australia. *Environment International* **27**, 173–80.

Ludwig, J. A., G. N. Bastin, R. W. Eager, et al. 2000. Monitoring Australian rangeland sites using landscape function indicators and ground- and remote-based techniques. *Environmental Monitoring and Assessment* **64**, 167–78.

Ludwig, J. A., R. W. Eager, G. N. Bastin, V. H. Chewings, and A. C. Liedloff. 2002. A leakiness index for assessing landscape function using remote-sensing. *Landscape Ecology* **17**, 157–71.

Ludwig, J. A., R. W. Eager, R. J. Williams, and L. M. Lowe. 1999. Declines in vegetation patches, plant diversity and grasshopper diversity near cattle watering points in the Victoria River District, northern Australia. *Rangeland Journal* **21**, 135–49.

Ludwig, J. A. and D. J. Tongway. 2002. Clearing savannas for use as rangelands in Queensland: altered landscapes and water-erosion processes. *Rangeland Journal* **24**, 83–95.

Ludwig, J. A., D. J. Tongway, G. N. Bastin, and C. James. 2004. Monitoring ecological indicators of rangeland functional integrity and their relationship to biodiversity at local to regional scales. *Austral Ecology* **29**, 108–20.

Mott, J. J., J. Williams, M. H. Andrew, and A. Gillison. 1985. Australian savanna ecosystems. Pages 56–82 in J. C. Tothill and J. J. Mott (eds.). *Ecology and Management of the World's Savannas.* Canberra: Australian Academy of Science.

Pickett, S. T. A. and P. S. White (eds.). 1985. *The Ecology of Natural Disturbance and Patch Dynamics.* New York: Academic Press.

Pickup, G. 1985. The erosion cell: a geomorphic approach to landscape classification in range assessment. *Australian Rangeland Journal* **7**, 114–21.

Pickup, G., G. N. Bastin, and V. H. Chewings. 1998. Identifying trends in land degradation in non-equilibrium rangelands. *Journal of Applied Ecology* **35**, 365–77.

Pringle, H. J. R. 2002. *Grazing Impacts in Rangelands: Assessment of Two Contrasting Landscape Types in Arid Western Australia from Different Land Management Practices.* PhD Thesis, Australian National University, Canberra.

Pringle, H. J. R. and K. L. Tinley. 2001. Ecological sustainability for pastoral management. *Journal of Agriculture* **42**, 30–5.

Prosser, I. P., I. D. Rutherford, J. M. Olley, et al. 2001. Large-scale patterns of erosion and sediment transport in river networks, with examples from Australia. *Marine and Freshwater Research* **52**, 81–99.

Scheffer, M., F. Westley, and W. Brock. 2003. Slow response of societies to new problems: causes and costs. *Ecosystems* **6**, 493–502.

Smyth, A. K. and C. D. James. 2004. Characteristics of Australia's rangelands and key design issues for monitoring biodiversity. *Austral Ecology* **29**, 3–15.

Strayer, D. L., R. E. Beighley, L. C. Thompson, et al. 2003. Effects of land cover on stream ecosystems: roles of empirical models and scaling issues. *Ecosystems* **6**, 407–23.

Tongway, D. J. and J. A. Ludwig. 1997. The conservation of water and nutrients within landscapes. Pages 13–22 in J. A. Ludwig, D. T. Tongway, D. O. Freudenberger, J. C. Noble, and K. C. Hodgkinson (eds.). *Landscape Ecology, Function and Management: Principles from Australia's Rangelands.* Melbourne: CSIRO Publishing.

TS-CRC. 2000. A review of pastoral monitoring programs and their real and potential contribution to biodiversity monitoring. In *Background Paper 2, Rangelands Monitoring: Developing an Analytical Framework for Monitoring Biodiversity in Australia's Rangelands.* Tropical Savannas Cooperative Research Centre (TS-CRC) for the National Land and Water Resources Audit. Canberra: Australian Government, Natural Heritage Trust Ministerial Board (http://audit.ea.gov.au/ANRA/rangelands/docs/change/BP02.pdf).

Turner, M. G., R. H. Gardner, and R. V. O'Neill. 2001. *Landscape Ecology in Theory and Practice: Pattern and Process.* New York: Springer-Verlag.

Wallace, J. F. and N. A. Campbell. 1998. *Evaluation of the Feasibility of Remote Sensing for Monitoring National State of the Environment Indicators.* State of the Environment Technical Paper Series. Canberra: Australian Government Department of the Environment.

Whitehead, P. J., J. Woinarski, P. Jacklyn, D. Fell, and D. Williams. 2000. *Defining and Measuring the Health of Savanna Landscapes: a North Australian Perspective*. Darwin: Tropical Savannas Cooperative Research Centre, Charles Darwin University.

Wilson, B. A., V. J. Neldner, and A. Accad. 2002. The extent and status of remnant vegetation in Queensland and its implications for statewide vegetation management and legislation. *Rangeland Journal* 24, 6–35.

Woinarski, J. C. Z., A. N. Andersen, T. B. Churchill, and A. J. Ash. 2002. Response of ant and terrestrial spider assemblages to pastoral and military land use, and to landscape position, in a tropical savanna woodland in northern Australia. *Austral Ecology* 27, 324–33.

Woinarski, J. C. Z. and A. Fisher. 2003. Conservation and the maintenance of biodiversity in the rangelands. *Rangeland Journal* 25, 157–71.

Wu, J. and R. Hobbs. 2002. Key issues and research priorities in landscape ecology: an idiosyncratic synthesis. *Landscape Ecology* 17, 355–65.

10

The preoccupation of landscape research with land use and land cover

10.1 Introduction

For most people, their initial contact with the landscape is by the observation of landform and land cover. Human-perception analysis evaluates what is observed in a holistic way and interprets simultaneously according to the available knowledge. Landscape can be approached in multiple ways (Muir 1999, Cosgrove 2003, Claval 2005) and similar concepts have subtle differences in meaning. In common language and disciplines related to policy and planning, the concepts of land use and land cover are sometimes erroneously used as synonyms, while scientific communities use clearly distinct definitions (Baulies and Szejwach 1997). An important conceptual difference also exists between landscape and land (Zonneveld 1995, Antrop 2001, 2003, Olwig 2004). Land is more associated with territory, terrain, soil, and land value, which depend on its utility. The landscape is considered as a perceivable expression of the dynamic interaction between natural processes and human activities in an area (Council of Europe 2000). Although land use and land cover are essential components in the characterization of the landscape, the concept of landscape is broader and encompasses social, economic, and symbolic aspects as well. The increasing magnitude and pace of the changes in land use and land cover have become of worldwide concern in policy-making (Fresco *et al.*, 1996), land management (Dale *et al.* 2000, Pontius *et al.* 2004), and modeling land-use changes (Veldkamp and Lambin 2001, Agarwal *et al.* 2002). Issues such as global warming, land degradation, and deforestation rarely are focused directly upon the landscape as a whole, integrating natural, cultural, and scenic values.

For landscape ecologists and geographers, land use and land cover are primary features of the landscape to be studied. The central paradigm is the

Key Topics in Landscape Ecology, ed. J. Wu and R. Hobbs.
Published by Cambridge University Press. © Cambridge University Press 2007.

continuous interaction between patterns formed by patches of land use and land cover and the processes that define the functioning of the landscape (Forman and Godron 1986). The study of changes in land use and land cover thus forms a key issue in landscape ecological research. A basic question, when applying landscape ecological principles in spatial planning, landscape management, and conservation, is: "What forms and spatial arrangements of land use can be suggested as being ecologically appropriate?" The mission statement and objectives of the International Association for Landscape Ecology (IALE) centers on interdisciplinary synergism involving all activities dealing with the landscape (http://www.landscape-ecology.org/). The increasingly faster changes of the landscape demand more comprehensive land-use policy and planning, and the members of IALE have an important role to play in this.

The use of landscape ecological knowledge in planning and management is still in its infancy (Dale *et al.* 2000). Landscape planning is a complex problem as multiple approaches are possible and many aspects need to be dealt with. Applying landscape ecological knowledge demands increasingly an inter- and transdisciplinary approach (Bastian 2001, Opdam *et al.* 2001, Tress *et al.* 2003). In highly dynamical complex landscapes, such as urbanized ones, the discussion focuses on issues of multifunctional land use (Fry 2001, Brandt and Vejre 2004), sustainability of landscapes (Haines-Young 2000), and designing future landscapes (Nassauer 1997, Steinitz 2001). Spatial planning, rural development, landscape conservation, and landscape design are directly involved in many aspects of land-use or land-cover changes. Changes in land use and land cover are important indicators of processes that act on different spatial scales ranging from action by local agents to global processes (Dale *et al.* 2000, Agarwal *et al.* 2002). Thus land use and land cover are essential data sources in landscape classification and typology (Mücher *et al.* 2003) as well as modeling changes and predictions (Baulies and Szejwach 1997, Veldkamp and Lambin 2001).

The purpose of this paper is to explore the general context in which the terms land use/land cover and land-use/land-cover changes occur in relation to landscape research, planning, and design activities, different landscape types, and possible causes and processes of change. The goal is to detect correspondences and differences between different approaches and activities dealing with the landscape and land use/land cover at a global scale. For this an Internet search-based approach has been used. The results reveal different patterns of associations in dealing with these concepts, indicating how disciplines in landscape research, planning, management, and design are involved differently. Thus, areas and topics for further integration of landscape ecological objectives can be formulated.

10.2 Method

As the aim of this analysis is to explore the global context for the use of the key concepts related to land use and land cover, the use of an Internet search seems a straightforward and appropriate approach. Google was chosen as it is a general search engine that allows searches on an equal basis in all domains and allows one to cover a broad variety of activities related to landscape, land use, and land-use change. Landscape and land-use/land-cover changes are the result of many factors combining natural processes and human activities (Dale *et al.* 2000). Decisions on land use involve agents at several scales and affect a wide range of spatial extent and duration (Agarwal *et al.* 2002). Consequently, it is important not to restrict this search to scientific databases alone, as many agents outside the scientific community are involved in the landscape and in land-use change.

Google uses the PageRank algorithm to define the significance and importance of the web pages that match the search, which reduces the sensitivity of the search for repeating and common words within the web pages. Page-Rank defines the importance of a web page by analyzing the link structure from and to the page (Brin and Page 1998), and is similar to a citation index (Ridings and Shishigin 2002). Google also uses stemming to find word and spelling variations, thus entering "land use" will also search for "land-use." Google automatically uses the Boolean AND-operator between keywords. Using the advanced search facilities, explicit word combinations can be searched for. A search of "landscape ecology" AND "land use" gave 34 200 hits and included "land-use" and "Landscape Ecology" as a consequence of the stemming and because Google is not case sensitive. The search for "landscape ecology" AND "land cover" gave 11 900 hits, for "landscape ecology" AND "land-use change" 6 640 hits, and "landscape ecology" AND "land-cover change" only 3 760 hits. The number of hits resulting from a search for a single keyword or keywords in specific combinations gives an indication of their absolute importance (magnitude) at the time of the search.

In the example above, "land use" was associated with "landscape ecology" two times more frequently than "land cover." The combination of "land use" AND "land-use change" is meaningless and will result in the same number of hits as "land-use change." The relative importance of certain keywords in association was expressed as a percentage. A search for "landscape ecology" resulted in 121 000 hits, while "land use" returned 5 660 000 hits and "land cover" 602 000 hits. Thus the combinations of "land use" and "land cover" with "landscape ecology" resulted in an correspondence of 28 percent of "land use" in "landscape ecology" and 10 percent of "land cover." In a similar way "land-use change" occurs only 3 percent in the total of "land use," and "land-cover change" occurs twice as much (6 percent) in the total of "land cover." Acronyms such as LULC, LU/LC for a combination of the terms "land use" and "land cover," and LUCC (Land Use Land Cover Change), the acronym of the important international and interdisciplinary project by the International Geosphere Biosphere Programme and the International Human Dimension Programme (Fresco *et al.* 1996)), were not used as search keywords, as many nonrelevant topics are included as well.

The large number of hits and the important differences between the results of different combinations of keywords were used to explore general correspondences at a certain moment. First, keywords were selected and grouped into different sets. The selection of the keywords was based upon their frequent use in various landscape disciplines. Common words were avoided by using combinations with land and landscape; so the search term "landscape management" was used instead of simply "management." These were analyzed according to their absolute response and to their relative occurrence within the different categorical sets. In the following, the exact formulation of the keywords as used in the searches is represented in italics: *landscape ecology* means a search with "landscape ecology" in Google.

The first set of keywords consisted of terms related to *landscape* (and equivalents in five other languages), *countryside*, *environment*, *land use* and *land cover*. Landscape types were searched using following keywords: *natural landscape*, *urban landscape*, *cultural landscape*, *rural landscape* and *countryside*. The second set explored the responses for different disciplines: *landscape architecture*, *landscape ecology*, *landscape history*, *landscape science*, *landscape geography*, and *land-use planning*. The third set referred to activities such as *landscaping*, *landscape design*, *landscape management*, *landscape protection*, *landscape conservation*, *landscape assessment*, *landscape evaluation*, and *landscape classification*. *Landscape planning* was considered here arbitrarily as a discipline while *land-use planning* as an activity. To make an integrated analysis easier, disciplines and activities are combined in one table. The next set consisted of keywords related to causes and processes, grouped in three subsets: (1) mainly natural causes and processes, (2) human-social factors and (3) economical factors. Finally, keywords related to *education* and *teaching* as well as the use of *landscape metrics* and *landscape indicators* were searched.

10.3 Results

The English term *landscape* is the most dominant compared to the other selected languages (German, Dutch, Spanish, French, and Portuguese). The relative importance of *countryside* compared to *landscape* should be noted

TABLE 10.1. *Hits of selected keywords related to landscape, land use, and land cover*

Keyword	Hits	Percentage
landscape	6 170 000	80.6
Landschaft	738 000	9.6
paisaje	373 000	4.9
paysage	185 000	2.4
landschap	177 000	2.3
paisage	10 800	0.1
		100.0
countryside	2 120 000	
land	44 800 000	
environment	39 700 000	
ecology	4 370 000	
land use	2 160 000	88.8
land cover	254 000	10.4
land use/land cover	19 300	0.8
		100.0
land-use change	95 600	87.5
land-cover change	13 700	12.5
		100.0
landscape change	12 600	

(Table 10.1). The keyword *land use* occurs approximately 8.5 times more than *land cover* and the combined use of both terms is very limited (Table 10.1). Table 10.2 gives the total hits of disciplines related to landscape research and the percentage of common occurrence. *Landscape architecture* is by far the most dominant, and *landscape ecology* offers about a quarter of all the hits on the selected disciplines. *Landscape architecture* has the most hits with *landscape planning* (43.1 percent), *landscape ecology* (29 percent) and *landscape history* (10.8 percent), while the correspondence between other pairs of disciplines is not significant. A recently introduced term, *landscape science*, obtains more hits than *landscape geography*. Table 10.3 summarizes the correspondences between the disciplines and three selected activities. Obviously *landscape design* and *landscaping* scores are high in *landscape architecture*, but also within *landscape history* and *landscape planning*. However, *landscape architecture* corresponds to only about 1 percent of all hits in *landscaping*, which is also more associated with other terms such as gardening. Both disciplines seem to be least involved in *landscape management*. *Landscape ecology*, *landscape science* and *landscape geography* have a similar correspondence to all three activities, but all score low in the total number

TABLE 10.2. *Total hits of disciplines and percentage of correspondence between disciplines*

	Total hits	Percentage	Landscape architecture	Landscape ecology	Landscape planning	Landscape history	Landscape science	Landscape geography
Landscape architecture	145 000	60	–	29.0	43.1	10.8	0.9	0.1
Landscape ecology	57 500	24		–	4.2	0.7	0.3	0.1
Landscape planning	27 000	11			–	3.2	5.3	2.8
Landscape history	9 260	4				–	0.2	0.2
Landscape science	1 560	<1					–	0.2
Landscape geography	620	<1						–
Sum	240 940	100						

TABLE 10.3. *The relative importance of selected activities and disciplines: percentage of all activity hits and percentage of all discipline hits*

	Percentage of activities			Percentage of disciplines		
	Landscaping	Landscape design	Landscape management	Landscaping	Landscape design	Landscape management
Landscape architecture	1	11	1	17	18	<1
Landscape ecology	<1	1	6	3	4	5
Landscape planning	<1	2	4	15	20	8
Landscape history	<1	1	1	15	21	3
Landscape science	<1	<1	<1	4	9	7
Landscape geography	<1	<1	<1	4	2	6

of hits on the discipline. *Landscape ecology* and *landscape planning* score highest of all disciplines within the total hits on *landscape management*. Comparing disciplines in relation to *countryside* (Table 10.4) shows the importance of *land-use planning* and *landscape history*, while the other disciplines, including *landscape planning*, are less correlated to the term *countryside*. Table 10.5 shows a clear difference in the use of the terms *landscape* and *countryside* in the different activity types; *countryside* and *landscape assessment* and *conservation/protection* have much in common.

TABLE 10.4. *Hits of countryside as percentage of the total hits per discipline*

Disciplines	Percentage of *countryside*
Land-use planning	17
Landscape history	12
Landscape geography	7
Landscape planning	6
Landscape science	5
Landscape architecture	4
Landscape ecology	3

TABLE 10.5. *Hits of activity types as percentage of the total hits of landscape and countryside*

Activities	Percentage of total *landscape* hits	Percentage of total *countryside* hits
Landscaping	28	2
Landscape design	3	2
Landscape management	1	6
Landscape protection	< 1	17
Landscape conservation	< 1	25
Landscape assessment	< 1	29
Landscape evaluation	< 1	15
Landscape classification	< 1	11

Table 10.6 shows the correspondences between *land use, land cover* and associations of these with *change* with the selected disciplines and activities. The use of the combined *land use/land cover* is slightly higher for the discipline *landscape ecology* and the activities *landscape evaluation* and *classification*. The ratio between the hits on *land use* and *land cover* differs a lot between disciplines and activities. *Landscape architecture* and related keywords use almost exclusively *land use* as a concept, and so do, to a lesser extent, activities related to *landscape management* and *conservation*. The term *land use* occurs only about four times more than *land cover* in *landscape ecology*. For the activities *land cover* scores clearly higher with *landscape assessment, evaluation,* and *classification*. In most cases the term *change* can be considered as an important issue as it matches one third to one half of all the hits in disciplines and in particular with the activities. The low scores of *landscape planning* (26 percent), *landscape architecture* (31 percent) and the activities *landscaping* (25 percent) and *landscape design* (19 percent) should be noted.

TABLE 10.6. *Correspondence between* land use, land cover *and* change *with disciplines and activities: the occurrence of the land use and land cover in combination with disciplines and activities is expressed as a percentage of the total hits on discipline or activity*

	Land use as percentage of discipline or activity	Land cover as percentage of discipline or activity	Change as percentage of hits for disciplines or activities
Disciplines			
Landscape architecture	14	1	31
Landscape ecology	26	7	40
Landscape planning	23	3	26
Landscape history	16	2	34
Landscape science	13	2	30
Landscape geography	20	4	42
Activities			
Landscaping	8	< 1	25
Landscape design	6	< 1	19
Landscape management	20	2	30
Landscape conservation	36	5	45
Landscape protection	33	2	48
Landscape assessment	40	9	50
Landscape evaluation	44	10	48
Landscape classification	43	20	49

Table 10.7 compares the main categories of landscape types in combination with the search terms *change, land use, land cover* and combinations, as well as their association with the disciplines *geography, history, ecology,* and *landscape ecology*. The most frequently occurring keyword is *countryside*, which is considered here as a landscape type. It has at least 30 times more hits than *natural landscape, cultural landscape,* and *urban landscape,* and even more than 50 times the number of hits on *rural landscape*. Again, the number of hits on *land use* is greater than on *land cover* or their combined use. However, the association of *land use* with *countryside* is only 4 percent, indicating that *countryside* covers much more than *land use*. For the other landscape types this association varies from 11 percent to 18 percent. The number of hits of *land cover* and the combined used of *land use/land cover* is rather similar. As a common term, the keyword *history* has the largest number of hits and is associated from 31 percent to 54 percent with the landscape types and in particular with *cultural* and *rural landscapes*. *Geography* scores relatively high with *cultural landscape* (14 percent) and *ecology* with *natural* (12 percent) and *cultural landscapes* (11 percent). *Landscape ecology* scores very low in relation to the landscapes types, indicating that the keywords of the

TABLE 10.7. *Main landscape types in relation to change, land use, land cover and the main disciplines (percentage of total hits of landscape type)*

	Countryside	Natural landscape	Urban landscape	Cultural landscape	Rural landscape
Total hits	2 200 000	72 900	66 900	66 300	41 400
Land use	4	18	11	13	19
Land cover	< 1	2	1	1	2
Land use/cover	< 1	2	1	1	2
History	31	32	37	48	54
Geography	4	7	9	14	8
Ecology	3	12	9	11	9
Landscape ecology	< 1	1	1	2	2

landscape types are associated with many more other things and the concept of landscape ecology is not well penetrated here.

Table 10.8 compares keywords related to causes and processes of *change, land use* or *land cover*, which are grouped in three categories: nature and environment, humans, and economy. For each of these groups, the keywords are sorted in descending order according to the number of hits on *land use*. The absolute numbers of hits for the different categories of causes and processes vary a lot, but within each group the percentage of *land use* and *land cover* is very similar. A similar observation can be made for the associations with *land-use change* and *land-cover change*. The ratio between the number of hits of *land use* and *land cover* shows the relative importance with respect to the associated causes and processes. In the group of natural and environmental causes and processes, *land use* and *land cover* are proportionally most important in relation to *nature, pollution, climate, fire* and *degradation*, and the other terms are significantly less represented. *Land use* is approximately 10 to 20 times more common than *land cover*, except for the category *global warming*. *Land use* and *land cover* are highly associated in the group "humans" with keywords such as *population* and *man*, but least of all with urban issues and with *landscaping* and *landscape design*. However, the keyword *land use* is almost dominant in the *landscaping* and *landscape design*. In the group of economical related causes and processes, the keyword *development* is the most important one, followed by the keywords *agriculture* and *economy*. The association varies between *land use* and *land cover* in a similar way, except that the term *land use* is relatively more used in relation to *agriculture* and *mobility* than *land cover*. *Land cover* occurs more associated with *forestry* in this group. The pattern given by the associations with *land-use change* and *land-cover change* is similar. The dominance of *land use* over *land cover* does not

TABLE 10.8. *The relative importance of* land use, land cover *and* change *in different groups of causes and processes*

Causes/processes	Land use	Land cover	Land-use change	Land-cover change
Nature/environment				
Nature	28	27	19	19
Pollution	22	13	14	12
Climate	20	29	29	27
Fire	19	13	10	12
Degradation	7	9	10	12
Desertification	2	3	4	5
Land degradation	1	2	3	4
Soil degradation	1	1	1	2
Saliniz(s)ation	0	1	1	1
Global warming	0	3	7	6
Sum	100	100	100	100
Humans				
Population	41	47	40	38
Man	20	19	13	13
Population growth	10	6	13	11
Landscape design, landscaping	8	2	4	1
Humans	8	8	9	11
Urbaniz(s)ation	5	7	11	12
Population density	3	6	5	7
Urban growth	3	3	4	5
Urban sprawl	2	2	0	3
Sum	100	100	100	100
Economy				
Development	38	46	40	35
Agriculture	21	27	20	22
Economy	16	5	11	9
Farming	8	5	7	6
Grazing	4	5	5	6
Accessibility	4	5	4	4
Mobility	4	1	2	2
Forestry	2	4	8	12
Globaliz(s)ation	1	1	2	3
Industrializ(s)ation	1	1	1	2
Sum	100	100	100	100

TABLE 10.9. *Percentage of the keywords* education, teaching, *and* courses *in the total hits per discipline, activity and for* land use *and* land cover

	Percentage of total hits
Landscape architecture	87
Landscape geography	53
Landscape history	49
Landscape science	47
Landscape ecology	41
Landscape conservation	46
Landscape management	43
Landscape assessment	43
Landscape evaluation	41
Landscape planning	37
Landscape classification	30
Land-use planning	56
Land use	48
Land cover	23
Land-use change	31
Land-cover change	41
Landscape change	44

exist except for the categories *landscaping* and *landscape design*. Also, *urbanization* shows a higher score with *land-use change* and *land-cover change* than the same keywords without the term "change." *Land-cover change* is also clearly more associated with *forestry*.

The keywords *education, teaching* and *courses* score almost half of all the hits of landscape disciplines, activities, land use and land cover, with an important exception for *landscape architecture*, which clearly uses the Internet for promoting education and training activities (Table 10.9).

In a similar way, Table 10.10 compares the keywords *metrics* as well as *indicator, index,* and *indices* with the disciplines, activities, and land use and land cover, in terms of the number of hits in each group. The terms *indicator, index,* or *indices* occur more frequently than *metrics*. Although the number of hits on *metrics* is much lower, the variation shows a clear pattern. *Metrics* are mainly used *in landscape classification, evaluation,* and *assessment* in relation to *landscape change* and in particular to *land cover*. Of all the disciplines considered, landscape ecology and landscape science refer most frequently to landscape metrics.

TABLE 10.10. *Correspondence between keywords related to metrics and indicators to disciplines, activities, and land-use and land-cover related concepts*

	Metrics as percentage of total hits	Indicators as percentage of total hits
Landscape	0.5	16
Landscape change	4.3	30
Landscape architecture	0.4	32
Landscape ecology	2.9	27
Landscape planning	0.9	18
Landscape history	0.3	29
Landscape science	1.2	29
Landscape geography	0.6	22
Landscaping	0.1	19
Landscape design	0.2	19
Landscape management	0.8	21
Landscape protection	0.2	20
Landscape conservation	1.0	29
Landscape assessment	2.5	29
Landscape evaluation	3.2	31
Landscape classification	5.1	41
Land use	0.5	28
Land cover	1.2	36
Land use/land cover	3.1	33
Land-use change	1.2	19
Land-cover change	4.7	37

10.4   Discussion

10.4.1   The Internet survey

Most Internet-based analyses relate to content analysis of texts (Miller and Riechert 1994, Popping 2000, West 2001) or to technical aspects of Internet searching and tools (Litkowski 1999, Ridings and Shishigin, 2002). Internet searches of scientific databases have been used in a comparative analysis of land-use change models (Agarwal *et al.* 2002), but do not cover all domains related to landscape, such as countryside and design. The Internet survey here is more general and explorative. Hits of associated keywords do not necessarily mean a reasonable relationship among them, and the nature of the relationship remains unknown. For example, a search for *history* and *countryside* will give results including topics such as "natural history," "countryside and gardening," and "history of the countryside." The PageRank algorithm, which is similar to a citation index, only reflects the importance of the web page in terms of linking (Ridings and Shishigin 2002), and thus is not useful in defining the significance and quality of the content. However, the majority of the results obtained in the study by refined searches using combinations of keywords showed meaningful associations. The large number of hits and the important differences between the outcomes contributes to the reliability. Besides the relative importance of hits, their absolute number was also indicative for the interpretation.

Other important limitations on the use of an Internet search do exist. First, the worldwide context is severely biased by the dominant use of the English language, as is demonstrated by the occurrence of the concepts related to *landscape* (Table 10.1). Second, the Internet is a highly dynamic medium. The results refer only to one moment. However, a test made at two different times (January and May 2003) showed that an overall growth in the number of hits could be noted, but that the correspondences between the keywords used remained rather stable. The results for the different search keywords obtained at two different times were highly correlated ($r=0.999$, $p<0.001$), and the average growth of the number of hits was 16 percent in this period of five months. Third, it appears that Google approximates and rounds the number of results differently according to the magnitude of the search results. This makes sophisticated statistical analyses unsuitable for comparing the number of hits.

10.4.2 Dealing with land use, land cover and change

The general observation is that the term *land use* occurs more frequently than *land cover* and is used in a wide variety of contexts. Variations in meanings of these terms cannot be detected directly by this survey, but the contexts in which they are used indicate such variations. *Land use* is an important issue in activities in the landscape related to evaluation, classification, assessment, conservation, protection, and to a lesser degree management. *Land cover* appears most in relation to landscape classification, evaluation, and assessment, thus the more technical aspects on which management and conservation should be based. Land use is a more important issue in landscape ecology, landscape planning, and landscape geography, but less in landscape architecture and landscape history, which is confirmed by the very low correspondences between land use and activities of landscape design, where the term land cover is also not significant. Landscape ecology and geography use both terms most equally. On the other hand, the term land cover is relatively equal to, or in some cases more important than, land use when compared to keywords that relate to causes and processes of change. This is clear in the association with agriculture and forestry. Apparently land cover is still associated more with vegetation, although the scientific definitions also include different forms of abiotic cover-age (Baulies and Szejwach 1997, Dale *et al.* 2000, Akbari *et al.* 2003). Landscape management is still poorly associated with landscape ecology, confirming the concerns raised by several landscape ecologists (Dale *et al.* 2000, Bastian 2001, Opdam *et al.* 2001).

Change is an important issue in all disciplines and activities, but remark-ably less important in disciplines that intentionally create changes such as landscape architecture and landscape planning and related activities such as design and landscaping. Land-cover and land-use change are proportionally more important in association with keywords related to causes and processes, such as *nature, pollution, climate, fire*, and *degradation* than with other natural and environmental causes and processes. They are also highly associated with key-words such as *population* and *man*, but least of all with urbanization issues, *land-scaping*, and *landscape design*, and *development, agriculture*, and *economy*. *Land use* is relatively more used than *land cover* in relation to *economy* and *mobility*, while *land cover* is more associated with *forestry*.

10.4.3 The context of land, landscape, and countryside

The different contexts in which land use and land cover are used indicate different approaches to the landscape. The fact that land cover is more associ-ated with vegetation and the more technical issues of classification, evaluation, and assessment makes it more an attribute or quality of the land and less asso-ciated with broader landscape values. Zonneveld (1995) discussed the differ-ences between land and landscape in relation to scientific disciplines. Land is more associated with soil, terrain, territory, and a series of qualities that give it a value. This value derives from an assessment of the potential land uses, and is often expressed in monetary terms. Land relates directly to land ownership. Landscape is more related to holistic and perceptive aspects of a territory, and is seen as the result of the interactions between natural processes and human activities (Council of Europe 2000). It also refers to common heritage values, both cultural and natural, belonging to a community or even to humankind, as is the case for the cultural landscape on the UNESCO World Heritage list. These values are often considered as soft values, and are rarely expressed in monetary terms.

The territory of a community is perceived through the landscape, which inte-grates the material qualities of the land with many nonmaterial values (Claval 2005). In rural areas this synthesis is expressed as the concept countryside, where history and culture are important aspects. To preserve these values, land-scape conservation and protection, and landscape assessment are important activities. Particularly in Europe, Landscape Character Assessment (LCA) has become popular. Land use and land cover are basic data sources here (Mücher *et al.* 2003). However, the actual land use is only partially expressed by the land cover (Dale *et al.* 2000), and can be described precisely only at fine scales. Studying land-cover change refers primarily to land changes, while land-use changes embrace a broader landscape context. Modeling and predicting land-use and land-cover changes often focuses upon changes in land qualities (Fresco *et al.* 1996, Agarwal *et al.* 2002, Pontius *et al.* 2004), and less upon changes in the cultural, social, and aesthetic values of the landscape.

10.4.4 Issues not covered by the Internet survey

This Internet-based analysis allows the defining of domains of common activity, but tells nothing about the nature or scale of this activity. The associations, however, indicate different scales of action linked to the action domains of the agents involved in landscape changes. Landscape architecture and landscape planning seem to have a domain most close to the real, material actions on the ground, mostly involving individual or local agents, but with relatively little interest in the effect of the changes in a more global context and in the long term. On the other hand, sciences such as landscape ecology have a focus on land use and land cover as well as on changes over a larger spatial extent and longer duration. The link between these two domains and scales seems to be rather weak. Landscape ecologists are aware of this, and insist on better integration of ecological knowledge at all levels of action that induce or control landscape changes. Landscape management seems to be the common ground between these two domains.

Similar observations have been made in a series of studies. The Ecological Society of America, for example, proposed five ecological principles to guide decisions in land-use change and formulated eight guidelines for this (Dale *et al.* 2000). The importance of landownership in land-use change was recognized, but the focus was mainly on the USA and ecological landscape values only. A key issue is the translation of general scientific knowledge to the local agent. Agarwal *et al.* (2002) indicated that decision-making is also an important dimension as they formulated a three-dimensional framework for assessing land-use change models, including space, time, and human decision-making. The importance of scale was defined for each of these dimensions. Spatial resolution and extent in the space dimension, and time step and duration in the time dimension were proposed as well as the equivalent concepts in human decision-making "agent" and "domain." Agent refers to an individual person, a landowner, a household, a company, as well as groups of these organized as a neighborhood, municipality, region, or state. Each of these has a specific domain of action characterized by a spatial extent and duration. Six levels of complexity are recognized in human decision-making as a criterion for the assessment of land-use change models. The Land Use and Cover Change projects (LUCC) place the main problems in defining appropriate land-use change models in the lack of data, in particular, social data such as property rights and economical data on globalization (Fresco *et al.* 1996, Baulies and Szejach 1997).

Hägerstrand (1995) analyzed the connections and interactions between the micro- and macro-scale aspects of management of the biosphere, in particular how abstract knowledge can be turned into actions on the ground that cause real land-use and landscape changes. He stressed the importance of what he called territorial competence as the combination of the freedom and limitation of the landowner in making choices and considered it the most important factor at the micro scale. The choices of this primary agent depend on legal constraints, technical ability, and knowledge. Higher-order domains of human decision-making have power over larger spatial extents, but do not act directly on the landscape. They have the spatial competence to regulate and control land-use and land-use changes, but this does not necessarily correspond to reality in the landscape. He stated: "Global change is after all not the outcome of a few human actors of an immense scale. It is the nearly incalculable number of small actions which pile up to major changes in space and over time" (Hägerstrand 1995). Clearly, the contribution of sciences in this framework is not only the improvement of technological tools, but also the knowledge transfer to the domains of decision-making adapted at all scales.

10.5   Conclusions: key issues for further integration in landscape ecology

Land use and land cover are basic concepts in many disciplines and activities related to landscape research and management. They are used in many different contexts and at very different scales, which causes inevitable subtle changes in meaning. In inter- and transdisciplinary landscape studies, a precise definition of the concepts in the appropriate context is essential (Tress *et al.* 2005). Some disciplines focus on one concept. For example, landscape architecture uses mostly the term land use, whereas landscape ecology uses mostly land cover. Land use and land cover are important concepts used both as a component for characterizing landscape types and an indicator of landscape changes. Although landscape change is becoming an increasingly important issue, landscape architecture and design show a rather poor association with issues related to change and landscape dynamics. Landscape ecology is most involved with change.

Land-cover and land-use change are important in activities related to landscape protection and conservation, as well as technical issues of classification and evaluation. Land-cover and land-use changes are also important environmental processes, involving degradation, fire, pollution, climatic change, population dynamics, economical development, and particularly agriculture and forestry. Landscape change is often expressed by changes in the land cover resulting from changing natural processes and human activities, such as other choices in land use. In intensively used landscapes, the overall change of the landscape is only rarely caused by vast natural calamities; mostly it is the result of numerous small changes in discrete patches, induced by numerous agents. For many of them, not only land-cover changes but also changes in many other nonmaterial values are important. These cultural, social, economical, and aesthetical values are not only associated with the concept of landscape in its general meaning, but even more with the concept of the countryside as well.

Landscape management is the activity where landscape architecture, landscape planning, and landscape ecology seem to meet, but clearly with different perspectives. Better communication and transfer of scientific knowledge to the planners and designers seems appropriate here for more integration. Inter- and transdisciplinary studies dealing with land-use, land-cover, and landscape change can be improved by building several bridges between disciplines and activities. Looking at land-use and land-cover changes at different scales and as the integrated result of both natural processes and social, economic, and cultural needs are some of these. Transferring knowledge about these changes and the processes that induce them from scientific observations to planners, designers, and managers is another. The focus should be on linking scientific research integrated at a global scale with decision-making of agents at the local scale.

References

Agarwal, C., G. M. Green, J. M. Grove, T. P. Evans, and C. M. Schweik. 2002. *A Review and Assessment of Land-Use Change Models: Dynamics of Space, Time, and Human Choice*. General Technical Report NE-297. Newtown Square, PA: USDA, Forest Service, Northeastern Research Station.

Akbari, H., L. Shea Rose, and H. Taha. 2003. Analyzing the land cover of an urban environment using high-resolution orthophotos. *Landscape and Urban Planning* 63, 1–14.

Antrop, M. 2001. The language of landscape ecologists and planners: a comparative content analysis of concepts used in landscape ecology. *Landscape and Urban Planning* 55, 163–73.

Antrop, M. 2003. Continuity and change in landscapes. Pages 1–14 in U. Mander and M. Antrop (eds.). *Multifunctional Landscapes* Vol. 3: *Continuity and Change*. Southampton: WIT Press.

Bastian, O. 2001. Landscape ecology: towards a unified discipline? *Landscape Ecology* 16, 757–66.

Baulies, X. and G. Szejach (eds.). 1997. *LUCC Data Requirements Workshop*. LUCC Report Series No.3, Barcelona: Institut Cartogràfic de Catalunya.

Brandt, J. and H. Vejre. 2004. Multifunctional landscapes: motives, concepts and perceptions. Pages 3–32 in J. Brandt and H. Vejre (eds.). *Multifunctional Landscapes: Theory, Values and History*. Vol. I. Southampton: WIT Press.

Brin, S. and L. Page. 1998. The anatomy of a large-scale hypertextual web search engine. *Computer Networks* **30**, 107–17.

Claval, P. 2005. Reading the rural landscapes. *Landscape and Urban Planning* **70**, 9–19.

Cosgrove, D. 2003. Landscape: ecology and semiosis. Pages 15–20 in H. Palang and G. Fry (eds.). *Landscape Interfaces: Cultural Heritage in Changing Landscapes*. Dordrecht: Kluwer Academic Publishers.

Council of Europe. 2000. European Landscape Convention. Firenze (http://www.coe.int/t/e/Cultural.Co-operation/Environment/Landscape/).

Dale, V.H., S. Brown, R.A. Haeuber, *et al.* 2000. Ecological principles and guidelines for managing the use of land. *Ecological Applications* **10**, 639–70.

Forman, R. and M. Godron. 1986. *Landscape Ecology*. New York: John Wiley & Sons, Inc.

Fresco, L., R. Leemans, B.L. Turner II, *et al.* (eds.). 1996. *Land Use and Cover Change (LUCC) Open Science Meeting Proceedings*. LUCC Report Series No.1., Amsterdam.

Fry, G. 2001. Multifunctional landscapes: towards transdisciplinary research. *Landscape and Urban Planning* **57**, 159–68.

Hägerstrand, T. 1995. *A Look at the Political Geography of Environmental Management*. Landscape and Life: Appropriate Scales for Sustainable Development, LASS Working Paper No.17. Dublin: University College Dublin.

Haines-Young, R. 2000. Sustainable development and sustainable landscapes: defining a new paradigm for landscape ecology. *Fennia* **178**, 7–14.

Litkowski, K.C. 1999. Towards a meaning-full comparison of lexical resources. In *Proceedings of the Association for Computational Linguistics Special Interest Group on the Lexicon, June 21–22, College Park, Maryland* (http://www.clres.com/Comparison.of.Lexical_Resources.html).

Miller, M.M. and B.P. Riechert. 1994. Identifying themes via concept mapping: a new method of content analysis. In *Communication Theory and Methodology Division of the Association for Education in Journalism and Mass Communication Annual Meeting, Atlanta, Georgia* (http://excellent.com.utk.edu/~mmmiller/pestmaps.txt).

Mücher, C.A., R.G.H. Bunce, R.H.G. Jongman, *et al.* 2003. *Identification and Characterisation of Environments and Landscapes in Europe*. Wageningen: Alterra-rapport 832.

Muir, R. 1999. *Approaches to Landscape*. London: MacMillan Press.

Nassauer, J.I. 1997. *Placing Nature: Culture and Landscape Ecology*. Washington, DC: Island Press.

Olwig, K.R. 2004. "This is not a landscape": circulating reference and land shaping. Pages 41–66 in H. Palang, H. Sooväli, M. Antrop, and S. Setten (eds.). *European Rural Landscapes: Persistence and Change in a Globalising Environment*. Dordrecht: Kluwer Academic Publishers.

Opdam, P., R. Foppen, and C. Vos. 2001. Bridging the gap between ecology and spatial planning in landscape ecology. *Landscape Ecology* **16**, 767–79.

Pontius, R.G., E. Shusas, and M. McEachern. 2004. Detecting important categorical land changes while accounting for persistence. *Agriculture, Ecosystems and Environment* **101**, 251–68.

Popping, R. 2000. *Computer-Assisted Text Analysis*. Thousand Oaks, CA: Sage.

Ridings, C. and M. Shishigin. 2002. *PageRank Uncovered* (http://www.texaswebdevelopers.com/docs/pagerank.pdf).

Steinitz, C. 2001. Landscape ecology and landscape planning: links and gaps and common dilemmas. Pages 48–50 in U. Mander, A. Printsmann, and H. Palang (eds.). *Development of European Landscapes*. Tartu: Publicationes Instituti Geographici Universitatis Tartuensis.

Tress, B., G. Tress, and G. Fry. 2003. *Interdisciplinary and Transdisciplinary Landscape Studies: Potential and Limitations*. Wageningen: Delta Series 2.

Tress, B., G. Tress, and G. Fry. 2005. Integrative studies on rural landscapes: policy expectations and research practice. *Landscape and Urban Planning* **70**, 177–91.

Veldkamp, A. and E. F. Lambin. 2001. Predicting land-use change. *Agriculture, Ecosystems and Environment* **85**, 1–6.

West, M. D. (ed.). 2001. *Applications of Computer Content Analysis*. Westport, CT: Ablex.

Zonneveld, I. S. 1995. *Land Ecology*. Amsterdam: SPB Academic Publishing.

BRENDAN G. MACKEY, MICHAEL E. SOULÉ, HENRY A. NIX,
HARRY F. RECHER, ROBERT G. LESSLIE, JANN E. WILLIAMS,
JOHN C.Z. WOINARSKI, RICHARD J. HOBBS, AND
HUGH P. POSSINGHAM

11

Applying landscape-ecological principles to regional conservation: the WildCountry Project in Australia

11.1 Introduction

One of the great challenges facing humanity in the twenty-first century is the conservation and restoration of biodiversity (Convention on Biodiversity 1992). In this chapter we present the landscape-ecological underpinnings of a new nongovernment organization (NGO)-driven conservation initiative in Australia, namely the WildCountry Project.

Global and national analyses highlight the extent of environmental degradation and the need for urgent protection and restoration of biodiversity (e.g., SEAC 1996, Environment Australia 2001, World Resources Institute 2001, NLWRA 2002). Such analyses also suggest that existing conservation strategies and plans are insufficient to prevent continuing losses.

The primary question, at the most general level, is: how can a conservation system be designed and implemented for Australia that is likely to maintain biodiversity for centuries to millennia? Dedicated protected areas are a core component of a nation's biodiversity conservation system. By our calculations (Fig. 11.1) only about 6 percent of Australia is in a secure protected area. There is no theoretical or empirical basis to the proposition that this level of reservation, while necessary, is sufficient for securing the conservation of Australia's biodiversity. In any case, protected area networks are largely the result of various historical contingencies rather than the principles of modern reserve design (Margules and Pressey 2000). We suggest that the percentage of Australia reserved in protected areas is unlikely to ever exceed 10–15 percent. Our calculations (Fig. 11.1) also show that about 84 percent of the Australian continent has a native vegetation cover, is outside a protected area, and is not used for agriculture or forestry. Of this 84 percent, about 56 percent

Key Topics in Landscape Ecology, ed. J. Wu and R. Hobbs.
Published by Cambridge University Press. © Cambridge University Press 2007.

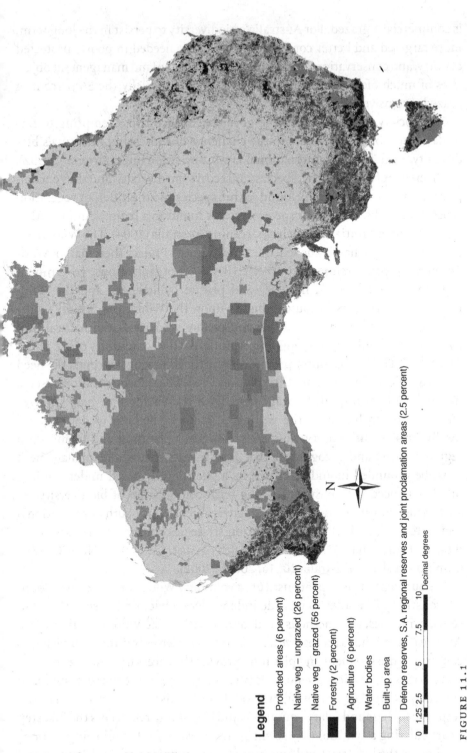

Legend

- Protected areas (6 percent)
- Native veg – ungrazed (26 percent)
- Native veg – grazed (56 percent)
- Forestry (2 percent)
- Agriculture (6 percent)
- Water bodies
- Built-up area
- Defence reserves, S.A. regional reserves and joint proclamation areas (2.5 percent)

0 1.25 2.5   5   7.5   10
Decimal degrees

N

FIGURE 11.1

Broad categories of land use and land cover for Australia that identify regions where different approaches are needed for implementing landscape-wide conservation assessment and planning that promotes ecological connectivity. The legend also indicates the percentage of the Australian continent covered by each class. (Source: *1996/97 Land Use of Australia*, Version 2, National Land and Water Resources Audit.) The boundaries of the reserves that permit grazing in S.A. were extracted from the NPWSA Property Boundaries Data set from the National Parks and Wildlife S.A., Department for Environment and Heritage. Note that the protected area boundaries for NSW are current as of 2003, whereas the boundaries for Victoria and W.A. are based on the best available land tenure data which were compiled *circa* 1997.

is commercially grazed. For Australia's biodiversity to persist in the long term, more targeted and better configured reserves are needed in poorly protected country, and conservation must be integrated into the land management objectives of much of the remaining 84 percent, and especially the 56 percent of grazed, extensive country.

Civil society has now joined the government sector in attempting to formulate appropriate responses to the challenge of conserving Australia's biodiversity. As defined by international law (Convention on Biodiversity 1992), biodiversity refers to genetic, species, and ecosystem diversity, and thus encompasses, *inter alia*, the diversity found within species and the different vegetation types, food webs, and landscape ecosystems found in a region. Amidst other nongovernment initiatives such as Greening Australia (2004), The Wilderness Society Australia has launched the WildCountry Project (hereinafter Wild-Country) in partnership with other civil society organizations, government at state and local levels, industry and private landowners, and the Wild-lands Project USA. WildCountry builds upon the Wilderness Society's mission, namely, "to protect, promote and restore wilderness and natural processes for the wellbeing and ongoing evolution of the community of life across Australia." The WildCountry project reflects the following concepts: the need for a significant improvement in the protected area network and off-reserve management, community engagement with stakeholders to help catalyze and sustain "coalitions of the willing" capable of helping to develop and locally implement conservation assessment and planning and action on a regional basis, and recognition that assessments, plans and management must be grounded in and informed by a scientifically based understanding of what is needed to ensure the long-term conservation of biodiversity. As such, WildCountry is consistent with government policy both at the national and state level, and related conservation strategies and programs (Commonwealth of Australia 1997, Commonwealth of Australia 1999, ANZECC 2001, Commonwealth of Australia 2001a, 2001b, 2002).

The authors of this paper constitute a voluntary WildCountry Science Council, constituted in order to provide independent advice on the scientific concepts, principles, and methods needed to underpin the WildCountry project. Are existing methods for reserve design adequate? Do prevailing approaches to conservation assessment and planning provide the necessary information? Are there critical ecological phenomena and processes not yet incorporated into currently existing conservation methodologies? This paper provides an initial response to these and related questions and in so doing represents the first step in articulating a WildCountry scientific framework. In the following sections we discuss the historical and conceptual underpinnings of WildCountry and the necessary scientific principles. We conclude by considering some implications of these for WildCountry implementation.

As noted above, WildCountry assumes that, for much of Australia, voluntary changes based on partnerships between stakeholders will be the way forward. NGOs such as the Wilderness Society are well placed to help such partnerships. Governments can be constrained by inertia, vested interests or prior policy decisions. NGOs, on the other hand, can have greater flexibility and, often, greater longevity, than governments. This approach to conservation will invariably need to mesh with other programs that aim at redesigning agricultural and pastoral systems to ensure sustainability (e.g., Landcare Australia 2004). In order to facilitate such a partnership approach, education of and engagement with local communities will be key components of a WildCountry framework. Whilst acknowledging the importance of these social dimensions to WildCountry, our focus in this chapter is on the necessary scientific components of a WildCountry framework – though the social dimensions are touched upon in those sections below that address broad-scale threatening processes and approaches to systematic planning.

11.2   Foundation principles

11.2.1   Core areas

It is axiomatic that dedicated core areas must be a key component in the WildCountry framework. These are areas, primarily managed for their conservation values, that contain relatively intact ecosystems (e.g., minimal broad-scale vegetation clearing) and that have low exposure to anthropogenically driven threatening processes (however, note the discussion below on management). At a regional scale, core areas should represent all major landscapes. Another key consideration in defining dedicated core areas is the long-term prospects for retaining or improving the quality of relative wildness. Dedicated core areas must be sufficiently large to have the capacity to "self-manage" through natural processes that include the dispersal of biota and their propagules, natural selection, species evolution, and biotic regulation of local biogeochemical and water cycles (Gorshkov *et al.* 2000). There is, however, no simple answer to the question of how large an area needs be to retain core-area characteristics. Given the extent of anthropogenic perturbation in Australia (particularly in the intensive land-use areas, Fig. 11.1), we can readily anticipate that in certain landscapes it will not be possible to find large areas that have not been subject to broad-scale clearing, overgrazing, large-scale disruption of hydrological regimes, and other intensive land uses. Thus, an emphasis on linking relatively intact habitat cores that represents "the best that is left," together with substantial ecological restoration, will be necessary.

Given the importance of core protected areas to WildCountry, a logical starting point in defining the components of an appropriate scientific framework is to consider the criteria developed for the Australian Regional Forest Agreement (RFA) process (AFFA 2003). Three main criteria were adopted for the RFA, namely: comprehensiveness, adequacy, and representativeness (CAR). Comprehensiveness refers to the extent to which the pre-European distributions of forest ecosystem types are captured by the protected-area network. Representativeness refers to how well the within-forest type variability is sampled by the protected-area network. Adequacy refers to the likelihood that the protected-area network will ensure the long-term viability of the biodiversity that resides therein. In practice, the criteria of adequacy and representativeness were not substantially applied in the RFA process, and targets were only set for the first criterion – "comprehensiveness." Thus, following extensive assessments, forest tenure was changed in each region so that a nominated percentage of the pre-European distribution of forest types ecosystems was included within the protected-area network. Targets were also set to ensure a percentage of the potential habitat of threatened and rare vertebrate animal and vascular plant species were captured within the protected-area network. Interestingly, wilderness targets were also prescribed but on the basis that wilderness quality reflects a social value of no biodiversity conservation relevance.

The RFA criteria, as applied to date, have been useful in helping to promote the implementation of explicit conservation criteria and systematic reserve design in Australia (e.g., GBRMPA 2003). While they remain relevant to WildCountry, it is equally important to appreciate their limitations. The RFA criteria ignore landscape condition and thus do not explicitly consider the impact of human land-use activity on ecosystem structure and function, and animal habitat. Furthermore, landscape variation in primary productivity was not considered. Thus, in identifying priority conservation areas the distinction was not necessarily made between heavily perturbed, low productivity and relatively intact, high productivity forests.

In practice, the setting of percentage targets for representation (i.e. the comprehensiveness criterion) proved to be a relatively arbitrary process without strong and explicit scientific foundation. In any case, it is arguable whether the concept of setting percentage targets for representation is relevant in intensively cleared landscapes where only fragments of native vegetation remain. In these circumstances it could be argued that all the remnant patches have conservation value. Similarly, experience gained from studying land degradation in southern Australia has yielded little by way of guidelines as to the ecologically permissible percentage of native vegetation that can be cleared within intact landscapes. In both these contexts, the risk with a CAR approach as applied in the RFA process is to promote ecologically and numerically minimalist conservation outcomes, whereas the WildCountry conservation objectives are expansive and long-term. Nonetheless, the CAR criteria as originally conceived remain useful and relevant to the problem of systematic reserve design, and as such are one set of inputs to a WildCountry scientific framework.

11.2.2 The Wildlands Project

Additional guidance was sought from the methodology and scientific principles underlying the Wildlands Project (hereinafter Wildlands) in North America (Foreman 1999). The vision of Wildlands is to protect and restore North America's ecological integrity. The project is creating an alternative, map-based land-use plan for the continent, with the emphasis on connectivity and the restoration of ecological interactions. Formed in 1991 by scientists and conservationists, Wildlands emphasizes maintaining, connecting, and buffering wild lands, repairing landscapes that have been compromised by such factors as habitat fragmentation and loss of species, maintaining natural disturbance regimes, and communicating the ecological values of wilderness, plants, and animals (Soulé and Terborgh 1999). The approach is to restore missing species and processes, and to anticipate climatic and landscape changes that might compromise natural values and society's opportunities for enlightened economies. This is called "rewilding" (Soulé and Noss 1998). Wildlands recognizes that the application of these broad conservation principles will vary depending on regional ecology, the history of disturbance, and existing land use.

A major component of rewilding in North America is the maintenance of ecologically effective populations of large mammalian carnivores and other highly interactive species, the loss of which initiates cascading or dissipative changes through the ecosystem (Soulé *et al.* 2003). There is persuasive scientific evidence that such strongly interacting species and processes are vitally important to healthy ecosystems. Because large predators require extensive space and connectivity, the modeling of their habitat requirements is a key tool in network design in North America. Reconciling this rewilding approach with the more traditional methods of biodiversity conservation has been one of the greatest challenges for Wildlands, but is also what distinguishes its approach from that of most other conservation groups (Soulé and Noss 1998).

Following the principles of systematic conservation planning (Margules and Pressey 2000), the Wildlands regional plans feature explicit goals, quantitative targets (based on defensible ecological calculations), rigorous methods for locating new reserves, and explicit criteria for implementing conservation action. Focal species analysis can complement the incorporation of special elements and representation of vegetation types by addressing questions concerning the size and configuration of reserves and other habitats necessary to maintain species diversity and ecological resilience over time.

Wildlands provides three key concepts that are potentially relevant to the WildCountry scientific framework in Australia, namely: (1) continental and regional connectivity of large core reserves as required to support the long-term conservation requirements of large carnivores and other spatially extensive ecological processes (Soulé and Terborgh 1999), (2) complementary land management in surrounding landscapes, and, (3) where necessary, restoration of natural processes and disturbance regimes, the control of invasive species, and the reintroduction of native species. Of particular interest was the first principle, regarding the need for conservation-area designs to reflect continental and regional connectivity, the pivot points of which are large core reserves. Is this principle of large-scale connectivity equally relevant to the Australian situation, or are there major differences in the ecologies of Australia and North America that require the concept to be revisited for WildCountry?

11.2.3 Connectivity revisited

As noted above, in a North American context, large-scale connectivity has been considered by the Wildlands project in terms of the maintenance of ecologically effective populations of large mammalian carnivores and other wide-ranging focal species. The absence of large predators often leads to numerical release (abnormally high abundances) and behavioral release (e.g., abnormal levels of foraging or predation) of herbivores and mesopredators, thereby changing community composition, dynamics, and the structure of vegetation. More generally, Wildlands emphasizes the need to maintain ecologically effective populations of keystone and other highly interactive species at the regional scale (Soulé and Noss 1998, Crooks and Soulé 1999, Terborgh *et al.* 1999, Soulé *et al.* 2003).

From this perspective, planning for connectivity means ensuring large core areas to be embedded within landscapes that include compatible-use areas and habitat linkages (Frankel and Soulé 1981, Noss and Cooperrider 1994, Hobbs 2002a). It is argued that a conservation-area design based on this principle is better able to sustain the long-term ecological viability of these large species compared to a conventional system of isolated parks and reserves. This approach requires working at spatial and temporal scales exceeding those normally employed to manage natural areas and natural resources.

There are major differences in the ecologies of Australia and North America that suggest the Wildlands principle of large-scale connectivity for large mammalian carnivores may not be as relevant to WildCountry. First, and most importantly, Australia lost its megafauna around 50000 years ago (Beck 1996). Thus, the long-term requirements of large predators might appear irrelevant in the framing of a continental conservation strategy for Australia. A second difference between Australian and North American ecology stems from the climatic systems that dominate these continents. Much of Australia is characterized by extreme variability in the distribution of rainfall as well as deeply weathered landscapes of low relief and low soil fertility. These dominating factors have generated distinctive ecological responses in the plants and animals everywhere, but particularly in the arid and semi-arid zones (Friedel *et al.* 1990, Morton *et al.* 1995).

Notwithstanding these differences, large-scale connectivity may still be an important conservation planning principle for Australia but primarily for different reasons than in North America. The following sections consider a set of ecological phenomena and processes that operate at large scales in both space and time. We argue that their ongoing functioning is necessary for the long-term resilience of landscape ecosystems, the maintenance and regeneration of habitat, and ultimately the viability of populations. Furthermore, we suggest that the landscape linkages necessary to maintain their functioning have yet to be substantially integrated into conservation assessment and planning.

11.3   Large-scale connectivity

Connectivity is generally considered in terms of wildlife corridors – narrow bands of native vegetation connecting core habitat areas (Lindemayer and Nix 1993). Here the word is used to draw attention to large-scale ecological phenomena and processes that require the maintenance of landscape linkages at regional to continental scales. The necessary landscape linkages may include core areas, comprise continuous habitat such as riparian corridors and appropriately spaced stepping-stones (Dobson *et al.* 1999, Roshier *et al.* 2001), or reflect some other kind of spatial "teleconnection."

11.3.1   Trophic relations and interactive species

Whilst Australia lacks the large mammalian carnivores of North America, species at any given trophic level can play a major role in regulating resource availability and population dynamics over species at other levels, e.g., large herbivores (Oksanen and Oksanen 2000), pollinators (honeyeaters; Paton *et al.* 2000) and mesopredators such as the dingo *Canis lupus dingo* (Caughley *et al.* 1980). Maintaining large-scale connectivity for such trophically interactive species (Soulé *et al.* 2003) is critical to consider in conservation planning.

The broader implications of maintaining and/or restoring trophic levels in a food web on a landscape-wide basis have generally not been used in Australia to guide conservation assessment and planning.

11.3.2 Hydroecology

The term hydroecology describes the role that vegetation plays in regulating surface and subsurface hydrological flows, and in turn the importance of water availability to plants and animals (Mackey *et al.* 2001). The significance in Australia of hydroecology is amplified by high year-to-year variability in rainfall (Hobbs *et al.* 1998). Hydroecological processes can be observed in all regions of Australia, including Cape York Peninsula (Horne 1995, Horne *et al.* 1995), the Southern Tablelands of NSW (Starr *et al.* 1999), the Central Highlands of Victoria (Vertessey *et al.* 1994), and inland Australia (Friedal *et al.* 1990, Stafford Smith and Morton 1990). Generally, our land management has not protected catchment-scale processes that affect groundwater recharge and discharge, although these are critical for maintaining perennial springs and water holes, river base flows, and perennial stream flow. Biodiversity conservation and planning must pay particular attention to such whole-of-catchment processes.

11.3.3 Long-distance biological movement

Both vertebrates and invertebrates can have stages in their life cycles that are associated with large-scale movement. A vast diversity of organisms and their propagules forage, disperse, and migrate (Cannon and Gardner 1999, Drake *et al.* 2001, Isard and Gage 2001). Examples of ecologically significant long-distance biotic dispersal include the use of rainforest patches by animals in Northern Australia (Palmer and Woinarski 1999, Shapcott 2000, Bach 2002), and dispersive avifauna in Australian woodlands and open-forest (Paton *et al.* 2000, NLWRA 2002). Thus, there is a need to maintain networks of suitable habitat for dispersive species over large regions. A conservation system is needed that is extensive enough to embrace the full breadth of continental variability in climate, productivity, and vegetation, and the resultant fauna dynamics (Nix 1974).

A type of biological movement of special conservation interest is dispersal to and from refugia – places where populations of a species can persist during a period of detrimental change occurring in the surrounding landscape. Thus, refugia are locations that provide refuge from threatening processes. They enable species to maintain their presence in landscapes and are potential sources for reestablishment. Refugia can be defined at a range of scales and with respect to various threatening processes, including inappropriate fire regimes (Mackey *et al.* 2002), global climate change (Lovejoy 1982), and drought (Stafford Smith and Morton 1990). Refugia are probably important in all ecosystems (thought not all movement associated with refugia is necessarily large scale) but only rarely has their significance been considered in conservation assessment and planning.

11.3.4 Ecologically appropriate fire regimes

Fire is a natural part of virtually all Australian landscapes and has an important influence on the biological productivity, composition, and landscape patterning of ecosystems (Reid *et al.* 1993, Williams *et al.* 1994, Whelan 1995, Bradstock *et al.* 2002, Catchpole 2002, Mackey *et al.* 2002). The conservation implications of ecologically inappropriate fire regimes can be substantial. In systems fragmented by human activity, broad landscape processes have been disrupted leading to altered fire regimes (Gill and Williams 1996, Hobbs 2002b). Remnant vegetation in agricultural areas may suffer from the absence of fire over long periods. In large core conservation areas, there may be an overriding need for deliberate and carefully planned fire management, allowing for large and/or high intensity wildfires. The role of Aboriginal burning practices demands special attention especially in Northern Australia (Price and Bowman 1994, Williams and Gill 1995, Bowman *et al.* 2001, Yibarbuk *et al.* 2001, Keith *et al.* 2002).

11.3.5 Climate change and variability

As a consequence of human-forced climate change (IPCC 2002), it is likely that Australian ecosystems will be exposed in the coming decades to an increase in the frequency of extreme weather events, higher average daily temperatures (especially higher minimum daily temperatures), and changes in the spatial and seasonal distribution of precipitation (CSIRO 2001). Such changes have direct and indirect impacts on all aspects of biodiversity, including species distributions, community structure, and ecosystem processes (Mackey and Sims 1993, Hannah and Lovejoy 2003, Thomas *et al.* 2004). Providing connectivity to promote biotic adaptation to climate change is a formidable challenge, but is central to continental- and regional-scaled conservation assessment and planning for the coming decades (NTK 2003).

11.3.6 Coastal zone fluxes

There are two perpendicular directions of flow in the coastal zone. One is the flux of matter and energy between sea and land; the other direction of flow is parallel to the coast, such as the migration of marine organisms, including shorebirds, and the movement of coastal currents. Connectivity of land/coastal-zone flows is particularly important given the concentration of Australia's population in coastal regions (Cosser 1997). Terrestrial conservation assessment and planning must include these important links with the marine environment. A landscape could have conservation value primarily because it contributes to ecosystem function in the adjacent coastal zone. Indeed, a "source to sea" planning framework is essential. A more comprehensive treatment of these connectivity processes will be published elsewhere.

11.4   Research and development issues

11.4.1 Dispersive fauna

Conservation planning for dispersive fauna requires data at landscape and continental scales on movements and the spatial and temporal distribution of habitat resources, including the dispersion of food resources in response to environmental variability. Meeting these information needs is conceptually tractable but logistically will require a significant investment in IT-based systems. Data from various sources (remotely sensed, field-survey records, digital maps) and themes (climate, topography, substrate, vegetation, wildlife, land use, land tenure) must be assimilated into usable formats at the best available resolutions across the continent. Advances in GIS, environmental modeling and remote sensing provide the capacity to describe, classify, and map landscapes in ways that are relevant to the assessment of fauna distributions and habitat requirements (Mackey *et al.* 1988, 1989, 2001, Lesslie 2001, Mackey and Lindenmayer 2001, Nix *et al.* 2001). They can also be used to directly track temporal variability in the distribution and availability of primary production and food resources. Critically, these analyses can now be undertaken at a continental scale with high spatial and temporal resolutions. Of particular interest are high-resolution digital elevation models (Hutchinson *et al.* 2000) and land-cover data derived from satellite-borne sensors such as MODIS ($\sim$250 m spatial resolution), Landsat TM ($\sim$25 m resolution) and JERS-1 SAR ($\sim$18 m resolution). Derived remotely sensed products now include various estimates of food resource production in response to environmental variability, including net primary productivity, above ground biomass, leaf-area index, and land-cover classes (Landsberg and Waring 1997, Austin *et al.* 2003, NASA 2003). These analytical capabilities add to existing technologies and aid in both identifying core protected areas and in designing the necessary buffers, corridors, linkages, and management changes in the surrounding landscape matrix.

11.4.2 Protected-area and off-reserve management

The design and establishment of core areas for biodiversity conservation can only be part of a WildCountry framework. Decisions must also be made about the ongoing management of such areas together with the necessary off-reserve management regimes. Management of core conservation areas will affect neighboring lands (and hence the regional community's attitude to WildCountry values and outcomes) and vice versa. In Australia, almost all lands, including protected areas, are affected by the increasing impact of feral animals and plants and altered disturbance regimes. Feral animals degrade the most remote deserts of central Australia, and feral animals and weeds transform the furthest reaches of central Arnhem Land. In the absence of preventative management, these threats drive the landscape and its natural values further into decay. It is an abrogation of responsibility to leave the conservation values of lands unprotected from the array of new elements that are altering these landscapes.

We noted above that an effective system of reserves requires high levels of connectivity either by managing the "matrix" (all areas that are not part of the network of lands and waters under some kind of biodiversity protection) to allow for the movement and dispersal of plants and animals or by creating linkages specifically for that purpose. It is unrealistic to assume that all essential connectivity can be contained within a system of reserves in isolation. It is more reasonable to assume that large areas of habitat (or landscape components that contribute to ecological function) will remain outside the reserve system. The way the matrix is managed will be critical for the long-term conservation of biodiversity (Hale and Lamb 1997, Lindenmayer and Recher 1998, Lindenmayer and Franklin 2002), including the effectiveness of the linkages needed to maintain the connectivity of large-scale ecological processes.

Off-reserve land can have a vital role to play in protecting and restoring hydrological relations, accommodating the impacts of long-term climate change, providing for the seasonal and episodic movements of animals, the dispersal of propagules, and the exchange of genetic material between core areas. For these reasons, the capacity to manage effectively will depend on the willingness of adjoining landowners and leasees to change management practices to enhance conservation outcomes. There is a growing number of examples where off-reserve conservation can serve as a key element in engaging landowners and other stakeholders in the conservation process, especially if the engagement includes the development of local capacity and understanding (e.g., Dilworth *et al.* 2000).

The challenges facing off-reserve land-use management vis-à-vis connectivity will vary depending on the environmental context, regional conservation objectives, land-use history, the degree of degradation of the habitat, and management regimes. In Australia, three broad categories of land use and land cover can be recognized (Fig. 11.1). First, there are extensive areas in the tropical north and arid, central and southern Australia that have suffered minimal clearing of native vegetation, but are now witnessing the loss of biodiversity as the result of introduced herbivores and predators, livestock, weeds, and altered fire regimes (Finlayson 1961, Morton 1990, Woinarski *et al.* 1992, Russell-Smith *et al.* 1998, Franklin 1999, Woinarski *et al.* 2001, Lewis 2002). However, this category retains the potential for effective connectivity. Second, there are landscapes dominated by agricultural production where the pre-European settlement vegetation has been largely removed, and only isolated and usually degraded remnants persist; examples include the sheep/wheat belts of southeast Australia and southwest Western Australia. The maintenance of ecological flows is far more challenging in such areas (Saunders and Hobbs 1991, Hobbs *et al.* 1993, McIntyre and Hobbs 1999). Third, there are areas that are dominated by native tree vegetation, but are subject to substantial resource extraction, in particular, forest ecosystems in southern and eastern Australia.

We cannot assume that the matrix is benign for native plants and animals. Indeed, the nature of the matrix will vary depending on the prevailing land use, and closer attention to the impact of different matrix types on species' movement and survival is needed (e.g., Davies *et al.* 2001). Some matrix areas will be ecological sinks, although species will respond differently to different kinds and degrees of disturbance, pollution, and degradation. Other areas will retain some capacity to contribute to biodiversity and the maintenance of ecosystem processes. Within the categories of landscapes just described there are significant differences in the management practices needed to restore and buffer core areas, promote ecological connectivity, protect off-reserve biodiversity, and protect on-reserve biodiversity from off-reserve hazards. Identifying the appropriate mix of complementary management practices remains an ongoing research challenge.

11.4.3 Fire regime management and social values

Management of landscapes for biodiversity conservation is not only about remedial or preventative work on invasive organisms. Effective management demands good relationships with the human communities that inhabit these landscapes. While the livelihoods of all communities in regional Australia are coupled to access to land, for Aboriginal Australians, lands cut off from people are considered "lands without life." It follows that conservation planning in the areas of Australia that are legally recognized as Aboriginal land (about 13 percent of the continent, largely but not exclusively in central and northern Australia) cannot be separated in practice from issues related to the social and economic aspirations of Aboriginal Australians. To our knowledge, Aboriginal Australians did not generally engage in broad-scale clearing or silviculture. Rather, fire was the most important component of Aboriginal land management. Substantial parts of the Australian landscape probably still reflect the impact of past Aboriginal fire management practices. In some areas, the management system persists. Understanding past and present fire regimes is a critical research task for integrating fire management into large-scale conservation planning in Australia. The challenge of integrated fire management for biodiversity conservation is no less complex when considering the management systems, values, aspirations, and rights of nonindigenous pastoralists in regional Australia.

11.4.4 Whole-of-landscape conservation planning

Significant advances have been made in identifying networks of dedicated reserves that represent some kind of optima with respect to representativeness of biodiversity at a regional scale, their spatial configuration, and the potential impact of removing land from other land uses. Systematic reserve design usually also incorporates information generated from population viability analysis undertaken for target species. The whole-of-landscape approach promoted by WildCountry suggests a similar, but more complex planning process. "Landscape viability analysis" is needed, which enables the entire landscape to be evaluated and the optimum set identified of dedicated reserves, areas of connectivity, and off-reserve management requirements.

If the problem of how to optimally allocate conservation effort can be properly formulated as a decision-theory problem then decision theory-algorithms can help solve the problem efficiently (Possingham *et al.* 2001). It is important in this context to separate the following three parts of conservation planning:

(1) *Defining the problem* in terms of the objectives and constraints – this is where the conservation values (and related socioeconomic values) that the planning is intended to promote or protect are quantified using some kind of mathematical formulation.
(2) *Describing the system state and its dynamics* – as per the target components of biodiversity and the large-scale processes discussed in this paper. This means answering such questions as: What and where are the habitat/ecosystem types? How do different activities (zoning into reserves or other uses) affect the viability of species? What are the

consequences of zoning decisions on ecological processes? And what are the consequences of spatial relationships of different human activities for ecological processes and species viability? The system state and its dynamics can include socioeconomic variables and sub-models.

(3) *Applying an algorithm* used to generate planning options. If the problem is properly defined and the system state and dynamics are adequately accounted for, then algorithms can be applied that find the best or some good solutions that aid or initiate the decision-making process. The algorithm often needs ancillary software to present alternatives and facilitate the use of potential solutions in the decision-making process. Ultimately the algorithm is no more than a decision support tool that uses computers to see possibilities that we may miss.

Traditionally the "reserve design" problem has been defined such that the objective is to minimize costs given a suite of conservation targets. However, there has been little analytical consideration of the connectivity issues discussed here. More recently, the Marxan family of software (www.ecology.uq.edu.au/marxan.htm) have been applied to solve spatial problems where the objective has a spatial component (minimize boundary length, minimum reserve size) and the targets attempt to deliver adequacy (Possingham *et al.* 2000, Possingham *et al.* 2001, Noss *et al.* 2002; also note the work of Andelman *et al.* 1999, Singleton *et al.* 2001). If issues of connectivity can be clearly defined, they can be incorporated into the algorithms. The challenge is to articulate the connectivity process issues discussed here so that they can be formulated mathematically. The existing Marxan algorithms then need to be modified to accommodate the required new kinds of objectives and constraints.

A computer-based planning tool is needed that draws upon these modified algorithms, accesses the spatial information base, and that can be used to prepare information and options for stakeholders interested in advancing biodiversity conservation in their region. As landscape viability is of equal concern for all users of the land resource, such a planning tool should be generally welcomed as a tool for meshing production and conservation objectives. Nevertheless, the difficulties with this approach should not be underestimated, as in many areas we lack basic information with which to guide landscape management, and we cannot always wait for complete information to make decisions. Simpler approaches that base decisions on partial information may stimulate activity and enthusiasm within local communities (e.g., Lambeck 1997, Dilworth *et al.* 2000). While these approaches can be criticized (e.g., Lindenmayer *et al.* 2002), they may form a useful kernel on which to build greater scientific sophistication that leads to action.

11.5 Conclusion

In summary, the WildCountry scientific framework draws from landscape ecology principles, which include the following main elements:

- Core protected-area networks must be based on systematic reserve design principles that build upon the criteria of comprehensiveness, adequacy, and representativeness, complimented by, among others, criteria related to primary productivity and landscape condition.
- Biodiversity conservation assessment and planning (including protected-area design) must move beyond traditional conservation design principles by aiming for the maintenance and restoration of large-scale (in space and time) ecological and evolutionary processes over the entire landscape. Assessments and plans must reflect the landscape linkages necessary to maintain large-scale ecological phenomena and processes related to trophic relations and interactive species, hydro-ecology, long-distance biological movement, refugia from threatening processes, ecological fire regimes, climate change and variability, and coastal zone fluxes.
- Proximity of the reserve system to sources of disturbance requires, as a minimum, buffering and consideration of complementary land uses and management. Whole-of-landscape conservation assessment and planning will be unavoidable; recognizing that the entire landscape (protected areas, leasehold land, Aboriginal land, unallocated crown land, private land) within which protected areas are embedded must be better managed to promote biodiversity conservation.

Regional planning must therefore include management guidelines and prescriptions for, among other things, broad-scale threatening processes including feral animals, weeds and ecologically inappropriate fire regimes, both in protected and unprotected areas. Ecological restoration in degraded landscapes will be necessary, particularly in the intensive land-use areas of Australia (Fig. 11.1). Restoration objectives should reflect the need to restore the identified large-scale connectivity processes. Landscape viability analysis will enable the entire landscape to be evaluated and the optimum set identified of dedicated reserves, areas of connectivity, and matrix (off-reserve) management requirements.

Management for biodiversity conservation that facilitates long-term ecological connectivity will remain an ongoing research and development challenge. It must be recognized that the matrix is never static, and it may be impossible to predict the quantity and quality (intensity) of development that could eventually occur on any specific parcel in any given region. Thus, the conservation utility of the matrix must be considered with caution and be recognized as complementary to dedicated core areas. In fact, it may be prudent to assume that the matrix will change, and, in the worst-case scenarios, lose all positive conservation values over time.

It must be noted that the emerging WildCountry framework described here goes beyond current reserve-based assessment and management, and hence needs a step-up in activity and funding. We acknowledge that it has proven difficult to maintain current levels of conservation management, with many agencies facing reduced budgets and having to deal with increasing threats. The challenge then is to recognize the full extent of the actions needed and convince land managers, communities, governments, and relevant agencies of the need for a broadly based landscape approach.

Acknowledgments

Thanks to Kathryn Edwards for technical assistance with Fig. 11.1.

References

AFFA. 2003. Regional Forest Agreements web site. Australian Government Department of Agriculture, Fisheries and Forestry (http://www.affa.gov.au/).

Andelman, S.J., I. Ball, F. W. Davis, and D. M. Stoms. 1999. *SITES V 1.0: an Analytical Toolbox for Designing Ecoregional Conservation Portfolios*. Unpublished manual prepared for the nature conservancy. (available at http://www.biogeog.ucsb.edu/projects/tnc/toolbox.html).

ANZECC. 2001. *Australian and New Zealand Environment and Conservation: Review of the National Strategy of the Conservation of Australia's Biological Diversity*. Canberra: Commonwealth of Australia.

Austin, J. M., B. G. Mackey, and K. P. van Niel. 2003. Estimating forest biomass using satellite radar: an exploratory study in a temperate Australian Eucalyptus forest. *Forest Ecology and Management* **176**, 575–83.

Bach, C. S. 2002. Phenological patterns in monsoon rainforests in the Northern Territory, Australia. *Austral Ecology* **27**, 477–89.

Beck, M. W. 1996. On discerning the case of late Pleistocene megafaunal extinctions. *Paleobiology* **22**, 91–103.

Bowman, D. M. J. S., O. Price, P. J. Whitehead, and A. Walsh. 2001. The "wilderness effect" and the decline of *Callitris intratropica* on the Arnhem Land Plateau, northern Australia. *Australian Journal of Botany* **49**, 665–72.

Bradstock, R. A., J. E. Williams, and A. M. Gill (eds.). 2002. *Flammable Australia: The Fire Regimes and Biodiversity of a Continent*. Cambridge: Cambridge University Press.

Cannon, R. and M. G. Gardner. 1999. Assessing the risk of windborne spread of foot and mouth disease in Australia. *Environment International* **25**, 713–23.

Catchpole, W. 2002. Fire properties and burn patterns in heterogeneous landscapes. Pages 49–75 in R. A. Bradstock, J. E. Williams, and A. M. Gill (eds.). *Flammable Australia: The Fire Regimes and Biodiversity of a Continent*. Cambridge: Cambridge University Press.

Caughley, G., G. Grigg, J. Caughley, and G. J. E. Hill. 1980. Does dingo predation control the densities of kangaroos and emus? *Australian Wildlife Research* **7**, 1–12.

Commonwealth of Australia. 1997. *Nationally Agreed Criteria for the Establishment of a Comprehensive, Adequate and Representative Reserve System for Forests in Australia*. Canberra: Government Printer.

Commonwealth of Australia. 1999. *National Forest Policy Statement*. (http://www.affa.gov.au/content/publications.cfm?ObjectID=CDA4CAF9-D118-4E13-AAC472AC7603EBE5).

Commonwealth of Australia. 2001a. *National Objectives and Targets for Biodiversity Conservation 2001–2005*. Canberra: Government Printer.

Commonwealth of Australia. 2001b. *Australian Native Vegetation Assessment 2001*. Canberra: National Land and Water Resources Audit.

Commonwealth of Australia. 2002. *The Regional Forest Agreements* (http://www.rfa.gov.au).

Convention on Biodiversity. 1992. Website of the Secretariat on the Convention on Biodiversity (http://www.biodiv.org).

Cosser, P. 1997. *Nutrients in Marine and Estuarine Environments*. Canberra: Australian Government Department of Environment and Heritage.

Crooks, K. R. and M. E. Soulé. 1999. Mesopredator release and avifaunal extinctions in a fragmented system. *Nature* **400**, 563–6.

CSIRO. 2001. *Climate Change Projections for Australia*. Melbourne: CSIRO Atmospheric Research (http://www.dar.csiro.au/publications/projections2001.pdf).

Davies, K. F., B. A. Melbourne, and C. R. Margules. 2001. Effects of within- and between-patch processes on community dynamics in a fragmentation experiment. *Ecology* **82**, 1830–46.

Dilworth, R., T. Gowdie, and T. Rowley. 2000. Living landscapes: the future landscapes of the Western Australian wheatbelt? *Ecological Management and Restoration* **1**, 165–74.

Dobson, D., K. Ralls, M. Foster, *et al.* 1999. Reconnecting fragmented landscapes. Pages 129–70 in M. E. Soulé and J. Terborgh (eds.). *Continental Conservation: Scientific Foundations for Regional Conservation Networks*. Washington, DC: Island Press.

Drake, V. A., P. C. Gregg, I. T. Harman, *et al.* 2001. Characterizing insect migration systems in inland Australia with novel and traditional methodologies. Pages 207–34 in I. P. Woiwood, D. R. Reynolds, and C. D. Thomas (eds.). *Insect Movement: Mechanisms and Consequences*. Wallingford: CABI Publishing.

Environment Australia. 2001. *State of the Environment Report*. Canberra: Commonwealth of Australia (http://www.ea.gov.au/soe/).

Finlayson, H. H. 1961. On central Australian mammals. *Records of the South Australian Museum* **14**, 141–91.

Foreman, D. 1999. The Wildlands Project and the rewilding of North America. *Denver University Law Review* **77**, 535–53.

Frankel, O. H. and M. E. Soulé 1981. *Conservation and Evolution*. Cambridge: Cambridge University Press.

Franklin, D. 1999. Evidence of disarray amongst granivorous bird assemblages in the savannas of northern Australia, a region of sparse human settlement. *Biological Conservation* **90**, 53–68.

Friedel, M. H., B. D. Foran, and D. M. Stafford Smith. 1990. Where the creeks run dry or ten feet high: pastoral management in arid Australia. *Proceedings of the Ecological Society of Australia* **16**, 185–94.

GBRMPA. 2003. *Explanatory Statement, Great Barrier Reef Marine Park Zoning Plan 2003*. Australian Government, Great Barrier Reef Marine Park Authority (http://www.reefed.edu.au/rap/pdf/ES.25-11-03.pdf).

Gill, A. M. and J. E. Williams. 1996. Fire regimes and biodiversity: the effects of fragmentation of southeastern eucalypt forests by urbanization, agriculture and pine plantations. *Forest Ecology and Management* **85**, 261–78.

Gorshkov, G. V., V. V. Gorshkov, and A. M. Makarieva, 2000. *Biotic Regulation of the Environment*. Chichester: Spinger-Praxis.

Greening Australia. 2004. http://www.greeningaustralia.org.au/GA/NAT/.

Hale, P. and D. Lamb (eds.). 1997. *Conservation Outside Nature Reserves*. Brisbane: Centre for Conservation Biology, University of Queensland.

Hannah L. and T. E. Lovejoy (eds.). 2003. *Climate Change and Biodiversity: Synergistic Impacts*. Washington, DC: Conservation International.

Hobbs, R. J. 2002a. Habitat networks and biological conservation. Pages 150–70 in K. J. Gutzwiller (ed.). *Applying Landscape Ecology in Biological Conservation*. New York: Springer.

Hobbs, R. J. 2002b. Fire regimes and their effects in Australian temperate woodlands. Pages 305–26 in R. Bradstock, J. E. Williams, and A. M. Gill (eds.). *Flammable Australia: Fire Regimes and the Biodiversity of a Continent*. Cambridge: Cambridge University Press.

Hobbs, J. E., J. A. Lindesay, and H. A. Bridgman. 1998. *Climates of the Southern Continents: Past, Present and Future*. Chichester: John Wiley & Sons Ltd.

Hobbs, R. J., D. A. Saunders, and A. R. Main. 1993. Conservation management in fragmented systems. Pages 279–96 in R. J. Hobbs and D. A. Saunders (eds.). *Reintegrating Fragmented Landscapes: Towards Sustainable Production and Nature Conservation*. New York: Springer.

Horn, A. M. 1995. *Surface Water Resources of Cape York Peninsula*. Canberra: CYPLUS – Queensland and Commonwealth Governments.

Horn, A. M., E. A. Derrington, G. C. Herbert, R. W. Lait, and J. R. Miller. 1995. *Groundwater Resources of Cape York Peninsula*. Canberra: CYPLUS – Queensland and Commonwealth Governments.

Hutchinson, M. F., J. A. Stein, and J. L. Stein. 2000. *Upgrade of the 9 Second Australian Digital Elevation Model*. CRES, The Australian National University (http://cres.anu.edu.au/dem).

IPCC. 2002. *Climate Change 2001: The Scientific Basis*. Intergovernmental Panel on Climate Change (http://www.ipcc.ch/).

Isard, S. A. and S. H. Gage. 2001. *Flow of Life in the Atmosphere: an Airscape Approach to Understanding Invasive Organisms*. East Lansing: Michigan State University Press.

Keith, D., J. E. Williams, and J. Woinarski. 2002. Fire management and biodiversity conservation – key approaches and principles. Pages 401–428 in R. A. Bradstock, J. E. Williams, and A. M. Gill (eds.). *Flammable Australia: The Fire Regimes and Biodiversity of a Continent*. Cambridge: Cambridge University Press.

Lambeck, R. J. 1997. Focal species: a multi-species umbrella for nature conservation. *Conservation Biology* 11, 849–56.

Landcare Australia. 2004. http://www.landcareaustralia.com.au/.

Landsberg, J. J. and R. H. Waring. 1997. A generalized model of forest productivity using simplified concepts of radiation-use efficiency, carbon balance and partitioning. *Forest Ecology and Management* 95, 209–28.

Lesslie, R. G. 2001. Landscape classification and strategic assessment for conservation: an analysis of native cover loss in far southeast Australia. *Biodiversity and Conservation* 10, 427–42.

Lewis, D. 2002. *Slower Than the Eye Can See: Environmental Change in Northern Australia's Cattle Lands: A Case Study from the Victoria River District, Northern Territory*. Darwin: Tropical Savannas Cooperative Research Centre.

Lindenmayer, D. B. and J. F. Franklin. 2002. *Conserving Forest Biodiversity: A Comprehensive Multiscaled Approach*. Washington DC: Island Press.

Lindenmayer, D. B. and H. A. Nix. 1993. Ecological principles for the design of wildlife corridors. *Conservation Biology* 7, 627–30.

Lindenmayer, D. B. and H. F. Recher. 1998. Aspects of ecologically sustainable forestry in temperate eucalypt forests – beyond an expanded reserve system. *Pacific Conservation Biology* 4, 4–10.

Lovejoy, T. E. 1982. Designing refugia for tomorrow. Pages 673–80 in G. T. Prance (ed.). *Biological Diversification in the Tropics*. New York: Columbia University Press.

Mackey, B. G. and D. B. Lindenmayer. 2001. Towards a hierarchical framework for modeling the spatial distribution of animals. *Journal of Biogeography* 28, 1147–66.

Mackey, B.G., D.B. Lindenmayer, M. Gill, M. McCarthy, and J. Lindesay. 2002. *Wildlife, Fire and Future Climate: A Forest Ecosystem Analysis*. Melbourne: CSIRO Publishing.

Mackey, B.G., H.A. Nix, and P. Hitchcock. 2001. *The Natural Heritage Significance of Cape York Peninsula*. Canberra: ANU Tech P/L.

Mackey, B.G., H.A. Nix, M.F. Hutchinson, J.P. McMahon, and P.M. Fleming. 1988. Assessing representativeness of places for conservation reservation and heritage listing. *Environmental Management* **12**, 501–14.

Mackey, B.G., H.A. Nix, J. Stein, E. Cork, and F.T. Bullen. 1989. Assessing the representativeness of the Wet Tropics of Queensland World Heritage Property. *Biological Conservation* **50**, 279–303.

Mackey, B.G. and R. Sims. 1993. A climatic analysis of selected boreal tree species and potential responses to global climate change. *World Resources Review* **5**, 469–87.

Margules, C.R. and R.L. Pressey, 2000. Systematic conservation planning. *Nature* **405**, 243–53.

McIntyre, S. and R.J. Hobbs. 1999. A framework for conceptualizing human impacts on landscapes and its relevance to management and research. *Conservation Biology* **13**, 1282–92.

Morton, S.R. 1990. The impact of European settlement on the vertebrate animals of arid Australia: a conceptual model. *Proceedings of the Ecological Society of Australia* **16**, 201–13.

Morton, S.R., D.M. Stafford Smith, M.H. Friedel, G.F. Griffin, and G. Pickup. 1995. The stewardship of arid Australia: ecology and landscape management. *Journal of Environmental Management* **43**, 195–218.

NASA. 2003. *The Moderate Resolution Imaging Spectroradiometer (MODIS)* (http://modis.gsfc.nasa.gov/).

Nix, H.A. 1974. Environmental control of breeding, post-breeding dispersal and migration of birds in the Australian region. Pages 272–305 in H.J. Frith and J.A. Calaby (eds.). *Proceedings of the XVI International Ornithological Congress*. Canberra: Australian Academy of Sciences.

Nix, H.A., D.P. Faith, M.F. Hutchinson, *et al.* 2001. *The Biorap Toolbox: A National Study of Biodiversity Assessment and Planning for Papua New Guinea*. Canberra: The Centre for Resource and Environmental Studies, The Australian National University.

NLWRA. 2002. *Australian Terrestrial Biodiversity Assessment*. Canberra: National Land and Water Resources Audit.

Noss, R.F., C. Carroll, K. Vance-Borland, and G. Wuerthner. 2002. A multicriteria assessment of the irreplaceability and vulnerability of sites in the Greater Yellowstone Ecosystem. *Conservation Biology* **16**, 895–908.

Noss, R.F. and A.Y. Cooperrider. 1994. *Saving Nature's Legacy*. Washington, DC: Island Press.

NTK. 2003. *Developing a National Biodiversity and Climate Change Action Plan*. National Task Group on the Management of Climate Change Impacts on Biodiversity. Canberra: Australian Government, Department of Environment and Heritage.

Oksanen, L. and T. Oksanen. 2000. The logic and realism of the hypothesis of exploitation ecosystems. *The American Naturalist* **155**, 703–23.

Palmer, C. and J.C.Z. Woinarski. 1999. Seasonal roosts and foraging movements of the black flying fox *Pteropus alecto* in the Northern Territory: resource tracking in a landscape mosaic. *Wildlife Research* **26**, 823–38.

Paton, D.C., A.M. Prescott, R.J. Davies, and L.M. Heard. 2000. The distribution, status and threats to temperate woodlands in South Australia. Pages 57–85 in R.J. Hobbs and C.J. Yates (eds.). *Temperate Eucalypt Woodlands in Australia: Biology, Conservation, Management and Restoration*. Chipping Norton, NSW: Surrey Beatty & Sons.

Possingham, H.P., Andelman, S.J., Noon, B.R., Trombulak, S. and Pulliam, H.R. 2001. Making smart conservation decisions. Pages 225–44 in G. Orians and M. Soule (eds.). *Research Priorities for Conservation Biology*. Covelo: Island Press.

Possingham, H.P., I.R. Ball, and S. Andelman. 2000. Mathematical methods for identifying representative reserve networks. Pages 291–306 in S. Ferson and M. Burgman (eds.). *Quantitative Methods for Conservation Biology*. New York: Springer.

Price, O. and D. M. J. S. Bowman. 1994. Fire-stick forestry: a matrix model in support of skillful fire management of *Callitris intratropica R.T. Baker* by north Australian Aborigines. *Journal of Biogeography* **21**, 573–80.

Reid, J. R. W., A. Kerle, and S. R. Morton. 1993. *Uluru Fauna: the Distribution and Abundance of Vertebrate Fauna of Uluru (Ayers Rock – Mount Olga) National Park*. Canberra: N. T. ANPWS.

Roshier, D. A., P. H. Whetton, R. J. Allan, and A. I. Robertson. 2001. Distribution and persistence of temporary wetland habitats in arid Australia in relation to climate. *Austral Ecology* **26**, 371–84.

Russell-Smith, J., P. G. Ryan, D. Klessa, G. Waight, and R. Harwood. 1998. Fire regimes, fire-sensitive vegetation and fire management of the sandstone Arnhem Plateau, monsoonal northern Australia. *Journal of Applied Ecology* **35**, 829–46.

Saunders, D. A. and R. J. Hobbs (eds.). 1991. *Nature Conservation 2: The Role of Corridors*. Chipping Norton, NSW: Surrey Beatty and Sons.

SEAC. 1996. *Australia: State of the Environment 1996*. Collingwood: CSIRO Publications. (http://www.deh.gov.au/soe/).

Shapcott, A. 2000. Conservation and genetics in the fragmented monsoon rainforest in the Northern Territory, Australia: a case study of three frugivore dispersed species. *Australian Journal of Botany* **48**, 397–407.

Singleton, P. H., W. Gaines, and J. F. Lehmkuhl. 2001. Using weighted distance and least-cost corridor analysis to evaluate regional-scale large carnivore habitat connectivity in Washington. Pages 583–94 in *The Proceedings of the International Conference on Ecology and Transportation, Keystone, CO. September 24–27.*

Soulé, M. E., J. Estes, J. Berger, and C. Martinez del Rio. 2003. Ecological effectiveness: conservation goals for interactive species. *Conservation Biology* **17**, 1238–50.

Soulé, M. E. and R. F. Noss. 1998. Rewilding and biodiversity: Complementary goals for continental conservation. *Wild Earth* **8**, 18–28.

Soulé, M. E. and J. Terborgh (eds.). 1999. *Continental Conservation: Scientific Foundations of Regional Reserve Networks*. Washington DC: Island Press.

Stafford Smith, D. M. and S. R. Morton. 1990. A framework for the ecology of arid Australia. *Journal of Arid Environment* **18**, 225–78.

Starr, B. J., R. J. Wasson, and G. Caitcheon. 1999. *Soil Erosion, Phosphorous and Dryland Salinity in the Upper Murrumbidgee: Past Changes and Current Findings*. "Pine Gully," Wagga, NSW, Australia: Murrumbidgee Catchment Management Committee.

Terborgh, J., J. A. Estes, P. C. Paquet, *et al.* 1999. Role of top carnivores in regulating terrestrial ecosystems. Pages 39–64 in M. E. Soulé and J. Terborgh (eds.). *Continental Conservation: Design and Management Principles for Long-term, Regional Conservation Networks*. Washington DC: Island Press.

Thomas, C. D., A. Cameron, R. E. Green, *et al.* 2004. Extinction risk from climate change. *Nature* **427**, 145–8.

Vertessey, R. A., R. Benyon, and S. Haydon. 1994. Melbourne's forest catchment: effect of age on water yield. *Water* **21**, 17–20

Whelan, R. J. 1995. *The Ecology of Fire*. Cambridge: Cambridge University Press.

Williams, J. E. and A. M. Gill. 1995. *The Impact of Fire Regimes on Native Forests in Eastern NSW*. Sydney: NSW National Parks and Wildlife Service.

Williams, J. E., R. J. Whelan, and A. M. Gill. 1994. Fire and environmental heterogeneity in southern temperate forest ecosystems: implications for management. *Australian Journal of Botany* **42**, 125–37.

Woinarski, J., P. Whitehead, D. Bowman, and J. Russell-Smith. 1992. Conservation of mobile species in a variable environment: the problem of reserve design in the Northern Territory, Australia. *Global Ecology and Biogeography Letters* **2**, 1–10.

Woinarski, J. C. Z., D. J. Milne, and G. Wanganeen. 2001. Changes in mammal populations in relatively intact landscapes of Kakadu National Park, Northern Territory, Australia. *Austral Ecology* **26**, 360–70.

World Resources Institute. 2001. *World Resources 2000–2001. People and Ecosystems*. United Nations Development Programme, United Nations Environment Programme, World Bank, and World Resources Institute.

Yibarbuk, D., P. J. Whitehead, J. Russell-Smith, *et al.* 2001. Fire ecology and Aboriginal land management in central Arnhem Land, northern Australia: a tradition of ecosystem management. *Journal of Biogeography* **28**, 325–44.

12

Using landscape ecology to make sense of Australia's last frontier

12.1 Introduction

Just as the nineteenth century was a period of great biological discovery, driven by exploration and worldwide expansion of Western culture, there is no doubt that the dramatic global environment changes, driven by exploitation and pollution of the biosphere, will characterize the twenty-first century. A spin-off of the expansion of industrial civilization, that is driving the planetary environmental crisis, is the development and widespread availability of powerful digital technologies, such as geographic information systems, global positioning systems, digital aerial photography, and satellite imagery. These technologies provide unique insights into the rate and scale of environmental disturbances at the landscape-scale, which in aggregate drive global change. Natural resource managers and decision-makers tasked to achieve ecological sustainability necessarily focus on the landscape scale. Let us call the science that examines the ecological interaction between humans and landscapes *landscape ecology* (Naveh and Lieberman 1984). This discipline has the advantage of building on numerous other disciplines, including pure and applied physical and biological sciences and the more ambiguous, nuanced, and subtler fields in the humanities that have a stake in landscapes, including anthropology, environmental history, and various themes of human geography (Head 2001). Such a polyglot and young science is inherently vulnerable to bouts of introspection and anxiety about the conceptual bounds of the discipline and its philosophical roots (Wu and Hobbs 2002). I submit that the strength and utility of the transdisciplinary perspectives for making sense of and responding to global change is provided by landscape ecology. Further, these strengths are most apparent on a development frontier, such as the Australian monsoon

Key Topics in Landscape Ecology, ed. J. Wu and R. Hobbs.
Published by Cambridge University Press. © Cambridge University Press 2007.

tropics. Here, there is a remarkable juxtaposition of a technologically advanced society and a traditional indigenous population, set within a great expanse of minimally developed and biologically diverse tropical and arid landscapes. This globally unique situation also contrasts with the southern half of the Australian continent that has been transformed by European colonization in a mere 200 years of settlement.

In this chapter I draw on my experience in working on the north Australian frontier, reflecting on the potential of landscape ecology to contribute to the quest for sustainability in a time of tremendous environmental change. In such a culturally contested and rapidly changing region, the holistic and integrative approaches of landscape ecology are clearly apparent. So too is the power of story telling. Indeed, I explore these ideas by telling a number of "stories" about northern Australia, and my impressions of the practice of landscape ecology.

12.2   The north Australian frontier

The extraordinary diversity of endemic plant and animal species that so astonished and perplexed European colonists and observers such as Charles Darwin has rendered the keyword "Australia" synonymous with biological exceptions to the global rule, or at least the "normal" northern hemisphere rule. It is not surprising, therefore, that the study of Australian ecology has developed some independent traditions relative to global trends in ecology and evolution (Attiwill and Wilson 2003). The sense of studying the "exceptions" is amplified for those who work in northern Australia, a vast tropical frontier which, while sharing many biological similarities with southern Australia, has a number of salient physical and cultural features that make it different from the rest of the continent (Haynes *et al.* 1991). First, the climate of the north is controlled by the Australian summer monsoon. Relative to southern Australia, the north has back-to-front seasons: during the austral summer months (December to February), when the south of the continent regularly experiences high temperatures and oven-dry winds, the north has a hot, humid wet season characterized by week-long deluges that cause widespread flooding, and frequent tropical cyclones that wrack coastal regions. Conversely, during the austral winter months (June to August), the north enjoys a hot, rain-free austral "winter" while the southern days are short, chilly, and often wet.

In a mere 150 years there has been a remarkable confluence of a 40000 year-old culture with the Western tradition, associated with the last wave of colonization by the British Empire. In addition to perceived nineteenth-century geopolitical strategic imperatives, the economic drivers of northern settlement were exploitation of the endless landscapes by extensive cattle ranching and localized mining (Powell 1996). Infertile soils, labor shortages and isolation from markets stymied intensive agricultural development, so the northern Australian savannas experienced an insignificant degree of land clearing compared to southern Australia where agricultural development has seriously fragmented the native vegetation. Indeed, the north has the largest expanse of intact savannas of any region in the world, although this may change given developments in agricultural technologies and increasing global demand for food and fiber. Despite the social upheavals caused by European settlement, Aboriginal people that remain on their tribal lands have maintained one of the most ancient connections between humans and landscapes anywhere on Earth (Mulvaney and Kamminga 1999). It is increasingly clear that Aboriginal fire and game management has molded vegetation such that apparently "natural" landscapes have a profound "cultural" imprint (Yibarbuk *et al.* 2001).

For the first half of the twentieth century the north remained an exotic frontier, so far over the horizon that it was beyond serious consideration in national, let alone international, thinking. Sustained attacks by the Japanese during the Second World War, however, forced the Australian government to consider more seriously the future and potential of the north. The post-war period saw great investment by the Australian government in examining the economic potential of the vast "empty" landscapes using, for their time, advanced scientific and technological approaches, particularly aerial photographic mosaics. Indeed, these federally funded resource surveys prompted the development of "land system mapping," an often-overlooked, pioneering approach to "landscape ecology" (Christian and Stewart 1953). Land system mapping sought to characterize the edaphic, topographic, and biological resources as integrated mapping units.

In the last two decades of last century, the focus of landscape-scale research has shifted from exploring development potential to moderating the impact of developments and land management practices. There is a real risk that development will destroy the heterogeneity of the savanna habitat mosaic or disrupt the capacity of highly mobile wildlife assemblages to track resources in time and space (Bowman 1991, Woinarski *et al.* 1992). Examples of some of these research programs are the mapping of vegetation and fire activity across northern Australia (Fox *et al.* 2001, Russell-Smith *et al.* 2004), and the design of entire systems of conservation reserves using biological databases (Woinarski 1996, Woinarski *et al.* 1996, Parks and Wildlife Commission of the Northern Territory undated). These ambitious landscape-scale projects have triggered or provided important context for more focused research such as explaining why soils, landforms, and vegetation should be so closely coupled (e.g., Bowman and Minchin 1987), predicting the negative consequences of loss of naturally occurring rainforest isolates on other rainforest isolates (e.g., Price *et al.* 1995), or discerning the landscape pattern of Aboriginal fire usage (e.g., Bowman *et al.* 2004).

Clearly, the research priorities in the north contrast starkly with those undertaken in fragmented landscapes that characterize much of the developed world. The latter studies demand consideration of a finer spatial scale with an emphasis on corridors and fragments of habitats, and ecological restoration (Egan and Howell 2001, Liu and Taylor 2002). Nonetheless, there can be no doubt that quantitative landscape-scale analyses are a basic prerequisite for comprehending and formulating management of natural resources in a frontier like the north of Australia.

12.3   This is not a landscape

While the products of quantitative landscape ecology are of tremendous importance in framing and disciplining thinking about landscape, it is easy to overlook that they are at best a crude analog of an infinitely complex landscape. This devastatingly simple philosophical point was made by the surrealist painter Rene Magritte in this famous painting *The Treachery of Images*. The painting depicts a tobacco pipe, with a caption that reads "*Ceci n' est pas une pipe*" [this is not a pipe] (Foucault 1973). Magritte was making the point that the image is a representation of the thing, not the thing itself. Treating an abstraction as if it were a material thing is a philosophical fallacy known as reification.

Quantitative landscape ecologists routinely reify because it is not possible to providing a rigorous and unambiguous definition of landscape. This problem arises because of the fractal, multidimensional, and dynamic nature of landscapes and their ecological complexity and biological diversity. I suggest that the conceptual ambiguity and overwhelming complexity of "landscape" has resulted in two related tendencies in quantitative landscape ecology: the search for some absolute mathematical expression of landscape attributes – the so-called metrics – and the construction of deductive arguments about speculative "landscape processes."

12.4   The quadrat is dead

The search for metrics to mathematically describe "landscape" mirrors phytosociologists' quixotic quest in the second half of the twentieth century for the best numerical methodology to objectively define plant communities (Mueller-Dombois and Ellenberg 1974). In the course of this methodological development, the quadrat underwent a conceptual transformation from being a pragmatic device to help focus sampling effort to having an ill-defined "essence" that disconnected the observer from the reality and inherent complexity of vegetated landscapes. Great intellectual effort yielded rigorous sampling designs and numerical procedures to analyze quadrat data, and these undoubtedly revolutionized plant ecology. However, the outputs of the most sophisticated analyses (Jongman *et al.* 1995) remain what they always were – abstract and imperfect descriptions of vegetation based on small samples located in complex and dynamic landscapes. Similarly, there are no inherently right metrics; rather the choice of metric depends upon the purpose and context of the research.

12.5   Landscape models: but "there is no there there"

Another consequence of reification in quantitative landscape ecology is the overemphasis on deductive reasoning such as hypothetical models or flow diagrams of putative ecological processes. There is no doubt that such approaches are of great heuristic value. For example, recognition of the importance of localized fertile patches within tracts of infertile savanna has contributed to important insights into the mechanisms of widespread population declines and local extinctions of wildlife (Hobbs 2005). However, discussion of these models and idealized systems can overlook the fact that the components of these systems do not exist as discrete elements in the real world, or to use Gertrude Stein's famous phrase, "there is no there there." The operational definition of theoretical landscape elements is fraught, whether in the field or from remote-sensing data. A good example of this problem concerns the definition and delineation of the small naturally occurring fragments of "rainforest" that occur embedded in the north Australian savanna matrix. While there is agreement that such vegetation exists and is of great importance for the ecological function of the savanna matrix, there is no agreement as to how to define such vegetation (e.g., Lynch and Neldner 2000, Bowman 2001a). To avoid any pretence of absolute definitions and to sidestep a spiral of endless disputation, I suggest that landscape elements should be explicitly and operationally defined at the outset of a research program. This pragmatic approach requires one to take into account the environmental context, spatial scale, and purpose of the research. I believe that both theoretical and methodological development in quantitative landscape ecology must be literally grounded by comprehensive field checking and experimentation. An example of this approach was the recent calibration of four different methodologies to map fire scars using a landscape-scale field experiment (Bowman *et al.* 2003).

On a frontier, there is close engagement with landscapes and with land managers. Consequently, there is strong "selection pressure" to use quantitative techniques to document change rather than to invent, refine, and perfect techniques in isolation from pressing demands. Nonetheless, a real danger with such pragmatism is losing sight of the inherent value of theoretical research and becoming increasing intellectually isolated from debates, innovations, and developments of the rest of the discipline. The extraordinary opportunities to observe a rapidly changing environment combined with the tension between the pure and the applied aspects of landscape ecology makes the research on a frontier stimulating and challenging if researchers are adaptable. But such adaptability may take quantitative researchers into the qualitative realm.

12.6   Longing and belonging

Despite the intractable difficulties in neatly defining, quantifying, or even agreeing about the essential nature of "landscape," there can be no doubt that this concept is vitally important for land managers and the broader community. Discussion about "landscape" may act as a lightning rod for profound political and cultural debates about identity, place, and belonging. Both frontier and post-colonial societies typically have a keenly felt need for a sense of belonging to the landscapes they violently appropriated from indigenous people (Head 2000). Nature conservation and national-park movements signal a philosophical shift from the initial frontier mentality to a view that landscapes need to be cherished and preserved (Bonyhardy 2000). Aldo Leopold (1949), widely regarded as the father of the contemporary conservation movement, argued that sustainable settlement requires the development of a deep appreciation of where the settlers live. He argued that such environmental awareness involved an appreciation of geographical and historical contexts such as understanding how landscapes have evolved and how they change in response to seasonal cycles. For example, people need an appreciation of where landscapes fit in relationship to each other, from whence the winds blow, rivers flow, and migratory birds come and where they go. Leopold argued that the reciprocal interaction of land and people created a profound sense of belonging because humans and land had a closely shared history. He also realized that these interactions must be moderated by ethical restrictions on the scale and nature of resource usage.

Ironically, Leopold, like so many settlers, failed to grasp that such a holistic frame of landscape also underpinned the indigenous cultures that had been so dramatically and negatively transformed by settlement. For example, Australian Aborigines living on their "country" have an encyclopedic geographic and ecological knowledge, including complex mythologies about how landscapes were shaped and formed in the distant past during the so-called Dreamtime. Aboriginal people have profound physical and spiritual interconnections with landscapes that are formed and maintained through everyday use of natural resources such as hunting and gathering. Far from being "abstract," Aboriginal art often explicitly depicts complex practical and mythological knowledge of landscapes, but unlocking this knowledge requires an appropriate cultural frame (Watson 1993).

There can be no doubt that quantitative landscape ecology can play a pivotal role in the development of a sense of connectedness between settlers and their adopted landscapes. Furthermore, the discipline can be used to moderate the impacts of settlement. For example, policy-makers and land managers can anticipate areas of resource conflict and identify the geographic bounds of landscapes that are vulnerable to overexploitation or that have significant cultural and ecological values. But how does quantitative landscape ecology meet the challenge of communicating the great complexity, diversity, and dynamics of landscapes? If the discipline's outputs are often ambiguous, uncertain, and implicitly steeped in human values and perceptions, how can it meet the expectations of technocrats and policy-makers that want neat, black-and-white "answers"?

12.7 Tell me a story

Environmental historians have demonstrably succeeded in meeting some of the expectations, or at least capturing the attention, of land managers and policy-makers, notwithstanding limited field experience and a humanities background. How is it that they have succeeded in being so attuned to the public imagination? The answer, I believe, is that they have mastered the craft of telling stories about nature (Bowman 2001b and references therein). A good example of this genre is Stephen Pyne's (1998) essay about the dramatic shift in perceptions of the Grand Canyon by European colonists. In 1540 the Spanish explorers, who were the first Europeans to encounter the Canyon, dismissed it as hostile and worthless and nothing more than a geographic obstruction, yet 400 years later it is regarded as a natural wonder and part of the core of modern American identity.

According to the American environmental historian William Cronon (1992), environmental histories are nonfictional narratives that are steeped in human values and therefore interesting to people. These narratives are grounded in ecological fact but not limited by them. Environmental historians are unafraid of making much with little by breathing human interest into disjointed and incomplete facts. Environmental activists have also understood the power of stories and emotion in their campaigns to sway public opinion. This modus operandi often destabilizes and perplexes natural scientists who are uncomfortable with acknowledging and integrating human values into their thinking and quantitative analyses because it confounds the core principle of scientific "objectivity." This restrictive worldview, however, can be a great handicap in building a constituency to support research and to transform research outputs into real-world outcomes. This is particularly so for the media that is interested in presenting a "story" told by animated, interesting individuals.

Just as landscapes are open to many interpretations, so too are stories about landscapes. For example, the gross generalization that Australia is an arid, isolated, infertile chunk of Gondwana has been used to bolster claims to restrict the growth of human populations (Flannery 1994). However, these geographic facts have also been used to argue that "outback" Australia is a prime location for the storage of nuclear waste (Bowman 2004). Another danger is that a particular story can become orthodoxy because it is psychologically satisfying or fits a particular ascendant ideology, even though it may be based on slender evidence. It may then be difficult to subject the story to critical analysis, and it may be misused to achieve political ends. A good example of the power of stories relates to the massive clearing of native vegetation in central Queensland, Australia, where sketchy evidence of increases in the density of tree cover in some landscapes was used to justify wholesale destruction of native vegetation (see Bowman 2001b and references therein).

Landscape ecologists have the ability and the tools to record new data and to both critically assess the hypotheses that underpin existing stories and to create entirely new stories about landscapes. A concrete example is the use of aerial photography to detect the spatial extent and temporal pattern of landscape-scale vegetation cover change (e.g., Fensham and Fairfax 2003). This research has been used to bolster the hypothesis proposed by Fensham (2000) that landscape-scale changes in tree cover are better explained by drought cycles than by overgrazing or changed fire regimes. This "story" has far-reaching implications for attitudes and policy formulation about native-vegetation clearance.

Landscape ecologists often unwittingly tell stories about landscapes in their academic writings. If they are to influence land management, they must become comfortable with the power of story telling as a tool to reach a broader audience and encode information that can be readily comprehended by managers and policy-makers. Rather than making specific predictions about what will happen, Cronon (1993) suggested it is wiser to "offer *parables* about how to interpret what *may* happen" [original emphasis]. The stories landscape ecologists tell should not disguise the uncertainty that surrounds them, and their stories should be continually reevaluated in the light of both new research and practical experience. Indeed, environmental histories can be seen as hypotheses for active adaptive management systems (Walters and Holling 1990). The tools of landscape ecology can be employed to examine such hypotheses in a more or less rigorous – or at least quantitative – way (Whitehead *et al.* 2002).

12.8   Unexpected insights: confessions of an empiricist

As outlined in the following two vignettes, I have gained unexpected and, for me at least, profound insights in the course of undertaking field research for quantitative landscape ecology programs in northern Australia. These insights may be more important than the quantitative papers that I have produced, because they collapse my academic research into a compressed, emotionally charged, and insightful story that lives within me and continues to animate me. While the power of stories is self-evident for journalists, educators, advertisers, politicians, and scholars with a humanities background, their importance has only recently being explicitly acknowledged by applied scientists. For example, a recent paper in the *British Medical Journal* used "stories" to illustrate the relevance of complexity science to medical practitioners (Plsek and Greenhalgh 2001).

12.8.1   Shooting sacred buffalo

The original purpose for my field research in central Arnhem Land was to study the landscape-scale pattern of Aboriginal fire use (Bowman *et al.* 2004). My research agenda was opportunistically modified to include a study of buffalo (*Bubalus bubalis*) hunting (Bowman and Robinson 2002). This addition came about because I quickly realized that buffalo hunting was a potent social and (in nonmarket terms) economically productive activity that could bridge the cultural divide between me (a scientist) and the Aboriginal owners of the land upon which I was working. Buffalo hunting enabled us to experience the same landscape together. On one hunting trip, after killing a buffalo, I sat resting under a shady tree in a dry creek bed with my linguist collaborator Murray Garde and my Aboriginal informant and friend Joshua Rostrum. At my request, Murray asked Joshua, in Guene language, "What is a buffalo?" He was confidently told "the rainbow serpent." In that instant I grasped the ongoing complex and confusing relationship between humans and feral buffalo in northern Australia.

12.8.2   This is my land

I am currently examining changes in the distribution of mulga shrublands (*Acacia anuera*) and spinifex hummock grassland (*Triodia* sp.) in the Tanami Desert region of northern Australia. These vegetation types are powerful bio-indicators of landscape change in response to Aboriginal fire management practices. Long-term changes are being assessed through the examination of carbon isotope composition ($\delta^{13}C$) and radiometric dating (^{14}C) of soil organic matter (Witt, 2002). Medium-term changes and a landscape-scale context are being provided by a Geographic Information System (GIS) analysis of fire, land management, and vegetation community boundary changes using satellite and historical aerial photography. Shorter-term changes are being described using transects across vegetation boundaries stratified in respect of fire history, land tenure, and soil type. Recently I accompanied a group of government pastoral officers to arrange for permission to continue working on a pastoral property. We had chosen a bad day – the pastoralists had just received word that, due to a family company dispute, their lease had been sold and they were going to be evicted. The owner had been raised on the property and their children born there. The family's distress was palpable; they were going to lose more than an enterprise because their identity was strongly tied to that landscape. On the way to the pastoral property we had refueled at an Aboriginal community. The evidence of chronic social dysfunction was depressing and confronting. Momentarily, I held these two confronting scenes in my mind – each in their own way demonstrated the potent nexus between land and human identity. At that moment I comprehended the human dimension of the landscape change I was quantifying.

12.9 Conclusion

The great advantage of working on a frontier is that everything is new and change is the norm. Traditions in scientific thinking can only be made, not broken. My thinking about landscapes has been grounded by two decades of fieldwork in the great, uncompromising landscapes of northern Australia. This experience has shaped my attitudes towards landscape ecology in ways that practitioners from beyond the frontier might consider iconoclastic. I have learnt that there is no single approach to studying landscape ecology. Rather, context and purpose should determine choice of methodology, and an historical perspective is critical. Meeting the challenge of comprehending landscapes and formulating ecologically sustainable land management practices demands creativity and collaboration with scholars from a diversity of fields. In a frontier setting, where the consequences of the historical dispossession of indigenous people of their land by settlers suffuses all political discourse, it is impossible and counter-productive to deny that human values profoundly influence the way landscape is conceptualized and experienced. The fundamental importance of human values in landscape ecology favors the framing of stories or the formulation of parables rather than encouraging the quixotic quest for an "objective" representation of landscape. I have come to accept that it is impossible to ever comprehensively describe and truly understand landscape: the only truth is the landscape itself.

Acknowledgments

I am most grateful to Richard Hobbs. for providing me with the opportunity to present these ideas to the International Association Landscape Ecology Darwin World Congress in 2003. Fay Johnston, Don Franklin, Peter Whitehead, and two reviewers provided invaluable help in refining my thoughts.

References

Attiwill, P.M. and B. Wilson. 2003. *Ecology: An Australian Perspective*. Oxford: Oxford University Press.

Bonyhardy, T. 2000. *The Colonial Earth*. Melbourne: The Miegunyah Press.

Bowman, D.M.J.S. 1991. How short do you cut the string? Biogeography, development and conservation in northern Australia. *Global Ecology and Biogeography Letters* 1, 2–4.

Bowman, D.M.J.S. 2001a. Future eating and country keeping: what role has environmental history in the management of biodiversity? *Journal of Biogeography* 28, 549–64.

Bowman, D.M.J.S. 2001b. On the elusive definition of "Australian rainforest": response to Lynch and Neldner (2000). *Australian Journal of Botany* 49, 785–7.

Bowman, D.M.J.S. 2004. Painful choices. *Nature Australia* 28, 84.

Bowman, D.M.J.S. and P.R. Minchin. 1987. Environmental relationships of woody vegetation patterns in the Australian monsoon tropics. *Australian Journal of Botany* 35, 151–69.

Bowman, D.M.J.S. and C.J. Robinson. 2002. The getting of the Nganabbarru: observations and reflections on Aboriginal buffalo (*Bubalus bubalis*) hunting in Northern Australia. *Australian Geographer* 33, 191–206.

Bowman, D.M.J.S., A. Walsh, and L.D. Prior. 2004. Landscape analysis of Aboriginal fire management in Central Arnhem Land, north Australia. *Journal of Biogeography* 31, 207–23.

Bowman, D.M.J.S., Y. Zhang, A. Walsh, and R.J. Williams. 2003. Experimental comparison of four remote sensing techniques to map tropical savanna fire-scars using Landsat-TM imagery. *International Journal of Wildland Fire* 12, 341–8.

Christian, C.S. and G.A. Stewart. 1953. *General Report on Survey of Katherine-Darwin Region, 1946*. CSIRO Land Research Series No. 1. Melbourne: Commonwealth Scientific and Industrial Research Organization.

Cronon, W. 1992. A place for stories: nature, history and narrative. *The Journal of American History* 78, 1347–76.

Cronon, W. 1993. The uses of environmental history. *Environmental History Review* 17, 1–22.

Egan, D. and E.A. Howell. 2001. *The Historical Ecology Handbook: A Restorationist's Guide to Reference Ecosystems*. Washington, DC: Island Press.

Fensham, R.J. 2000. *Nature's Bulldozer: Tree Dieback in the Savannas* (http://savanna.ntu.edu.au/publications/savanna.links14/tree.dieback.html).

Fensham, R.J. and R.J. Fairfax. 2003. Assessing woody vegetation cover change in northwest Australian savannas using aerial photography. *International Journal of Wildland Fire* 12, 359–67.

Flannery, T.F. 1994. *The Future Eaters: An Ecological History of the Australasian Lands and People*. Sydney: Read Books.

Foucault, M. 1973. *This is Not a Pipe* (trans. James Harkness). Berkeley: University of California Press.

Fox, I.D., V.J. Neldner, G.W. Wilson, and P.J. Bannink. 2001. *The Vegetation of the Australian Tropical Savannas*. Brisbane: Environment Protection Agency.

Haynes, C.D., M.G. Ridpath, and M.A.J. Williams. 1991. *Monsoonal Australia: Landscapes, Ecology and Man in the Northern Lowlands*. Rotterdam: Balkema.

Head, L. 2000. *Second Nature: The History and Implications of Australia as Aboriginal Landscape (Space, Place, and Society)*. Syracuse: Syracuse University Press.

Head, L. 2001. *Cultural Landscapes and Environmental Change*. London: Arnold.

Hobbs, R.J. 2005. Landscapes, ecology and wildlife management in highly modified environments – An Australian perspective. *Wildlife Research* **32**, 389–98.

Jongman, R.H.G., C.J.F. ter Braak, and O.F.R. van Tongeren. 1995. *Data Analysis in Community and Landscape Ecology*. Cambridge: Cambridge University Press.

Leopold, A. 1949. *A Sand County Almanac, and Sketches Here and There*. 1989 Commemorative Edition. New York: Oxford University Press.

Liu, J. and W.W. Taylor. 2002. *Integrating Landscape Ecology into Natural Resource Management*. Cambridge: Cambridge University Press.

Lynch, A.J.J. and V.J. Neldner. 2000. Problems of placing boundaries on ecological continua: options for a workable national rainforest definition in Australia. *Australian Journal of Botany* **48**, 511–30.

Mueller-Dombois, D. and H. Ellenberg. 1974. *Aims and Methods of Vegetation Ecology*. New York: John Wiley & Sons, Inc.

Mulvaney, J. and J. Kamminga. 1999. *Prehistory of Australia*. Sydney: Allen and Unwin.

Naveh, Z. and A.S. Lieberman. 1984. *Landscape Ecology: Theory and Application*. New York: Springer.

Parks and Wildlife Commission of the Northern Territory (undated). *Northern Territory Parks Master Plan: Towards a Secure Future*. Darwin: Parks and Wildlife Commission of the Northern Territory.

Plsek, P.E. and T. Greenhalgh. 2001. Complexity science: the challenge of complexity in health care. *British Medical Journal* **323**, 625–8.

Powell, A. 1996. *Far Country: A Short History of the Northern Territory*. Melbourne: Melbourne University Press.

Price, O., J.C.Z. Woinarski, D. Liddle, and J. Russell-Smith. 1995. Patterns of species composition and reserve design for a fragmented estate: monsoon rainforests in the Northern Territory, Australia. *Biological Conservation* **74**, 9–19.

Pyne, S.J. 1998. *How the Canyon Became Grand*. New York: Penguin Books.

Russell-Smith, J., C. Yates, A. Edwards, *et al.* 2004. Contemporary fire regimes of northern Australia, 1997–2001: change since Aboriginal occupancy, challenges for sustainable management. *International Journal of Wildland Fire* **12**, 283–97.

Walters, C.J. and C.S. Hollings. 1990. Large-scale management experiments and learning by doing. *Ecology* **71**, 2060–8.

Watson, H. 1993. Aboriginal: Australian Maps. Pages 28–36 in *Maps Are Territories: Science Is an Atlas*. Chicago: The University of Chicago Press.

Whitehead, P.J., J.C.Z. Woinarski, D. Franklin, and O. Price. 2002. Landscape ecology, wildlife management and conservation in northern Australia: linking policy, practice and capability in regional planning. Pages 227–59 in J. Bissonette and I. Storch (eds.). *Landscape Ecology and Resource Management: Linking Theory with Practice*. New York: Island Press.

Witt, B. 2002. Century-scale environmental reconstruction by using stable carbon isotopes: just one method from the big bag of tricks. *Australian Journal of Botany* **50**, 441–54.

Woinarski, J.C.Z. 1996. Application of a taxon priority system for conservation planning by selecting areas which are most distinct from environments already reserved. *Biological Conservation* **76**, 147–59.

Woinarski, J.C.Z., G. Connors, and B. Oliver. 1996. The reservation status of plant species and vegetation types in the Northern Territory. *Australian Journal of Botany* **44**, 673–89.

Woinarski, J. C. Z., P. J. Whitehead, D. M. J. S. Bowman, and J. Russell-Smith. 1992. Conservation of mobile species in a variable environment: the problem of reserve design in the Northern Territory, Australia. *Global Ecology and Biogeography Letters* **2**, 1–10.

Wu, J. and R. Hobbs. 2002. Key issues and research priorities in landscape ecology: an idiosyncratic synthesis. *Landscape Ecology* **17**, 355–65.

Yibarbuk, D. M., P. J. Whitehead, J. Russell-Smith, *et al.* 2001. Fire ecology and Aboriginal land management in central Arnhem Land, northern Australia: a tradition of ecosystem management. *Journal of Biogeography* **28**, 325–43.

CLAIRE C. VOS, PAUL OPDAM, EVELIENE G. STEINGRÖVER,
AND RIEN REIJNEN

13

Transferring ecological knowledge to landscape planning: a design method for robust corridors

13.1 Introduction

There is still a big gap to cross between ecology and planning (Moss 2000, Opdam *et al.* 2002). This lack of integration is a problem in several ways. If a regional development plan projects a spatial pattern of ecosystems not sustaining the key ecological processes to serve the nature conservation objectives, it is by definition ecologically unsustainable. Moreover, for landscape ecology as an applied problem-solving science, its future and its justification (Moss 2000) is at stake, if landscape ecological knowledge is unable to provide a sound scientific basis for the planning of landscapes. In this chapter, we present an approach for the transfer of knowledge on population ecology to planning and design procedures. The method is based on two assumptions: regional stakeholders determine conservation targets as well as landscape design, and such decision-making is based on the principles of ecological sustainability. We developed this method in the context of the planning of robust corridors in the Netherlands.

Why should regional development plans be ecologically sustainable? Sustainable development is a widely accepted strategic framework in decision-making concerning land use now and in the future (IUCN 1992). It demands that landscape planning aims for "a condition of stability in physical and social systems, achieved by accommodating the needs of the present without compromising the ability of future generations to meet their needs" (WCED 1987, Ahern 2002). This implies that in decision-making about a future landscape a balance is achieved, in the short and long term and among ecological, cultural, and economic functions (Linehan and Gross 1998, Ahern 1999). It further implies that the spatial organization of the landscape sustains these functions in the long term. For species diversity, for example, this would imply that

Key Topics in Landscape Ecology, ed. J. Wu and R. Hobbs.
Published by Cambridge University Press. © Cambridge University Press 2007.

the pattern of ecosystems provides populations of species with the capacity to recover from local and regional disturbances.

However, ecological sustainability is not well developed in landscape planning for two reasons. The first one is that the explicit inclusion of ecological principles in landscape planning is quite a recent advancement (Ahern 2002). The second reason is that ecological sustainability has not been properly defined in a spatially explicit context. We will use the definition proposed by Opdam *et al.* (2005): a landscape is ecologically sustainable if: (1) the landscape structure supports the required ecological processes, so that the landscape can deliver its natural goods and services to present and future generations, (2) the landscape can change over time without losing its key resources, and (3) stakeholders are involved in decision-making about landscape functions and patterns. The first and second conditions require the understanding of the interdependence between landscape pattern and ecological and abiotic processes (Wu and Hobbs 2002). However, often landscape plans are not based on landscape ecological concepts (see Bergen Jensen *et al.* 2000, Steiner 2000, Tress *et al.* 2003). The third condition requires that nonecologists can handle information based on ecological knowledge, and this information is not presented as a standard recipe, but is open to decision-making and integration in a complex, multi-actor planning process.

The definition can be criticized for not being explicit about how many ecological services and goods are needed for present and future generations. We believe that defining ecosystem and biodiversity targets for a region is a context-dependent democratic activity, in which international as well as national conservation targets are considered in the regional context. Stakeholders are primarily responsible for setting ambition levels, with science in the role of providing a basis for decision-making. It is outside the scope of this chapter to discuss this issue, and we simplify it here by taking biodiversity as a resource that should be conserved in the landscape for future generations (Luck *et al.* 2003). This general notion can be transformed into the goal that populations of species are persistent in the long term. However, the required conservation effort increases with the amount of species in a planning area occuring in persistent populations (Opdam *et al.* 2005). This implies that planning for ecological sustainability always demands that a level of ambition is defined. We consider this as a political and societal question rather than a scientific one. This means that goal-setting for sustainable planning requires the involvement of stakeholders. Because conservation of biodiversity may not be compatible with other land-use functions, and requires money, goal-setting is part of a negotiation process. Ecological knowledge must be in a form that is suitable for such a process.

Our approach allows integration of landscape ecological knowledge into landscape planning and design. It entails three essential steps to bridge the gap between basic landscape ecological research and its application in landscape planning (also see Opdam *et al.* 2002, 2005):

- Step 1 – the translation of basic ecological knowledge of individuals and populations of a species into minimum levels of habitat quality, area, and configuration at a defined spatial scale.
- Step 2 – the integration of these conditions from single species into multi-species or ecosystem conditions.
- Step 3 – the transfer of that knowledge for use in the various steps of the planning process.

We focus on the total area and spatial distribution of habitat for a species in a region, assuming that abiotic fluxes and patterns allow for good-quality habitat, and that management of ecosystems is adequate. By this focus the chapter becomes less complicated, while it does not affect the message we want to get across. Moreover, we believe that in planning there is already quite some attention on the relation between ecosystems and abiotic conditions (Steiner 2000), whereas the significance of the spatial pattern of ecosystems to biodiversity is often neglected. A further simplification is that we neglect the response time of populations to changing ecosystem patterns. Though this is not realistic, this simplification does not alter the method we present, but it may alter the minimum area threshold of an ecosystem network.

We have developed our approach in the context of a case study: the recent introduction of a new concept in Dutch conservation policy: "robust corridors" (explained below). The aim of this chapter is to discuss the development of planning guidelines for effective corridors, based on the best available ecological knowledge, and the effective implementation of these guidelines in a complex multi-actor planning process. We will analyze the different steps in the implementation process from basic landscape ecology to on-the-ground application, and identify different types of knowledge and skills required in each phase. We are aware that no single "best method" exists for the implementation of landscape ecology in landscape planning. Strategies are, for instance, strongly dependent on the nature policy and planning traditions in various countries. However, we do think that our approach might be helpful to better understand the prerequisites for the integration of landscape ecology in planning.

13.2   Context of the case study

In the 1990s the Dutch government launched a far-reaching landscape ecological concept in nature policy: the National Ecological Network (NEN; MANFS 1990). The NEN was conceived as a structure of existing nature areas that was made more robust and cohesive by enlarging existing areas, developing new nature areas, and establishing local ecological corridors. The NEN was an answer to habitat loss and fragmentation, which were considered the prime causes for the loss of biodiversity. Habitat networks were thought to offer conditions for long-term conservation where individual areas were no longer large enough for persistent populations (Opdam *et al.* 1995, Hanski 1999, Hobbs 2002, Opdam 2002). Long-term survival of target species in habitat networks could be achieved when requirements were met concerning: total network size, habitat quality, network density, and connectivity of the network (so-called "network cohesion," Opdam *et al.* 2003). The biodiversity objectives of nature policy were defined in target ecosystem types and target species (Bal *et al.* 2001). In 2000, halfway through the implementation process, an evaluation took place to predict whether the NEN would indeed effectively protect these target species. It proved that the expected spatial cohesion after implementation of the NEN would still be insufficient for the sustainable conservation of the target species. Three main causes were identified:

(1) The nature areas were still too small. The autonomic process of obtaining land for nature development was only moderately successful in enlarging existing areas (Nature Policy Agency 1999).

(2) The connectivity was too low. The development of ecological corridors was too slow (Beentjes and Koopmans 2000) and about 50 percent of the planned corridors would be ineffective due to insufficient design (Bal and Reijnen 1997).

(3) Barriers caused by infrastructure were insufficiently solved (Reijnen *et al.* 2000).

A project in which both scientists and policy-makers participated resulted in a new landscape ecological concept, "robust corridors," which was chosen as the best solution for the lack of spatial cohesion (Pelk *et al.* 1999). Robust corridors connect the important nature regions in the Netherlands, over a distance of several tens of kilometers. Robust corridors consist of wide dispersal corridors and large new nature reserves. The concept was adopted by the national government (MANFS 2001), and it was decided to expand the NEN with supraregional robust corridors (Fig. 13.1). Subsequently, the governments of the 12 provinces were asked to explore the possibility of developing such robust corridors. In fact, this was the beginning of a negotiation process between the national government and the provincial governments about implementing the national policy at the provincial level. Because provinces were still in the process of implementing the original plan from 1991, and differed in the progress they had made and in the regional support by farmers, the new ambition of the national government was not received with enthusiasm everywhere. Moreover, the initial idea allocated unequal amounts of extra hectares to provinces. The

FIGURE 13.1
Robust corridors (dark lines) added to the National Ecological Network in the
Netherlands to connect important nature regions (grey areas)

final outcome of this first step in the planning process was the decision by the
national government about which robust corridors should be developed, the
total area needed, and the distribution over the provinces.

We supported this negotiation process during the explorative phase with a
framework for planning and design of ecologically effective robust corridors.
To develop spatial conditions for robust corridors, we went through step 1 –
"the translation of basic ecological knowledge of individuals and populations
of a species into general spatial conditions of the landscape," and step 2 – "the
integration of knowledge from single species to multi-species or ecosystem
conditions." Then we transferred these guidelines into a planning and design

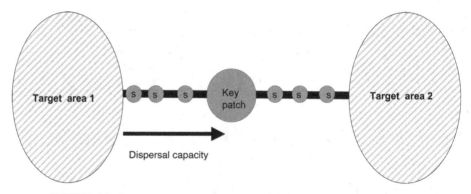

FIGURE 13.2
Translation of basic landscape ecological knowledge into spatial conditions for single species ecological corridors (step 1)

procedure to guide the feasibility studies carried out by the provinces. Here step 3 came into focus: "the transfer of knowledge for use in the various steps of the planning process." In the following sections we will describe the development of the guidelines for robust corridors and the implementation of design rules for robust corridors in the planning process.

13.3 The development of robust corridors and the implementation in the planning process

13.3.1 Step 1: the translation of basic landscape ecological knowledge into guidelines for single-species corridors

Robust corridors should facilitate the exchange of species between target areas through unsuitable landscapes over distances that will often exceed their dispersal capacity by far. Therefore the corridor consists of a combination of new habitat patches, where a species can establish and maintain a population, the so-called key patch, and measurements that will facilitate dispersal through an inhospitable matrix, so-called dispersal corridor and stepping stones (Fig. 13.2). A large body of basic landscape ecological research has formed the basis for guidelines for corridor conditions of single species. Empirical studies provide information on dispersal distances and whether a species needs special habitat elements (a dispersal corridor) to cross agricultural or urbanized landscapes (e.g., Bennett 1999, Rickets 2001, Ray *et al.* 2002, Tewksbury *et al.* 2002, Vos *et al.* 2002). Metapopulation models provide guidelines on the size of and distance between habitat patches, depending on the dispersal capacity and individual area requirements of target species (e.g., Verboom *et al.* 2001, Etienne *et al.* 2004, Ovaskainen and Hanski 2004,

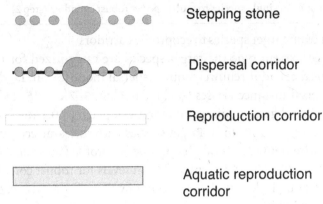

Stepping stone

Dispersal corridor

Reproduction corridor

Aquatic reproduction corridor

FIGURE 13.3
Four basic corridor types based on the dispersal capacity and mode of dispersal of the target species

Fahrig, Chapter 5, this volume). Rules for the required patch size of key patches were derived from metapopulation modeling and knowledge on individual area requirements of species (Verboom *et al.* 2001). Rules for the distance between habitat patches were based on knowledge about the dispersal capacity of species. Empirical research on movements of species in heterogeneous landscapes in combination with movement modeling provided knowledge of whether species avoid or prefer specific habitat types during dispersal (Vos *et al.* 2002). Based on species-specific dispersal characteristics, this knowledge was extrapolated into rules for dispersal corridors, measures at infrastructural barriers, and stepping stones (Fig. 13.2). Four basic corridor design patterns were developed, related to different modes of dispersal (Fig. 13.3).

13.3.2 Step 2: integration from single species to multi-species robust corridors

Empirical studies on the functioning of species in heterogeneous landscapes are often based on single species. Nature conservation goals are not. Thus landscape planners need integrated corridor guidelines, where requirements for single species are integrated into a multi-species design. The integration was achieved in the following way (see Box 1 for more details). Dutch nature policy has formulated explicit biodiversity aims in a list of target species that are representative for ecosystem types (Bal *et al.* 2001). A robust corridor that connects two target areas (for instance two marshlands) should facilitate the exchange of all (marshland) target species. As a first step towards integration for each ecosystem type the target species were grouped according to their dispersal capacity, dispersal mode, and individual area requirements (Box 13.1). Target species with similar requirements were combined in

Box 13.1.   Integration from single species to multi-species robust corridors (step 2)

---

(1) Integration from target species to ecoprofile corridors

For each ecosystem type the target species are categorized for relevant ecological corridor requirements: dispersal mode (by air, land, or water), dispersal distance classes ($\leq 1\,$km, 1–3$\,$km, 3–7$\,$km, 15–15$\,$km, and $> 35\,$km), and area requirements for a key patch ($\leq 0.1\,$km^2, 0.1–1$\,$km^2, 1–5$\,$km^2, 50–150$\,$km^2). Target species with the same ecological corridor requirements are combined in one ecoprofile (Vos *et al.* 2001, Pouwels *et al.* 2002). The different ambition levels for robust corridors (see Table 13.1) are linked to the ecoprofiles by their dispersal capacity. For the lowest ambition level the robust corridor functions for species with a dispersal capacity larger that 15$\,$km. The medium ambition level incorporates also less mobile species (dispersal capacity $> 3\,$km), while on the highest ambition level all ecoprofiles are included.

(2) Integration from ecoprofile corridors to a multi-species robust corridor

In the last integration step the ecoprofile corridors from a particular ecosystem type and ambition level are integrated into one robust corridor that is suitable for all ecoprofiles. In the example (Fig. 13.4) a robust corridor for a marsh ecosystem at the lowest ambition level consists of four ecoprofiles: "beaver," "great reed warbler," "otter" and "bittern." Rules for integration are:

(i) distance between patches is determined by the ecoprofile with the lowest dispersal capacity ("beaver"), and

(ii) area of patches is determined by the species with the largest area requirements ("bittern"). This integration of ecoprofiles leads to an area reduction of 40 percent compared to the required area for all separate ecoprofile corridors.

(3) Finally a robust corridor might consist of several ecosystem types (see Fig. 13.4).

"ecoprofiles" (Vos *et al.* 2001, Pouwels *et al.* 2002). For instance, the amphibian species tree frog (*Hyla arborea*), the common spadefoot (*Pelobatus fuscus*), and the pool frog (*Rana lessonae*) have comparable habitat requirements, dispersal distances, and individual area requirements and therefore were classified into one ecoprofile. This resulted in a reduction of 398 target species to 138 ecoprofiles (Broekmeyer and Steingröver 2001). Subsequently the requirements of a list of ecoprofiles of the same ecosystem type were integrated into one robust corridor that fits requirements for all ecoprofiles. In this integration step the distance between key patches is determined by the ecoprofile with the smallest dispersal capacity, while the size of key patches is determined by the ecoprofile with the largest individual area requirements. As is explained in Box 13.1 (Fig. 13.4),

TABLE 13.1. *Robust corridors with low, medium, and high level of ecological ambition*

Ecological ambitions	Low ambition level (1)	Medium ambition level (2)	High ambition level (3)
Create sustainable networks on national scale	Very mobile species (dispersal capacity > 15 km), with large area requirements	–	+
Connect regional networks to increase biodiversity in all suitable habitat		Medium mobile species (dispersal capacity 3–15 km), with intermediate area requirements	+
Connect networks for the whole ecosystem: the spreading of risks			Species low mobility (dispersal capacity < 3 km), with small area requirements

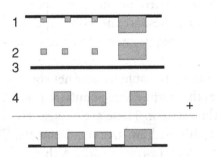

Ecoprofiles	Distance between key patches	Area key patch, stepping stone	Dispersal corridor (width)
1 "Beaver"	20 km	300 ha, 30 ha	50 m
2 "Great reed warbler"	20 km	300 ha, 30 ha	-
3 "Otter"	50 km	not needed	50 m
4 "Bittern"	30 km	750 ha, 75 ha	-
Total	Robust corridor Marshland, ambition level 1, length 25 km		

FIGURE 13.4
Integrating ecoprofiles into one robust corridor for a marshland ecosystem on the lowest ambition level: suitable for relatively mobile species with large area requirements (dispersal capacity > 15 km)

an important reduction of the total required area is reached, by integrating all corridor-requirements of the ecoprofiles into one robust corridor.

13.3.3   Step 3: developing tools for the implementation of flexible design rules in the planning process

It was important that feasibility studies were carried out in the same way at the national and provincial levels, and that an analysis of cost-effectiveness produced a ranking of corridors. The provinces were asked to analyze the cost-effectiveness for different scenarios. In a handbook, examples of robust corridors were worked out (Broekmeyer and Steingröver 2001). For a flexible application of the design rules in the planning process a CD-ROM was produced providing information on all robust corridors per ecosystem type and per ambition level. Different scenarios could be generated by varying the ambition level or the number of ecosystem types that were incorporated in the corridor. During the feasibility studies, planners were searching for the most effective, best accepted, and economically most stable corridor options. To facilitate this negotiation process between the national and provincial governments, we decided to maximize flexibility in the planning and design guidelines. The following options allowed such flexibility: (1) defining the ambition level, (2) defining the ecosystem type(s) to be included, (3) finding the preferred location, (4) defining the sequence of corridor elements, and (5) combining other land-use functions.

13.3.3.1 *Defining the ambition level*

Conservation targets were divided in three ambition levels, depending on the number of species that should use the corridor (Table 13.1). At the lowest ambition level (i.e., level 1), robust corridors are created for only those species that require habitat networks on a national scale. These are the mobile species (dispersal capacity > 15 km) with very large area requirements, such as otter (*Lutra lutra*) and bittern (*Botaurus stellaris*). These species need enough cohesion between all large Dutch nature areas to create sustainable habitat networks. At ambition level 2, requirements for species that form viable population networks on a regional level are included in the corridor. These are moderately mobile species (dispersal capacity 3–15 km) with medium area requirements, such as grass snake (*Natrix natrix*) and blue throat (*Luscinia svecica*). Although these species are able to survive in networks on a regional level, the effectiveness of the NEN as a whole will improve by making exchange between these regions possible for moderately mobile species. Thus target species will be able to reach suitable habitat in regions where they would otherwise be missing, increasing species number per nature reserve. At the highest ambition level (3) the robust corridors will be suitable for the whole ecosystem and exchange between regions for all species should be possible. These corridors were believed necessary to allow species to respond to large-scale disturbances, such as climate change. When favorable habitat conditions shift as a result of climatic change, all species should be able to follow these changes within the Netherlands (Opdam and Wascher 2004). For this strategy to be effective, robust corridors are needed not only in the Netherlands but also in the whole European ecological network (so-called "Natura 2000" reserves; Jongman *et al.* 2003).

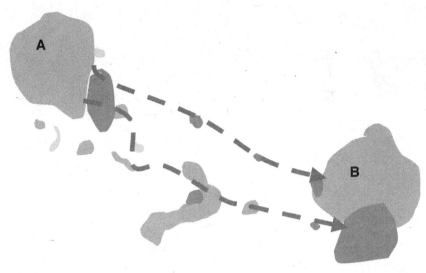

FIGURE 13.5
Two scenarios for a robust corridor between target areas A and B. The grey tones indicate different ecosystem types. The dashed lines are two scenarios for the location of the robust corridor. The top line would be the shortest route between A and B. However, this route would require large investments to develop new nature areas. The bottom line is longer, but incorporates more existing nature areas

13.3.3.2 Defining the number of ecosystem types in the robust corridor

Natural regions to be connected by robust corridors often consist of different ecosystem types, arranged in mosaics of woods, marshes, dry heath, moors, and grasslands. Therefore a robust corridor design should consider all ecosystem types present in a region. For instance, if one wants to enhance exchange of forest species over large distances, one would need a forested corridor and create new forested nature reserves. For heath species, the corridor needs to be extended with a heath corridor, and new heath areas, etc. One may, however, decide to develop the corridor zone for only one or two of the ecosystem types present, which would require less area and thus be easier to accomplish.

13.3.3.3 Finding the preferred location

Because the CD-ROM linked the scenarios directly to the number of target species that would benefit from the corridor, the potential benefit between scenarios could be compared. The best location for the corridor was determined by analyzing the cost-effectiveness of different options. The trade-off was compared between the shortest route between target areas and the route that incorporated most of the existing nature areas or encountered least barriers (Fig. 13.5). The optimal allocation was found by producing alternative trajectories of the corridor, and choosing between the options can be based on costs (amount of additional area required, number of barriers to be crossed) as opposed to the number of species that will benefit from the corridor.

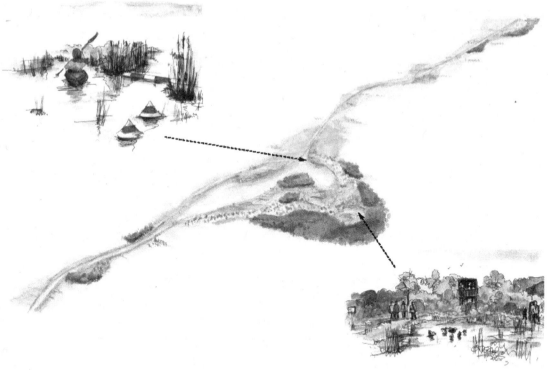

FIGURE 13.6
Visualization of function combinations. A robust marshland corridor can be combined, to some extend, with recreation (artist impression by Karel Hulstein; step 3). The illustration presents the robust corridor, consisting of a marshland dispersal-corridor and new marshland areas. The target areas that are to be connected by the robust corridor are not in the illustration (from Broekmeyer and Steingröver 2001)

13.3.3.4 *Defining the sequence of corridor elements*

Guidelines were incorporated as to how the sequence of corridor elements could be altered without losing the ecological effectiveness. By creating room for different design options within one ambition level, we intended to create flexibility in the design process, and thereby to increase the probability that a proper design could be negotiated between parties.

13.3.3.5 *Combining other functions*

Finally the handbook discusses the possibilities to combine the ecological function of the corridor with other functions of the landscape: water management, agriculture, and recreation. For instance, when recreation is stimulated by these corridors, this will be beneficial for the regional economy and contribute to gaining support for the plans (Fig. 13.6). However, recreation is also considered as a potential disturbance for animal populations, with the effect of decreasing the carrying capacity of the habitat. We reviewed existing knowledge on the impact of recreation on species, (e.g., Hill *et al.* 1997, Henkens 1998, Miller *et al.* 1998, Vos *et al.* 2003) and filled knowledge gaps with best expert guesses on disturbance distances. This knowledge was summarized in a series of design principles for recreation facilities, including minimum distances and multiplying factors for minimum area to compensate for a potential decrease in habitat quality. Similar procedures were followed for the integration of particular forms of agriculture, cultural landscapes values, and water retention within these corridors.

13.4 Discussion

13.4.1 Contribution to key issues

Wu and Hobbs (2002) proposed six key issues of landscape ecology. In this chapter, we have mainly addressed three issues: integration between basic research and application, transdisciplinarity, and cooperation with policy-makers and stakeholders. Following Wu's and Hobbs' claim that effective communication requires willingness, desire, and commitment on the landscape ecologists' part, we have developed an interactive design approach with two main achievements: it links the required ecological functioning to the selected conservation goal, and it offers opportunities for decision-making. Therefore, it allows a context-oriented design of corridors, decisions by nonecologists that are still ecologically sustainable, and room for negotiation between parties in a planning and design process. We have integrated the three key issues, thereby facilitating interactive sustainable planning. Being "interactive" is important because the ecological researcher is not the decision-maker: his or her role is clarifying the boundary conditions for landscape patterns given a target chosen by stakeholders and policy-makers. The researcher also develops methods to facilitate the setting of feasible (nature conservation) targets. In addition to what we have developed in our case study, such methods may include the determination of ecosystem functions to be ensured in the region for future generations. In that role the researcher demarcates the playing field for the game the stakeholders have chosen, and warns where boundaries are crossed. If during the planning process it turns out that the required conditions are not attainable, the researcher facilitates the reformulation of goals. Methods like the design approach discussed here can provide the decision process with a scientific foundation, because they make the planning process systematic, transparent, and reproducible. Such methods can be peer-reviewed and discussed in scientific meetings, and thereby become acceptable for society as a general basis for decision-making.

Critical to this role of the researcher is that the method allows the actors to move and switch between alternatives in the negotiation process. As Theobald *et al.* (2000) pointed out, "Planning is a process that uses scientific data, but ultimately depends on the expression of human values." The actors explore the domain of sustainable options to find the solution that best fit their preferences. The best landscape design is not found on the basis of ecological criteria only. It may not be the cheapest option either, but the one with the best balance among ecological, social, and economic profits. It is a compromise rather than an optimal design from the point of view of ecological criteria (Haines-Young 2000). We have argued in this chapter that the compromise should not be found in accepting ecologically less sustainable conditions, but in lowering the ambition level of the objectives, requiring less area or a more feasible spatial pattern for the corridor.

Our approach does not use formal optimization methods (e.g., Cabeza and Moilanen 2001, McDonnell *et al.* 2002, Rothley 2002, Hof and Flather, Chapter 8, this volume). We believe that such methods are useful for generating a series of theoretical scenarios for the location of nature reserves, which could be useful in landscapes with a much lower intensity of current land use. The practical significance of such research increases if they could generate the domain of sustainable configurations, rather than the ecologically best solution, and if they could be made suitable for planning with interactive stakeholder involvement.

13.4.2 Further development of the corridor design method

13.4.2.1 *Step 1: translating basic species ecology into spatial conditions*

Future landscape ecological research will enhance our understanding of the functioning of species in fragmented landscapes. This will provide new insights about sustainable spatial conditions for species survival and consequently influence the design rules for robust corridors. Thus it is important that the structure of design rules is flexible, so that new knowledge can be incorporated and distributed easily (for instance by CD-ROM). As the implementation of the corridors in the landscape will be a long-term process, there is time to incorporate improved design rules. On the other hand, monitoring the effectiveness of the robust corridors might also generate basic knowledge on species functioning. As Golly and Bellot (1991) put it, "Landscape plans are actually hypotheses of how a proposed landscape structure will influence landscape processes." Thus the implementation of robust corridors in the landscape generates landscape experiments that in themselves form the basis for new knowledge on species functioning. Study questions, for instance, are whether the corridors are actually used by the target species and whether species presence in the target areas is increasing. In particular, monitoring the (genetic) species diversity in the target areas before and after implementation will generate strong evidence of its effectiveness.

13.4.2.2 Step 2: knowledge integration

Species-specific knowledge is not effective in decision-making by stakeholders who cannot handle the large variation in spatial scales and habitat requirements. We proposed a method in which species data were integrated in a framework for landscape design, and in which we tried to balance between loss of detail and necessary simplification and generalization. We combined corridor requirements for species that have roughly similar reactions to scale and configuration of habitat pattern into so-called ecoprofiles (Vos *et al.* 2001). An alternative approach is to identify the most critical species per landscape characteristic, for instance, area-limited and dispersal-limited species (Lambeck 1999). The integration problem has been tackled in various ways (e.g., Lambeck and Hobbs 2002), including the introduction of focal or umbrella species (Cox *et al.* 1994, Noss and Cooperrider 1994) that are expected to provide protection for other species. Knowledge integration is dependent on sound species-specific knowledge that comes from basic studies. Also, the development of integrative frameworks needs specific research input and the development of new approaches. Too often integrative methods are not published in scientific papers. This is detrimental to the development of conservation planning as a science because, when methods are not made transparent and repeatable, the effectiveness cannot be compared.

A relatively new challenge for basic landscape ecology is the development of spatial conditions for multifunctional ecosystem networks. In our approach, we gave some indications based on expert judgment on whether robust corridors could be combined with recreation, water management, and agriculture. However, the level of quantitative knowledge on the interrelationship between nature and other functions is quite poor. If ecological functions are impaired by other functions, how can ecosystem networks be designed to compensate for that loss? For instance, it is known that recreation pressure lowers the nest density and reproduction success for bird species (Yalden and Yalden 1990). Incorporating this decline in habitat quality will require larger habitat networks for viable bird populations (Vos *et al.* 2003). We urgently need more quantitative studies on the interaction between land-use functions allocated within ecosystem networks.

13.4.2.3 Step 3: flexible design rules

Flexibility is important if ecological structures are to be inserted into a specific multifunctional landscape context. End users of the corridor design approach could vary the ambition level by varying the type of ecosystems to be included in the corridor, as well as the number of species for which the corridor could serve as a functional connection. These choices have consequences for the amount of space and the abiotic conditions required. The designer can also shift the chain of corridor elements in the landscape to find the most appropriate locations with the least amount of area to be acquired. Then, as an additional degree of freedom, we built in the possibility to shift the order of elements, to optimize the fit of the corridor at the local level. Finally, we gave guidelines to use the corridor for water storage and recreation, including consequences for the design and the area required for ecological functioning. We hypothesize that such degrees of flexibility in ecological design rules increase the effectiveness of ecological knowledge in improving the ecological sustainability of landscape development. This aspect of the relation between ecology and planning deserves much more research, including formal testing of the hypothesis.

13.4.3 Impact on the planning process

We have shown how we were actively involved in the explorative phase of planning robust corridors. But how has our method affected the planning process? As Theobald *et al.* (2000) pointed out, the ultimate question is how ecological information has altered the decision-making process. We have not been able to measure this effect for our method, because we did not attempt any systematic comparison with other planning approaches. We do know, however, that the 12 provinces involved in the explorative studies all worked with the handbook and CD-ROM. Possibly because this converged their thinking and because the corridors were crossing provincial borders, we witnessed, for the first time in the ten-year history of the National Ecological Network, that all the provinces really worked together on common goals. They also readily accepted the priority ranking of the proposed corridors by the national government. We assume that our method and the role we played in facilitating the goal-setting improved the decision-making process and the acceptance of the new concept by the provinces. We were asked to evaluate whether in these explorative studies the handbook was appropriately applied. Although most users were positive about the value of the handbook, we found big differences in the interpretation of the steps of the method, possibly partly attributable to the very short time in which the explorations had to be made. We suspect that some users had insufficient basic knowledge of ecological processes, and no time to acquire all the background information presented in the handbook. We take this as an indication that the design tool still needs further adjustments, based on the experiences of users. This can only be achieved by a profound understanding of the planning process, and requires the interactive participation of landscape ecologists in landscape planning.

References

Ahern, J. 1999. Spatial concepts: planning strategies and future scenarios: a framework method for integrating landscape ecology and landscape planning. Pages 175–201 in J. Klopatek and R. Gardner (eds.). *Landscape Ecological Analysis*. New York: Springer.

Ahern, J. 2002. *Greenways as Strategic Landscape Planning: Theory and Application*. Ph.D. Thesis. Wageningen University, Wageningen.

Bal, D., H.M. Beije, M. Fellingern, *et al.* 2001. *Handbook Nature Target Types*. LNV Expertise Centre Report Number 2001/020, Wageningen.

Bal, D. and R. Reijnen. 1997. *Nature Policy Practice: Efforts, Effects, Expectations and Chances*. Wageningen: Expertise Centrum LNV.

Beentjes, R.A. and J.C.M. Koopman. 2000. *Pulsing Veins: Giving an Impulse to the Realisation of Ecological Corridors in the Netherlands*. Den Haag: Projectgroup Ecological Corridors.

Bennett, A.F. 1999. *Linkages in the Landscape*. Gland, Switzerland and Cambridge, United Kingdom: The World Conservation Union (IUCN) Forest Conservation Programme.

Bergen Jensen, M., B. Persson, S. Guldager, U. Reeh, and K. Nilsson. 2000. Green structure and sustainability – developing a tool for local planning. *Landscape and Urban Planning* 52, 117–33.

Broekmeyer, M. and E. Steingrover (eds.). 2001. *Handbook of Robust Corridors and Ecological Prerequisites*. Wageningen: Alterra.

Cabeza, M. and A. Moilanen. 2001. Design of reserve networks and the persistence of biodiversity. *TREE* 16, 242–8.

Cox, J., R. Kautz, M. MacLaughlin, and T. Gilbert. 1994. *Closing the Gaps in Florida's Wildlife Habitat Conservation System*. Tallahassee: Florida Game and Freshwater Fish Commission.

Etienne, R.S., C.J.F. Ter Braak, and C.C. Vos. 2004. Application of stochastic patch occupancy models to real metapopulations. Pages 105–32 in I. Hanski and O.E. Gaggiiotti (eds.). *Ecology, Genetics, and Evolution of Metapopulations*. San Diego: Elsevier Academic Press.

Golly, F.B. and J. Bellot. 1991. Interactions of landscape ecology, planning and design. *Landscape and Urban Planning* 21, 3–11.

Haines-Young, R. 2000. Sustainable development and sustainable landscapes: defining a new paradigm for landscape ecology. *Fennia* 178, 7–14.

Hanski, I. 1999. Habitat connectivity, habitat continuity, and metapopulations in dynamic landscapes. *Oikos* 87, 209–19.

Henkens, R.J.H.G. 1998. *Ecological Capacity of Ecosystem Types I: The Effect of Outdoor Recreation on Breeding Birds*. IBN-report 363. Wageningen: IBN.

Hill, D., D. Hockin, D. Price, *et al.* 1997. Bird disturbance: improving the quality and utility of disturbance research. *Journal of Applied Ecology* 43, 275–88.

Hobbs, R.J. 2002. Habitat networks and biological conservation. Pages 150–70 in K.J. Gutzwiller (ed.). *Applying Landscape Ecology in Biological Conservation*. New York: Springer Verlag.

IUCN. 1992. *The Rio Declaration on the Environment*. Gland: IUCN, UNEP, WWF.

Jongman, R.H.G., M. Külvik, and I. Kristiansen. 2003. European ecological networks and greenways. *Landscape and Urban Planning* 68, 305–19.

Lambeck, R.J. 1999. Landscape planning for biodiversity conservation in agricultural regions. In *Biodiversity Technical Paper Number 2*. Canberra: Environment Australia.

Lambeck, R.J. and R.J. Hobbs. 2002. Landscape and regional planning for conservation. Pages 360–80 in K.J. Gutzwiller (ed.). *Applying Landscape Ecology in Biological Conservation*. New York: Springer.

Linehan, J.R. and M. Gross. 1998. Back to the future, back to basics: the social ecology of landscapes and the future of landscape planning. *Landscape and Urban Planning* 42, 207–24.

Luck, G.W., G.C. Daily, and P. Ehrlich. 2003. Population diversity and ecosystem services. *TREE* **18**, 331–6.

MANFS. 1990. *Nature Policy Plan 1990*. Den Haag: Ministry of Agriculture, Nature and Food Safety.

MANFS. 2001. *Nature Policy Plan 2001: Nature for People, People for Nature*. Den Haag: Ministry of Agriculture, Nature and Food Safety.

McDonnell, M.-D., H.P. Possingham, I.R. Ball, and E.A. Cousins. 2002. Mathematical methods for spatially cohesive reserve design. *Environmental Modelling and Assessment* **7**, 107–14.

Miller, S.G., R.L. Knight, and C.K. Miller. 1998. Influence of recreational trials on breeding birds communities. *Ecological Application* **8**, 162–9.

Moss, M. 2000. Interdisciplinarity, landscape ecology and the "Transformation of Agricultural Landscapes". *Landscape Ecology* **15**, 303–11.

Nature Policy Agency. 1999. *Nature Balance 1999*. Samsom H.D. Tjeenk Willink, Alphen aan den Rijn, RIVM.

Noss, R.F. and A. Cooperrider. 1994. *Saving Nature's Legacy: Protecting and Restoring Biodiversity*. Washington, DC: Defenders of Wildlife and Island Press.

Opdam, P. 2002. Assessing the conservation potential of habitat networks. Pages 381–404 in K.J. Gutzwiller (ed.). *Applying Landscape Ecology in Biological Conservation*. New York: Springer-Verlag.

Opdam, P., F. Foppen, and C.C. Vos. 2002. Bridging the gap between empirical knowledge and spatial planning in landscape ecology. *Landscape Ecology* **16**, 767–79.

Opdam, P., R. Foppen, R. Reijnen, and A. Schotman. 1995. The landscape ecological approach in bird conservation, integrating the metapopulation concept into spatial planning. *Ibis* **137**, 139–46.

Opdam, P., E. Steingröver, and S. van Rooij. 2005. Ecological networks: a spatial concept for multi-actor planning of sustainable landscapes. *Landscape and Urban Planning* (in press).

Opdam, P., J. Verboom, and R. Pouwels. 2003. Landscape cohesion: an index for the conservation potential of landscapes for biodiversity. *Landscape Ecology* **18**, 113–26.

Opdam, P. and D. Wascher. 2004. Climate change meets habitat fragmentation: linking landscape and biogeographical scale level in research and conservation. *Biological Conservation* **117**, 285–97.

Ovaskainen, O. and I. Hanski. 2004. Metapopulation dynamics in highly fragmented landscapes. Pages 73–104 in I. Hanski and O.E. Gaggiiotti (eds.). *Ecology, Genetics, and Evolution of Metapopulations*. San Diego: Elsevier Academic Press.

Pelk, M., B. Heijkers, R. van Ettiger, *et al.* 1999. *Quality by Connectivity: Why, Where and How*. Wageningen: Ministry of Agriculture, Nature and Food Safety.

Pouwels, R., M.J.S.M. Reijnen, J.T.R. Kalkhoven, and J. Dirksen. 2002. *Ecoprofiles for Species Analysis of Spatial Cohesion with LARCH*. Wageningen: Alterra. Alterra-Report 493.

Ray, N., A. Lehmann, and P. Joly. 2002. Modelling spatial distribution of amphibian populations: a GIS approach based on habitat matrix permeability. *Biodiversity and Conservation* **11**, 2143–65.

Reijnen, R.E., E. van der Grift, M. van der Veen, *et al.* 2000. *The Road to Least Resistance: Priority List of to Be Removed Barriers*. Wageningen: Alterra and Expertise Centrum LNV.

Ricketts, T.H. 2001. The matrix matters: effective isolation in fragmented landscapes. *American Naturalist* **158**, 87–99.

Rothley, K. 2002. Dynamically based criteria for the identification of optimal bioreserve networks. *Environmental Modelling and Assessment* **7**, 123–8.

Steiner, F. 2000. *The Living Landscape: An Ecological Approach to Landscape Planning*. New York: McGraw Hill.

Tewksbury, J.J., D.J. Levey, N.M. Haddad, *et al.* 2002. Corridors affect plants, animals, and their interactions in fragmented landscapes. *PNAS* **99**, 12 923–6.

Theobald, D.M., N.T. Hobbs, T. Bearly, *et al.* 2000. Incorporating biological information in local land-use decision-making: designing a system for conservation planning. *Landscape Ecology* **5**, 35–45.

Tress, B., G. Tress, A. Van der Valk, and G. Fry. 2003. *Interdisciplinary and Transdisciplinary Landscape Studies: Potentials and Limitations*. Wageningen: Delta Series 2.

Verboom, J., R. Foppen, P. Chardon, P. Opdam, and P. Luttikhuizen. 2001. Introducing the key-patch approach for habitat networks with persistent populations: an example for marshland birds. *Biological Conservation* **100**, 89–101.

Vos, C. C., H. Baceco, and C. J. Grashof-Bokdam. 2002. Corridors and species dispersal. Pages 84–104 in K. J. Gutzwiller (ed.). *Applying Landscape Ecology in Biological Conservation*. New York: Springer.

Vos, C. C., P. Opdam, and R. Pouwels. 2003. Recreation and biodiversity: a landscape approach. *Landschap* **20**, 3–14.

Vos, C. C., J. Verboom, P. F. M. Opdam, and C. J. F. Ter Braak. 2001. Towards ecologically scaled landscape indices. *American Naturalist* **157**, 24–51.

WCED – World Commission on Environment and Development. 1987. *Our Common Future*. Oxford: Oxford University Press.

Wu, J. and R. Hobbs. 2002. Key issues and research priorities in landscape ecology: an idiosyncratic synthesis. *Landscape Ecology* **17**, 355–65.

Yalden, P. E. and D. W. Yalden. 1990. Recreational disturbance of breeding golden plovers (*Pluvialis apricarius*). *Biological Conservation* **51**, 243–62.

14

Integrative landscape research: facts and challenges

14.1 Introduction

There are many tensions in landscape management at spatial scales from individual fields to regions and upwards to global environmental change (Dalgaard *et al.* 2003). Farmers are under increasing pressure to produce non-food products including recreational opportunities, attractive landscapes, and habitats for wildlife. The many different forms of agri-environmental payment schemes are witness to these pressures. In urban landscapes we see a new emphasis on urban green space, urban green structures, and greenways fulfilling multiple goals (Fábos 2004, Gobster and Westphal 2004).

One of the trends in the funding of landscape research over the last 20 years has been the rapid growth of large-scale integrative projects (Höll and Nilsson 1999, Tress *et al.* 2005a). This trend must be seen against the background of environmental concerns that have placed greater demands on the way landscapes are managed and the widening range of objectives they should fulfil. This has fuelled the demand for new research tools to address these problems. Since the problems are complex and span several disciplines, it was natural to consider integrative forms of research as the way forward (Balsiger 2004). In this chapter, we explore several of the major concepts associated with integrative research modes, what funding bodies and researchers expect from such research, and what is being delivered. We discuss the organisational barriers to integration, merit system, and ways to improve the theory base. Finally, we present education and training needs for integrative research and recommend measures to enhance integrative landscape research.

Key Topics in Landscape Ecology, ed. J. Wu and R. Hobbs.
Published by Cambridge University Press. © Cambridge University Press 2007.

14.2 Methods

This chapter is based on results of the INTELS study investigating Interdisciplinarity and Transdisciplinarity in European Landscape Studies (http://www.intels.cc). To provide a framework for discussing integrative projects and their products, we present data gathered from 19 qualitative interviews with funding bodies, project leaders, and participants involved in integrative projects on European landscapes, as well as contact with 156 journal editors and results from an international web-based survey of 150 researchers involved in integrative landscape research projects. All figures and tables in this chapter are based on the results of these investigations. Additionally, we gathered information from a literature review, reports, and descriptions of research programs. We reviewed the literature on interdisciplinarity and transdisciplinarity, especially theoretical and methodological papers, and those on the practical application of integrative approaches. We collected written material from large national research programs within Europe. The projects were screened to collect statements about the expectation of funding bodies towards integrative projects and, in general, about their understanding of the approaches. We have used this information to review the challenges arising from the rapidly changing field of integrative research with a focus on what we can realistically expect it to achieve. The overall aim of the project is the development of a code of good practice for integrative landscape research (Tress *et al.* 2003a, Tress *et al.* 2005a).

14.3 Defining integrative research approaches

Integrative research approaches, especially interdisciplinarity and transdisciplinarity, are widely used in landscape ecological research. This is true for many other fields of research related to resource management, especially at larger scales, e.g., from landscapes to regions. At these scales, there is a tendency to move away from specialist research disciplines and put a greater emphasis on integrating several, often conflicting, interests, values, and goals. Within landscape research there has been an increasing interest in integrative research approaches (Naveh and Lieberman 1994, Nassauer 1995, Zonneveld 1995, Hobbs 1997, Brandt 2000, Décamps 2000, Klijn and Vos 2000, Palang *et al.* 2000, Naveh 2001, Tress *et al.* 2001, Bastian 2002, Wu and Hobbs 2002).

A major driving force behind the increasing number of integrative research projects has been the emphasis given to integrative research in national and international research funding programs (Tress *et al.* 2005a). Nevertheless, there is much confusion regarding the terminology describing integrative research approaches (Tress *et al.* 2005b). Various terms are used to express interaction between disciplines in landscape research. Expressions include terms such as integrated, holistic, interactive, transepistemic, collaborative, cross-disciplinary or supradisciplinary. Yet most authors using these expressions do not clearly define the meaning of these research approaches. This not only complicates communication of the core concepts, such as interdisciplinarity, but can also make it difficult to match funding body expectations with research achievements. It is paramount to the success of integrative research that agreement over the scope and aims of the project are reached very early on in the project. We stress the need to clarify concepts in the field of integrative research and make definitions explicit. Therefore, we start this chapter with descriptions of how the main concepts are used in this paper; not as an attempt to provide an authoritative set of definitions but as an aid to communication and further the debate on integrative approaches.

- *Disciplinarity*: projects that take place within the bounds of a single, currently recognized academic discipline. We fully appreciate the artificial nature of subject boundaries and that they are dynamic. The project has a disciplinary goal-setting and aims at development of new disciplinary theory and knowledge.
- *Multidisciplinarity*: projects that involve several different academic disciplines researching one theme or problem, but with multiple disciplinary goals. Participants exchange knowledge, but do not aim to cross subject boundaries to create new knowledge and theory. The research process progresses as parallel disciplinary efforts without integration but usually with the aim to compare results. Theory development is discipline oriented.
- *Interdisciplinarity*: projects that involve several unrelated academic disciplines in a way that forces them to cross subject boundaries to create new integrative knowledge and theory and solve a common research goal. By unrelated, we mean that they have contrasting research paradigms. We might consider the differences between qualitative and quantitative approaches or between analytical and interpretative approaches that bring together disciplines from the humanities and the natural sciences.
- *Transdisciplinarity*: projects that both integrate academic researchers from different unrelated disciplines and nonacademic participants, such as land managers and the public, to research a common goal. Transdisciplinary projects create new integrative theory and knowledge among science and society. Transdisciplinarity combines interdisciplinarity with a participatory approach.

In Fig. 14.1, we summarize the key characteristics of nonintegrative (disciplinary and multidisciplinary) and integrative (interdisciplinary

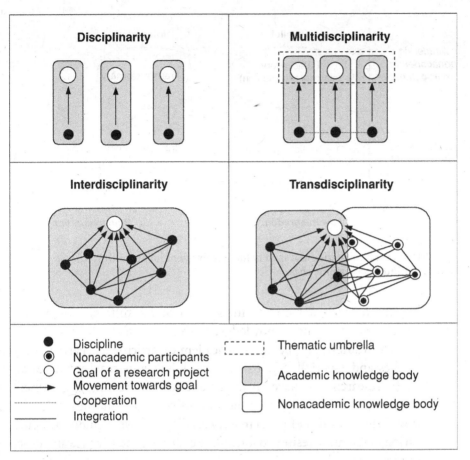

FIGURE 14.1
Main research modes referred to in this chapter showing the differences in degree of integration

and transdisciplinary) research concepts. We further define the concepts of integrative and participatory studies, because both will be necessary to understand the characteristics of the presented research concepts and both will be further used in this chapter. In Fig. 14.2, we visualize the different degrees of integration and stakeholder involvement of integrative and nonintegrative approaches. A detailed overview on all discussed concepts can be found in Tress *et al.* (2005b).

- *Integrative studies*: projects that either work in an interdisciplinary or a transdisciplinary way, in that new knowledge and theory emerges from the *integration* of disciplinary knowledge.
- *Participatory studies*: projects that involve academic researchers and nonacademic participants working together to solve a problem. Academic researchers and nonacademic participants exchange knowledge,

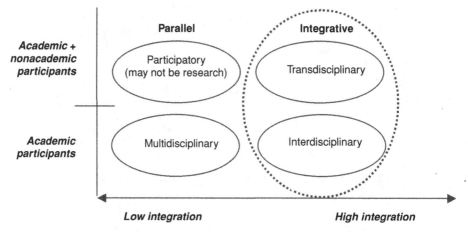

FIGURE 14.2
Degrees of integration and stakeholder involvement in integrative and nonintegrative approaches.

but the focus is not on the integration of the different knowledge cultures to create new knowledge. Both disciplinary and multidisciplinary studies may include nonacademic participants. Participatory studies and especially the use of local knowledge may not necessarily be research, but have an important role in creating engagement and empowerment in the application of scientific findings. It is also under the umbrella of participatory studies that we include the application of scientific results to formulate codes of good practice and other guidelines.

14.4   Motivations for integrative landscape studies

The motivation for participating in integrative studies has frequently arisen from other than academic needs. Individual scientists, project leaders, and research institutes claim that participation in integrative studies is often a response to the priorities of policy-makers and funding bodies. Researchers also claim that their project applications must be integrative to have any chance of winning large research grants in the field of natural resource management. Funding bodies claim that current societal and environmental problems cross policy sectors and disciplinary boundaries and thus call for a common effort. Besides problem-solving, increasing the interaction between science and society as well as building expertise are the key motivations for policy to promote integrative research (Tress *et al.* 2005a; Table 14.1). National research funds are the main driver of the interest in integrative landscape research. This source of funding finances the majority of large-scale integrative

TABLE 14.1. *Expectations of policy and funding bodies*

Expectations

(1) Solving environmental problems
  - research should provide and apply knowledge to solve pressing environmental problems

(2) Increase the social relevance of research
  - research should involve stakeholders in problem definition and solution
  - greater amount of research funding to be used at local level

(3) Scientific progress and expertise
  - long-term investment in high-quality research
  - investment in intellectual capital/competitive ability

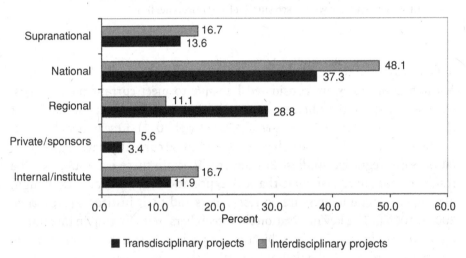

FIGURE 14.3
Sources of funding for interdisciplinary and transdisciplinary projects (data derived from web-based survey of 150 researchers)

studies (see Fig. 14.3). Only for transdisciplinary projects do regional research funds provide a significant proportion of project funding. Transdisciplinary projects frequently engage local or regional stakeholders, which explains the strong involvement of regional funding agencies.

Many of the problems facing landscape management are complex and may involve aspects of animal husbandry, economy, soil science, rural sociology, ecology, and cultural studies, etc. Researchers are expected to contribute to problem solving through examining natural resource management issues from several perspectives. On the other hand, research institutes claim this has made it difficult to fund pure research, even when this is to advance knowledge

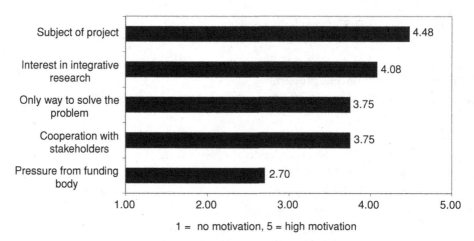

FIGURE 14.4
Top five motivating factors for researchers joining an integrative landscape study
(data derived from web-based survey of 150 researchers)

in fields essential to our understanding of sustainable land use. Disciplinary
research approaches are considered less able to meet current policy needs,
whereas integrative studies are expected to produce a greater proportion of
operational solutions (Gibbons *et al.* 1994, Ewel 2001). Researchers have little choice but to follow research policies, and these currently have high expectations of integrative studies. But how realistic are these expectations? The
results of our surveys suggest that the expectations are unrealistically high,
placing considerable pressure on researchers and their institutes (Tress *et al.*
2003a, 2005a). The key motivation for researchers participating in integrative
research is an interest in the subject of the project. Researcher interest in the
integration process or in stakeholder participation are lesser motivating factors
(Fig. 14.4). Researchers claim that pressure from funding bodies has only a low
effect on participation in integrative research. However, interviews with repre-
sentatives from funding bodies revealed a different picture. Funding bodies are
aware that they have a powerful steering role in integrative research and believe
that interest would be lower if funds were not used purposely to stimulate
integration.

However, the expectations of funding bodies place researchers in a difficult situation. The academic integration of disciplines is both very difficult
and can take longer than a single disciplinary research project or program.
We require, therefore, a more realistic understanding concerning the nature of
integrative research and especially of the limitations of any form of research to
solve natural resource management problems (van Asselt and Rijkens-Klomp
2002).

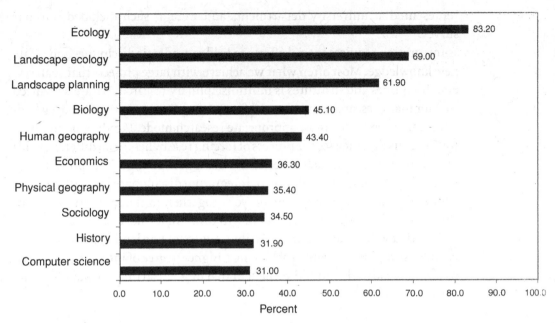

FIGURE 14.5
Top ten disciplines involved in integrative landscape research projects (data derived from web-based survey of 150 researchers)

14.5   What are we trying to integrate?

When conducting integrative research, it is necessary to understand the meaning of disciplines and their boundaries (Klein 1990, Lattuca 2001, Winder 2003) and to critically reflect on their current state and direction (Klein 2004). Disciplines are not static and are increasingly evolving into sub-disciplines with their own language and identity. New disciplines appear and old ones disappear, reflecting developments in knowledge cultures and academic institutions. From an epistemological perspective, some boundaries are harder to cross than others (Tress *et al.* 2005b). Integrating humanities and natural science perspectives is especially challenging. For many agricultural research projects, combining economic and ecological perspectives has been difficult, with areas of disagreement related to underlying model assumptions, timescales and what should and should not be taken into account. When undertaking integrative landscape research, such boundaries have to be identified and their nature understood before significant degrees of integration are possible. Disciplines that are most frequently involved in integrative landscape research include ecology, landscape ecology, landscape planning, biology, and human geography (Fig. 14.5). Many of these disciplines were considered as sub-disciplines or umbrella disciplines before, but are now independent disciplines represented at university departments and were as such included into our survey.

Integration requires special efforts to bridge academic disciplines and create new knowledge. Most often what we achieve with large projects that span several disciplines and institutes is multidisciplinary research. We are convinced that for many research purposes and for meeting the demands of funding bodies, it can often be the most appropriate research mode. Funding bodies have informed us that they see the process of forcing researchers from different fields to communicate with each other as the main goal of large-scale projects, not necessarily the more difficult task of integrating disciplinary knowledge. They believe that steering researchers to work together, in the same study area or through studying the same problem, will result in formal and informal interactions that will make valuable contributions to solving land-use management problems. Yet, any attempt at achieving a higher degree of integration of disciplines is considered as an add-on benefit that might lead to a new way of solving existing problems.

14.6   Organizational barriers to integration

Achieving a high level of integration between disciplines is difficult, and most projects struggle to realize their intended aim. Project organization, project design, and the day-to-day working environment of people working in large-scale projects can determine success or failure (Jakobsen *et al.* 2004). If institutional frameworks are unsupportive of the integration process as expressed through low resource allocation or cultural isolation, the barriers will be insurmountable. Coordinating the staff of large research teams in space and time is a major challenge of project organization. Spatial separation and infrequent or formal meetings will not help the process of integration. Supportive leadership and management styles combined with frequent and goal-oriented meetings are important factors. Projects that have no clear strategy on how to deal with integration issues often have difficulties in getting started or reaching successful outcomes. We have often observed that projects that set out to be integrative have neither integration goals nor a common problem definition.

The integration process may take longer, especially in defining common research goals, and thus needs more support in the early project phase. Participants should, as far as possible, have the opportunity for regular contact and spontaneous discussion to build the mutual trust and understanding needed to reach high levels of integration. To achieve this, it might be necessary to create temporary environments that physically bring interdisciplinary teams together across institutional boundaries. Research management can do much

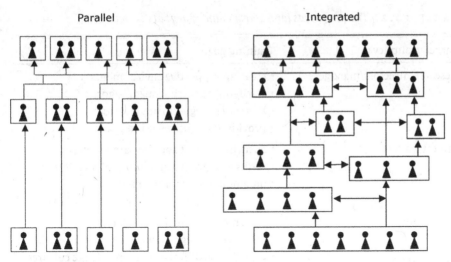

FIGURE 14.6
Fundamental differences in project organization between parallel and integrated
research design

to foster inter- and transdisciplinary studies. But this requires research managers to be sensitive to the needs of integrative research and how to create the special conditions and supportive environments needed by integrative research teams. We cannot over-emphasize the role of the project leader. Strong and committed leadership is essential in any new venture. The challenges of integrative research will call for leaders with highly developed interpersonal skills, research credibility, and the ability to maintain the motivation of the team, even when things go wrong.

As a result, many projects end as constellations of small independent (disciplinary) groups. During the course of a project, these groups work more or less independently of each other, only coming together at the end to integrate results. But this is seldom realized when it comes so late in the project process (Fry 2001). The design of integrative research should include measures that span the whole project period. The integration process should start at the beginning – at the stage of project formulation and application. At this stage we need to ask why we are integrating interests and what integration is expected to achieve. If we are unable to formulate answers to these questions, how are we going to know when we have been successful? If the integration does not start at the beginning of a project, this might result in a parallel instead of an integrated project design (Tress *et al.* 2005a; Fig. 14.6). The parallel project design allows for disciplines to carry out their own research agenda and come together at the end of the project to compare results, whereas the integrated design forces teams of mixed researchers to define and agree, at the start of the project, the research agenda needed to address a common research question.

TABLE 14.2. *Training needs for research management at different levels*

Target group/level	Training goal
Research directors/managers	• creating supportive environments • clarifying research strategies/goals • linking strategies to merit system • providing incentives to integrative research
Project leaders	• ensuring budget and time are adequate • facilitating formal and informal meetings • organizing seminars on methods • formulating integration implementation plan
Researchers	• integrating method and theory • organizing integration in daily project work • coping with differences in knowledge cultures • trust building and communication
Ph.D. students	• bridging the demands of a disciplinary Ph.D. and integrative work on a project

14.7   Education and training needs

All research requires a working knowledge of the accepted tools and methods that are integral to specific disciplines and knowledge cultures. Yet, we still observe that interdisciplinary and transdisciplinary studies often start without either participants or project leaders having a firm understanding of integrative research approaches. With no background training or experience in these approaches, researchers often have enormous problems in making the integration work and may return to the relative security of their disciplinary modes of research. To increase the success of integrative research, training is required on different levels and with different goal-settings as is shown in Table 14.2.

The special situation of Ph.D. students in integrative research projects demands special mention. It is common for large-scale integrative projects to involve several Ph.D. students. These students are often given responsibility for the task of achieving integration between the disciplines. Our surveys show that Ph.D. students in integrative projects take longer than average to complete their studies. This may be an especially acute problem for research students in transdisciplinary projects where the solving of a specific practical problem may not involve sufficient research activity or originality to qualify for a Ph.D. Training research students in the epistemological background to integrative research and in the social and psychological processes involved in working across subject and knowledge culture boundaries should be part of their

formal course work. It is very important for students to understand the dominant theoretical approaches of the different disciplines involved in a project if
they are to play a major role in the integration process (Klein 2004). Our belief
is that integrative research is less suited to Ph.D. students and that if they are
to tackle this work they will need significantly greater levels of support than is
current practice.

14.8   Improving the theory base

Despite high expectations for solving practical land-use problems, interdisciplinarity and transdisciplinarity are just alternative research approaches.
This implies that they have an underlying epistemological support, including integrative theories and concepts. However, these are only poorly developed in integrative landscape studies. Wu and Hobbs (2002), Moss (2000) and
Antrop (2001) have all pointed out that interdisciplinary work requires method
development, conceptual frameworks, and interdisciplinary theory. So far, little coherent interdisciplinary or transdisciplinary theory has emerged from
landscape research (Tress and Tress 2002). The same is true for the development
of inter- and transdisciplinary concepts and methodologies. One suggestion is
to increase efforts in support of a systematic collection of results and experiences of integrative studies in order to identify new generalizable knowledge
and to improve methods (Smoliner et al. 2001, Fry 2003). The implicit knowledge gained from practical experiences in integrative studies is only rarely
made explicit, and is, therefore, unavailable to the scientific community (Nonaka and Takeuchi 1995). This violates a basic academic tradition: to build on
existing knowledge. Instead, most integrative landscape studies suffer similar
starting problems, especially lack of common methods, and hence progress is
slow (Tress et al. 2003b).

There are several themes within landscape research that appear especially
interesting as starting points for integrative approaches (see Table 14.3).
Although different disciplines use their own jargon to describe these concepts,
they overlap to a great extent. Exploring the overlapping conceptual zones
offers a rich source for the development of common theory.

14.9   The merit system and the products of integrative research

A merit system gives scientists rewards for certain activities that institutes, universities, or the wider scientific community regard as important
achievements. Current academic merit systems are tailored for disciplinary
approaches and rely heavily on peer-reviewed publications in international
journals as the main criteria of success. Likewise, the career advancement of
scientists is still mainly based on disciplinary efforts. This is seen by some as a

TABLE 14.3. *Examples of concepts and themes that seem to have wide application to biodiversity, aesthetics and cultural aspects of landscapes (adapted from Fry 2003)*

Concept/theme	Description/application
Connectivity	Connectivity is one of the fundamental processes in landscape ecology. It relates to the functional linkages in a landscape and differs from connectedness, which refers to the physical connections between landscape elements. Connectivity is much more than being physically connected and may include the resistance to movement caused by barriers or by land-use types. Connectivity as a concept is increasingly important in cultural studies where the perception of time and space relate to the mode and characteristics of transport and how these have altered over time. Connectivity in landscapes is an important determinant of the ways in which animals or humans can navigate and move around in the landscape. It will have significance for resource availability and the frequency of cultural interaction and contact with other social groups. The concept involves not just the physical flows across landscapes, but also mental and physical barriers to movement. It also affects visual aspects by indicating accessibility.
Corridors	Corridors are linear landscape features that increase the flow of individuals, materials or energy between resource patches or suitable habitat/settlements and are important in defining the movement infrastructure of both animals and people (Dover 2000). Corridors are one of the most important ways of increasing landscape connectivity. They have been found to function for a wide range of animals and plants (via wind and water spreading of propagules or through seed vectors). The human parallel to landscape corridors is, of course, transport infrastructure. Traffic infrastructure is the most significant human corridor system, increasing access to and availability of physical and mental resources. An important research theme in geography is the way transport infrastructure and modes of transport affect our concepts of place and space.
Nodes	Nodes are intersections in movement corridors that result in important meeting places. They are especially of significance for determining the number of alternative routes individuals can take to move around a landscape. They are also important in visual orientation in built and natural landscapes. Nodes are therefore important in determining flows of species, nutrients, and energy around landscapes. If we examine a typical hedgerow network, we find nodes where hedges meet or cross. It is interesting to note that for a wide range of animal and plant taxa, the number of species that accumulate at nodes is richer than the surrounding landscape (Fry 1991, Sarlöv-Herlin and Fry 2000). The ecological interpretation of this phenomenon is that the diversity of species is a result of the favorable (food, shelter, and less disturbance) habitat at nodes and their significance for landscape-scale dispersal mechanisms. Movement models predict greater visitation rates to nodes than other sections of hedgerow networks. The importance of nodes to geographers and archaeologists is much older than the new-found relations in ecology. The role of nodes in transport infrastructure has long been a major explanatory variable in the locating of defences, settlements, and industry. Multiple infrastructure nodes such as waterways and roads have been especially important at different historical periods.

| Supplementation and Complementation | Supplementation and complementation are key landscape ecological concepts that reflect different strategies by which individuals and populations sustain themselves with essential resources in fragmented landscapes (Taylor et al. 1993). The process of supplementation relates to obtaining necessary resources from several small sources within accessible distance. We can think of an animal such as the wolf that has a large home range; it may be able to use several forest areas to meet all its needs as long as the forests are all accessible and within traveling distance. Complementation is a similar, but subtly different, concept that describes the acquisition of several quite different but essential resources, which must be "available" i.e., close enough to utilize without expending too much energy or being exposed to risks such as predation. "Availability" of resources in human terms is related to both physical and cultural/behavioral aspects. It is thus possible to discuss the theory of complementation in archaeological terms as the basis for many locational models (Fry et al. 2004). |
| Heterogeneity | Heterogeneity is a complex concept and much used in landscape analysis for a wide range of purposes. Heterogeneity sums up two aspects of landscapes: their grain size (the size of fields, forest patches etc.), and the variety that exists. Perhaps because of this dual influence on heterogeneity, it is not so easy to capture by numerical indices (Dramstad et al. 2001, Fjellstad et al. 2001). Despite difficulties in capturing heterogeneity, it ensures that a wide range of resources are available in a spatially restricted area, and is also one of the dominant visual aspects of landscapes and correlated to a wide range of animal and plant population and community characteristics. Because of this, landscape heterogeneity has been a central theme in the development of landscape indicators for biodiversity (Schneider and Fry 2001, Leitão and Ahern 2002, Olff and Richie 2002). It is also a major feature in human landscape preference studies and used in landscape characterization. In a paper based on the results of a multiple-interest landscape monitoring project, Dramstad et al. (2001) found that bird and plant diversity, the density of prehistoric grave mounds and human visual preferences all responded negatively to the grain size of the landscapes. In other words, the finer the scale of the landscape the higher the biodiversity, cultural heritage, and visual value of the landscape. The concept of heterogeneity is thus likely to provide an interesting starting point for theory generation across disciplinary boundaries in an attempt to explain its wide influence in humans and nature at the landscape level. |

TABLE 14.4. *Statements by researchers related to their experiences of trying to publish the results of integrative landscape research (Tress et al. 2005; Data derived from qualitative interviews)*

No.	Statement
1	For interdisciplinary or transdisciplinary studies you cannot find the right journals
2	Publications from interdisciplinary and transdisciplinary projects are not suited for journal publication
3	It becomes more difficult to publish in journals with high-impact factor, the more applied the study is

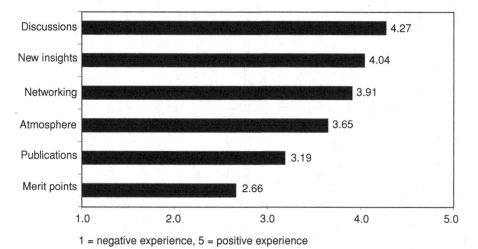

FIGURE 14.7
Some of the positive and negative experiences of researchers working in integrative landscape research (data derived from web-based survey of 150 researchers)

limitation to the development of integrative approaches. If scientists are to work with integrative approaches, their involvement should have an equal chance of being rewarded as disciplinary efforts. In Fig. 14.7, we present what researchers consider as negative and positive experiences of integrative research; merit points and publications are ranked more negatively than other effects. A merit system for integrative approaches may require academia to acknowledge a wider range of research products. Assessment of these products, however, will need the development of extensive, systematic, transparent, and fair systems of peer-reviewed achievement. There exists confusion over what integrative research can deliver. There is a wide belief among researchers that it is difficult to publish the results of integrative research (Table 14.4). To test this, we contacted more than 156 journals in landscapes, agriculture, forestry,

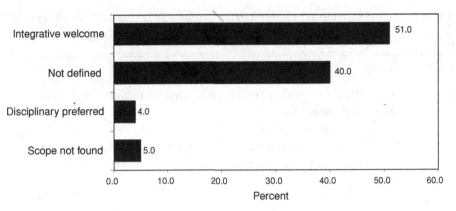

FIGURE 14.8
Editorial policies towards publishing integrative research papers of 156 journals
that publish landscape research; as extracted from their published instructions to
authors (n = 156)

ecology, planning, and cultural studies. All have published papers from landscape research. We asked the editors of these journals whether they would publish the results of integrative studies. Of the 97 that replied 96 said that they
would welcome such papers. Similarly, the instructions to authors of scientific
journals show that more than 50 percent actively seek papers from interdisciplinary research and a further 40 percent are neutral (Fig. 14.8). There may
be discrepancies between the declared publishing policy of journals and the
responses of reviewers to integrative papers, or there may be other reasons for
finding the results of integrative research difficult to publish. Researchers in
our survey mentioned the factors listed in Table 14.5.

An equally important product of integrative research is its ability to solve
environmental problems. However, we have found little evidence to suggest that integrative projects are more or less likely than single-disciplinary
research to provide solutions to environmental problems. Applied research is
more likely to focus on specific problems and their solutions, but whether integrative approaches result in more or better solutions is difficult to assess. There
is little empirical evidence either way. The long-term benefits of increasing
communication between disciplines are also difficult to assess.

14.10 Mapping the boundaries of research

At an abstract level, integrative or participatory research projects contain elements of knowledge creation, application, and reflection, as well as
feedback to science. These processes go hand in hand and mutually influence
each other. We have analyzed the process of knowledge creation in two steps to
clarify the boundary between research and consultancy/outreach activities
(Tress *et al.* 2003b; Fig. 14.9):

TABLE 14.5. *Statements by researchers concerning why integrative research may not get published (Tress et al. 2005; Data derived from qualitative interviews)*

No.	Reason
1	The work does not make a significant or novel contribution to the relevant bodies of knowledge
2	The work is too descriptive
3	It is difficult to write joint papers when co-authors belong to different research cultures – might take more time. Different knowledge cultures mean: • Different styles in writing • Different concepts of data • Different strategies for analyzing data • Lack of common theory base
4	The work is the application of existing knowledge, not original enough to be published
5	Large-scale interdisciplinary projects often lack replication or control or may study a single case making it difficult to generalize and be accepted by high-ranking journals

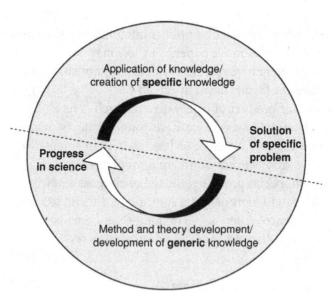

FIGURE 14.9
The process of knowledge creation – from problem solving to progress in science
(Tress *et al.* 2003b)

1. Existing knowledge is used to develop a solution to a specific problem. This knowledge may be derived from the collective expertise of the project team (which may include nonacademics as well as academics) or from the results of earlier research studies – part of the body of scientific knowledge. For a project to be considered as research demands that new knowledge has to be generated by the project team in order to solve the problem. This debate is very relevant to the increased consultancy and outreach activities of European research institutes.

2. The second part of the process of knowledge creation occurs when the focus is on the generation of *generic* knowledge. We also acknowledge that the systematic application of existing knowledge can be a form of hypothesis testing – leading to the production of generic knowledge. As science is interested in the nature and behavior of observable phenomena (Feynman 1998), it seeks knowledge that has relevance and validity beyond a specific context. This generic knowledge is fed back to science usually through the publication of a peer-reviewed scientific paper or book and is the main process through which progress in science takes place.

It would appear that many applied integrative projects only focus on the goal of gaining the knowledge needed for solving the specific problem defined by the funding agency. Once this has been achieved, there may be neither time nor money for more basic reflection on the knowledge created or how it relates to the wider scientific context. The focus of consultancy work is more on the application of existing knowledge than on the creation of generic knowledge and hence scientific advancement. Consultancy relies on the application of existing knowledge for the solution of a problem – the work is not usually considered as research even when that solution is contextual and unique. The difference between fundamental research and consultancy is also illustrated in Fig. 14.10. Research projects, in their intention to be applied or solve a specific problem, may transgress the border between research and consultancy.

14.11 Enhancing integrative landscape ecology research

What is a good interdisciplinary or transdisciplinary study? Attempts have been made to develop sets of evaluation criteria for integrative projects (Defila and Di Giulio 1998, Spaapen and Wamelink 1999, Balsiger 2004; see Table 14.6). There are, however, no widely recognized quality standards that could be used to evaluate projects through their various stages. Quality standards would have two main advantages. Firstly, agreed standards would make it easier for funding bodies to distinguish real interdisciplinary projects from

TABLE 14.6. *List of measures to increase the success of integrative research projects based on research of the INTELS project (Tress et al. 2003a, 2005)*

No.	Measure
1	Start the work early in the process with a plan for integration
2	Give participants opportunity for frequent formal and informal contact
3	Identify shared problems/challenges as work packages
4	Organize seminars specifically to communicate different research modes and approaches and to identify common theory and methods
5	Include a project plan for integrated products showing add-on value
6	In general small groups work better than large
7	The power of personal chemistry cannot be over-estimated
8	Good project leadership and management are essential to the success of large projects
9	Supportive institutional structures are required, these provide reward and identity
10	Training is required at the researcher, project leader, and institutional management level
11	Integrative projects may not be suitable for Ph.D. studies

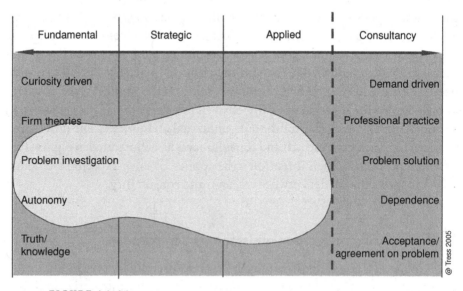

FIGURE 14.10
Defining the footprint of integrative research – from pure science to professional practice

those that only play with the name to improve their chance for getting support. Secondly, standards would also serve as guidelines for researchers when setting criteria and goals for project planning. Development of integrative research standards would contribute significantly to improving interdisciplinary and transdisciplinary projects. One of the most important of these criteria would be the degree of integration reached in a project and how this contributed to the add-on value of end products.

Good personal chemistry between researchers is also a key to success. Mutual trust, motivation, and pleasure in working together are important in any research team. However, the common ground is much smaller when contrasting disciplines are involved than when researchers all come from the same or closely related disciplines. Overcoming cultural barriers places high demands on the interpersonal relationships between members of interdisciplinary projects. As a result, smaller research teams are often better suited to crossing disciplinary boundaries than larger ones.

One characteristic of the large-scale research projects studied by the INTELS project was that most had no specific plan to reach integration. Yet, many of the other aims of these projects had clear objectives linked to specific methods and milestones to assess planned progress. We firmly believe that achieving integration should also be seen as a specific project aim with a full description of the planned progress of integration and how this will be achieved. Only in a few recently started initiatives, have we identified specific aims for improving knowledge and skills in integrative research among the aims of research programs.

14.12 Conclusion

Landscape ecology has undoubtedly been a major force in the development of integrative landscape research. This contribution to the development of integrative approaches is widely evidenced by the many landscape conferences that have held special sessions or had integration as a major theme as reported by Brandt (2000), Klijn and Vos (2000), Moss (2000), Tress *et al.* (2001), and Mander *et al.* (2004). The overarching concepts presented in Section 14.8 of this paper are but a few of those emerging within landscape ecology. Developing these concepts may lead to significant advances in the development of integrative landscape theory. We also urgently need new methods for use in integrative research. The ways different disciplines collect and analyze data are both different and often incompatible. Yet, here again we see new research initiatives that combine qualitative and quantitative approaches, and traditional knowledge with landscape metrics. The future will include a rich variety of such studies that will redefine the scope and direction of landscape ecology. We have seen the rapid increase in integrative landscape research projects. Many of these are large in terms of budget and personnel resources. The time is near when we can start to compile a meta-analysis of such projects to locate aspects of emerging theory and identify robust methodological approaches. The science of landscape ecology requires this academic growth as well as its contribution finding practical solutions to resource management issues.

To further improve the success rate of integrative landscape studies, the expectations of scientists and funding bodies need to reach a better balance and be made explicit. This would include greater reflection on and a more realistic appraisal of what integrative research approaches can achieve and what they cannot (see *Futures* **36**, issue (**4**), 2004). We need to acknowledge – against the tide of opinion – that integrative studies are not the solution to each and every land management problem nor will they always result in win–win resource-management situations. Working interdisciplinarily or transdisciplinarily will not prevent power struggles between interest groups and will not tell policy-makers what should be done. Integrative research approaches will, however, increasingly inform policy-makers of the consequences of different land-use scenarios for a wider range of landscape values and stakeholders and hence provide a better basis for decision-making.

In this chapter we have identified many of the challenges and difficulties facing integrative research. Despite the problems, we have also noted that it is possible to find successful interdisciplinary and transdisciplinary projects and that such projects increasingly succeed in publishing their results. In addition, participants from integrative studies report that their involvement gave them unexpected new insights into other fields of research and also into their own subjects (Kinzig 2001, Tress *et al.* 2005a). These insights sometimes fundamentally changed the way researchers perceived their own discipline or their methodological approach. To all those involved in integrative research from planning and policy levels to individual researchers, the future challenge is to further elaborate on the improvement and advancement of integrative research, to bring its full benefit to both academia and society.

References

Antrop, M. 2001. The language of landscape ecologists and planners: a comparative content analysis of concepts used in landscape ecology. *Landscape and Urban Planning* **55**, 163–73.

Balsiger, P. W. 2004. Supradisciplinary research practices: history, objectives and rationale. *Futures* **36**, 407–21.

Bastian, O. 2002. Landscape ecology: towards a unified discipline? *Landscape Ecology* **16**, 757–66.

Brandt, J. 2000. Editorial: the landscape of landscape ecologists. *Landscape Ecology* **15**, 181–5.

Dalgaard, T., N. J. Hutchings, and J. R. Porter. 2003. Agroecology, scaling and interdisciplinarity. *Agriculture Ecosystems and Environment* **100**, 39–51.

Décamps, H. 2000. Demanding more of landscape research (and researchers). *Landscape and Urban Planning* 47, 105–9.

Defila, R. and A. Di Giulio. 1998. Interdisziplinarität und Disziplinarität. Pages 111–37 in J.H. Olbertz (ed.). *Zwischen den Fächern – über den Dingen?* Opladen: Leske und Budrich.

Dover, J. 2000. Human, environmental and wildlife aspects of corridors with specific reference to UK planning practice. *Landscape Research* 25, 333–44.

Dramstad, W.E., G. Fry, W.J. Fjellstad, *et al.* 2001. Integrating landscape-based values. *Landscape and Urban Planning* 57, 257–68.

Ewel, K.C. 2001. Natural resource management: the need for interdisciplinary collaboration. *Ecosystems* 4, 716–22.

Fábos, J.G. 2004. Greenway planning in the United States: its origins and recent case studies. *Landscape and Urban Planning* 68, 321–42.

Feynman, R. 1998. *The Meaning of Everything.* London: Penguin Books.

Fjellstad, W.J., W.E. Dramstad, G.-H. Strand, and G.L.A. Fry. 2001. Heterogeneity as a measure of spatial pattern for monitoring agricultural landscapes. *Norwegian Journal of Geography* 55, 71–6.

Fry, G., B. Skar, G.B. Jerpåsen, V. Bakkestuen, and L. Erilestad. 2004. Predicting archaeological sites: a method based on landscape indicators. *Landscape and Urban Planning* 67, 97–107.

Fry, G.L.A. 1991. Conservation in agricultural ecosystems. Pages 415–43 in I.F. Spellerberg, F.B. Goldsmith, and M.G. Morris (eds.). *The Scientific Management of Temperate Communities for Nature Conservation.* London: Blackwell.

Fry, G.L.A. 2001. Multifunctional landscapes: towards transdisciplinary research. *Landscape and Urban Planning* 57, 159–68.

Fry, G.L.A. 2003. From objects to landscapes in natural and cultural heritage management: a role for landscape interfaces. Pages 237–53 in H. Palang and G. Fry (eds.). *Landscape Interfaces: Cultural Heritage in Changing Landscapes.* Dordrecht: Kluwer.

Gibbons, M., C. Limoges, H. Nowotny, *et al.* 1994. *The New Production of Knowledge: The Dynamics of Science and Research in Contemporary Societies.* London: Sage.

Gobster, P.H. and L.M. Westphal. 2004. The human dimensions of urban greenways: planning for recreation and related experiences. *Landscape and Urban Planning* 68, 147–65.

Hobbs, R. 1997. Future landscapes and the future of landscape ecology. *Landscape and Urban Planning* 37, 1–9.

Höll, A. and K. Nilsson. 1999. Cultural landscape as subject to national research programmes in Denmark. *Landscape and Urban Planning* 46, 15–27.

Jakobsen, C.H., T. Hels, and W.J. McLaughlin. 2004. Barriers and facilitators to integration among scientists in transdisciplinary landscape analysis: a cross country comparison. *Forest Policy and Economics* 6, 15–31.

Kinzig, A.P. 2001. Bridging disciplinary divides to address environmental and intellectual challenges. *Ecosystems* 4, 709–15.

Klein, J.T. 1990. *Interdisciplinarity: History, Theory and Practice.* Detroit: Wayne State University Press.

Klein, J.T. 2004. Prospects for transdisciplinarity. *Futures* 36, 515–26.

Klijn, J. and W. Vos, (eds.). 2000. *From Landscape Ecology to Landscape Science.* Dordrecht: Kluwer.

Lattuca, L.R. 2001. *Creating Interdisciplinarity: Interdisciplinary Research and Teaching among College and University Faculty.* Nashville: Vanderbilt University Press.

Leitão A.B. and J. Ahern. 2002. Applying landscape ecological concepts and metrics in sustainable landscape planning. *Landscape and Urban Planning* 59, 65–93.

Mander, Ü., H. Palang, and M. Ihse. 2004. Editorial: development of European landscapes. *Landscape and Urban Planning* 67, 1–8.

Moss, M. 2000. Interdisciplinarity, landscape ecology and the "Transformation of Agricultural Landscapes." *Landscape Ecology* 15, 303–11.

Nassauer, J.I. 1995. Culture and changing landscape structure. *Landscape Ecology* 10, 229–37.

Naveh, Z. 2001. Ten major premises for a holistic conception of multifunctional landscapes. *Landscape and Urban Planning* **57**, 269–84.

Naveh, Z. and A. Lieberman. 1994. *Landscape Ecology: Theory and Application*. 2nd edn. Berlin, Heidelberg: Springer.

Nonaka, I. and H. Takeuchi. 1995. *The Knowledge-creating Company*. Oxford: Oxford University Press.

Olff, H. and M. E. Richie. 2002. Fragmented nature: consequences for biodiversity. *Landscape and Urban Planning* **58**, 83–92.

Palang, H., Ü. Mander, and Z. Naveh. 2000. Holistic landscape ecology in action. *Landscape and Urban Planning* **50**, 1–6.

Sarlöv-Herlin, I. and G. Fry. 2000. Dispersal of woody plants in forest edges and hedgerows in a Southern Swedish agricultural area: the role of site and landscape structure. *Landscape Ecology* **15**, 229–42.

Schneider, C. and G. L. A. Fry. 2001. The influence of landscape grain size on butterfly diversity in grasslands. *Journal of Insect Conservation* **5**, 163–71.

Smoliner, C., R. Häberli, and M. Welti. 2001. Mainstreaming transdisciplinarity: a research-political campaign. Pages 263–71 in J. T. Klein, W. Grossenbacher-Mansu, R. Häberli, A. Bill, R. W. Scholz, and M. Welti (eds.). *Transdisciplinarity: Joint Problem Solving among Science, Technology, and Society*. Basel: Birkhäuser.

Spaapen, J. B. and F. J. M. Wamelink. 1999. *The Evaluation of University Research: A Method for the Incorporation of Societal Value of Research*. The Hague: NRLO-report 99/12.

Taylor, P. D., L. Fahrig, K. Henein, and G. Merriam. 1993. Connectivity is a vital element of landscape structure. *Oikos* **68**, 571–3.

Tress, B. and G. Tress. 2002. Disciplinary and meta-disciplinary approaches in landscape ecology. Pages 25–37 in O. Bastian and U. Steinhardt (eds.). *Development and Perspectives in Landscape Ecology*. Dordrecht: Kluwer.

Tress, B., G. Tress, H. Décamps, and A. d'Hauteserre. 2001. Bridging human and natural sciences in landscape research. *Landscape and Urban Planning* **57**, 137–41.

Tress, B., G. Tress, and G. Fry. 2005a. Integrative studies on rural landscapes: policy expectations and research practice. *Landscape and Urban Planning* **70**, 177–91.

Tress, G., B. Tress, and G. Fry. 2005b. Clarifying integrative research concepts in landscape ecology. *Landscape Ecology* **20**, 479–93.

Tress, B., G. Tress, A. Van der Valk, and G. Fry (eds.). 2003a. *Interdisciplinarity and Transdisciplinarity in Landscape Studies: Potential and Limitations*. Wageningen: Delta Series 2.

Tress, G., B. Tress, and G. Fry. 2003b. Knowledge creation and reflection in integrative and participatory projects. Pages 14–24 in G. Tress, B. Tress, and M. Bloemmen (eds.). *From Tacit to Explicit Knowledge in Integrative and Participatory Research*. Wageningen: Delta Series 3.

van Asselt, M. B. A. and N. Rijkens-Klomp. 2002. A look in the mirror: reflection on participation in integrated assessment from a methodological perspective. *Global Environmental Change* **12**, 167–84.

Winder, N. 2003. Successes and problems when conducting interdisciplinary or transdisciplinary (= integrative) research. Pages 74–90 in B. Tress, G. Tress, A. V. d. Valk, and G. Fry (eds.). *Interdisciplinarity and Transdisciplinarity in Landscape Studies: Potential and Limitations*. Wageningen: Delta Series 2.

Wu, J. and R. Hobbs. 2002. Key issues and research priorities in landscape ecology: an idiosyncratic synthesis. *Landscape Ecology* **17**, 355–65.

Zonnenveld, I. S. 1995. *Land Ecology: An Introduction to Landscape Ecology as a Base for Land Evaluation, Land Management and Conservation*. Amsterdam: SPB.

PART III

Synthesis

15

Landscape ecology: the state-of-the-science

15.1 Introduction

Good science starts with precise definitions because clearly defined terminology is a prerequisite for any fruitful scientific discourse. For rapidly developing interdisciplinary sciences like landscape ecology, unambiguous definitions are particularly important. Contemporary landscape ecology is characterized by a flux of ideas and perspectives that cut across a number of disciplines in both natural and social sciences, as evidenced in the previous chapters of this volume. On the one hand, after having experienced an unprecedented rapid development in theory and practice in the past two decades, landscape ecology has become a globally recognized scientific enterprise. On the other hand, more than 65 years after the term "landscape ecology" was first introduced, landscape ecologists are still debating on what constitutes a landscape and what landscape ecology really is (e.g., Wiens 1992, Hobbs 1997, Wiens and Moss 1999, Wu and Hobbs 2002).

Two major schools of thought in landscape ecology have widely been recognized: the European approach that is more humanistic and holistic and the North American approach that is more biophysical and analytical. To increase the synergies between the two approaches, not only do we need to appreciate the values of both approaches, but also to develop an appropriate framework in which different perspectives and methods are properly related. Toward this end, in this chapter we shall compare and contrast the European and North American approaches through several exemplary definitions (see Table 15.1). We shall argue that both approaches can be traced back to the original definition of landscape ecology, and that recent developments seem to show a tendency for unification of once diverging perspectives. Then, we shall propose a

Key Topics in Landscape Ecology, ed. J. Wu and R. Hobbs.
Published by Cambridge University Press. © Cambridge University Press 2007.

TABLE 15.1. *A list of exemplary definitions of landscape ecology*

Definition	Source
The German geographer Carl Troll coined the term "landscape ecology" in 1939, and defined it in 1968 as "the study of the main complex causal relationships between the life communities and their environment in a given section of a landscape. These relationships are expressed regionally in a definite distribution pattern (landscape mosaic, landscape pattern) and in a natural regionalization at various orders of magnitude" (Troll 1968; cited in Troll 1971).	• Troll, C. 1939. Luftbildplan and okologische bodenforschung. *Zeitschrift der Gesellschaft fur Erdkunde Zu Berlin* 241–98. • Troll, C. 1968. Landschaftsokologie. Pages 1–2 in *Pflanzensoziologie und Landschaftsokologie – Symposium Stolzenau.* Junk: The Hague. • Troll, C. 1971. Landscape ecology (geoecology) and biogeocenology – a terminological study. *Geoforum* 8, 43–6.
"Landscape ecology is an aspect of geographical study which considers the landscape as a holistic entity, made up of different elements, all influencing each other. This means that land is studied as the 'total character of a region', and not in terms of the separate aspects of its component elements" (Zonneveld 1972).	• Zonneveld, I. S. 1972. *Land Evaluation and Land(scape) Science.* Enschede, The Netherlands: International Institute for Aerial Survey and Earth Sciences.
"Landscape ecology is a young branch of modern ecology that deals with the interrelationship between man and his open and built-up landscapes" based on general systems theory, biocybernetics, and ecosystemology (Naveh and Liberman 1984). "Landscapes can be recognized as tangible and heterogeneous but closely interwoven natural and cultural entities of our total living space," and landscape ecology is "a holistic and transdisciplinary science of landscape study, appraisal, history, planning and management, conservation, and restoration" (Naveh and Liberman 1994).	• Naveh, Z. and A. S. Lieberman. 1984. *Landscape Ecology: Theory and Application.* New York: Springer-Verlag. • Naveh, Z. and A. S. Lieberman. 1994. *Landscape Ecology: Theory and Application,* 2nd edn. New York: Springer-Verlag.
"A landscape is a kilometers-wide area where a cluster of interacting stands or ecosystems is repeated in similar form; landscape ecology, thus, studies the structure, function and development of landscapes" (Forman 1981). Landscape structure refers to "the spatial relationships among the distinctive ecosystems;" landscape function refers to "the flows of energy, materials, and species among the component ecosystems;" and landscape change refers to "the alteration in the structure and function of the ecological mosaic over time" (Forman and Godron 1986).	• Forman, R. T. T. 1981. Interaction among landscape elements: a core of landscape ecology. Pages 35–48 in S. P. Tjallingii and A. A. de Veer (eds.), *Perspectives in Landscape Ecology.* Wageningen: Pudoc. • Forman, R. T. T. and M. Godron. 1986. *Landscape Ecology.* New York: John Wiley & Sons Inc.

"Landscape ecology focuses explicitly upon spatial pattern. Specifically, landscape ecology considers the development and dynamics of spatial heterogeneity, spatial and temporal interactions and exchanges across heterogeneous landscapes, influences of spatial heterogeneity on biotic and abiotic processes, and management of spatial heterogeneity" (Risser *et al.* 1984). "Landscape ecology is not a distinct discipline or simply a branch of ecology, but rather is the synthetic intersection of many related disciplines that focus on the spatial-temporal pattern of the landscape" (Risser *et al.* 1984).

"Landscape ecology emphasizes broad spatial scales and the ecological effects of the spatial patterning of ecosystems" (Turner 1989).

"Landscape ecology is the study of the reciprocal effects of the spatial pattern on ecological processes," and "concerns spatial dynamics (including fluxes of organisms, materials, and energy) and the ways in which fluxes are controlled within heterogeneous matrices" (Pickett and Cadenasso 1995).

"Landscape ecology investigates landscape structure and ecological function at a scale that encompasses the ordinary elements of human landscape experience: yards, forests, fields, streams, and streets" (Nassauer 1997).

Landscape ecology is "ecology that is spatially explicit or locational; it is the study of the structure and dynamics of spatial mosaics and their ecological causes and consequences" and "may apply to any level of an organizational hierarchy, or at any of a great many scales of resolution" (Wiens 1999).

"Landscape ecology emphasizes the interaction between spatial pattern and ecological process, that is, the causes and consequences of spatial heterogeneity across a range of scales" (Turner *et al.* 2001). "Two important aspects of landscape ecology . . . distinguish it from other subdisciplines within ecology": "First, landscape ecology explicitly addresses the importance of spatial configuration for ecological processes" and "second, landscape ecology often focuses on spatial extents that are much larger than those traditionally studied in ecology, often, the landscape as seen by a human observer" (Turner *et al.* 2001).

- Risser, P.G., J.R. Karr, and R.T.T. Forman. 1984. *Landscape Ecology: Directions and Approaches.* Special Publication 2. Champaign: Illinois Natural History Survey.

- Turner, M.G. 1989. Landscape ecology: the effect of pattern on process. *Annual Review of Ecology and Systematics* 20, 171–97.

- Pickett, S.T.A. and M.L. Cadenasso. 1995. Landscape ecology: spatial heterogeneity in ecological systems. *Science* 269, 331–4.

- Nassauer, J.I. 1997. Culture and landscape ecology: insights for action. Pages 1–11 in J.I. Nassauer (ed.). *Placing Nature: Culture and Landscape Ecology.* Washington, DC: Island Press.

- Wiens, J.A. 1999. Toward a unified landscape ecology. Pages 148–51 in J.A. Wiens and M.R. Moss (eds.). *Issues in Landscape Ecology.* Snowmass Village: International Association for Landscape Ecology.

- Turner, M.G., R.H. Gardner, and R.V. O'Neill. 2001. *Landscape Ecology in Theory and Practice: Pattern and Process.* New York: Springer-Verlag.

hierarchical and pluralistic cross-disciplinary framework for promoting interactions and synergies between different perspectives and methods. Finally, the
relevance of this framework to the admirable but elusive goal of unification will
be discussed.

15.2 Two dominant approaches to landscape ecology

15.2.1 The European approach

The term landscape ecology was coined by the German geographer,
Carl Troll (1939), who was inspired especially by the spatial patterns of landscapes captured by aerial photographs and the ecosystem concept put forward
by Arthur Tansley (1935). This new field of study was proposed to combine the
horizontal–geographical–structural approach with the vertical–ecological–
functional approach, in order to meet the needs for geography to acquire ecological knowledge of land units and for ecology to expand its analysis from
local sites to the region (Troll 1971). For example, information obtained from
local sites through ground-based work can be "extended areally by means of
knowledge of the distribution of the ecosystems derived from air photograph
study" (Troll 1971). From its very beginning, landscape ecology evidently had
a close conceptual relationship with ecosystem ecology. In a formal definition,
Troll (1968) described landscape ecology as "the study of the main complex
causal relationships between the life communities and their environment in
a given section of a landscape. These relationships are expressed regionally
in a definite distribution pattern (landscape mosaic, landscape pattern) and in
a natural regionalization at various orders of magnitude" (Troll 1968, 1971).
While the above definition seems semantically indistinguishable from that of
ecosystem ecology, Troll's explanation of the "complex causal relationships"
points to three important characteristics that distinguish landscape ecology
from ecosystem ecology: (1) broad spatial scales, (2) spatial pattern, and (3) multiplicity of scales.

In addition, a landscape as perceived by Troll (1939, 1971) includes humans
in addition to its physical and biological components, as does the ecosystem
by Tansley (1935). Like other holistic geographers in Europe and Russia of that
time, Troll considered a landscape as something of a *Gestalt* (a German word
referring to a configuration of elements or an integrated system organized in
such a way that the whole cannot be described merely as the sum of its parts).
Zonneveld (1972) further emphasized the holistic totality of the landscape
while defining landscape ecology as part of the applied science of land evaluation and planning (Table 15.1). Oddly, he claimed unequivocally that landscape
ecology was not part of the biological sciences, but a branch of geography. The

holistic landscape perspective culminated in Naveh's and Liberman's (1984, 1994) work which described a landscape as a biocybernetic subsystem of the so-called "Total Human Ecosystem" – "the highest level of co-evolutionary complexity in the global ecological hierarchy" (Naveh 2000). Naveh (1991) further stated that "Landscape ecology deals with landscapes as the total spatial and functional entity of natural and cultural living space. This requires the integration of the geosphere with the biosphere and the noospheric human-made artifacts of the technosphere." This is essentially what is called the "holistic landscape ecology," often described as a transdisciplinary environmental science (Naveh 2000).

In general, most landscape ecological studies in Europe since the 1930s have reflected more of the humanistic and holistic perspective, involving landscape mapping, evaluation, conservation, planning, design, and management (Zonneveld 1972, Naveh and Lieberman 1984, Schreiber 1990, Bastian and Steinhardt 2002). However, it should be pointed out that, influenced by geographic and socioeconomic settings as well as academic and cultural traditions, European landscape ecological studies do vary in terms of the research focus and methodology, ranging from tedious technical mapping of heavily populated areas and systematic land evaluation, to philosophical (and sometimes enigmatic) discourses of the wholeness of landscapes. Some of the fine traditions and exciting new developments in European landscape ecology are well reflected in several chapters of this volume (e.g., Antrop, Chapter 10, Voss *et al.*, Chapter 13, Fry *et al.*, Chapter 14).

15.2.2 The North American approach

Landscape ecology was introduced to North America in the early 1980s (Forman 1981, Risser *et al.* 1984, Forman and Godron 1986), more than 40 years after it had been practiced in central Europe, focusing on the human–land systems. In the following decade, landscape ecology quickly flourished in North America with a stream of new perspectives and methods (Forman 1990, Turner 2005; also see Iverson, Chapter 2 of this volume for an interesting and personable account of the early days of North American Landscape Ecology). Consequently, landscape ecology became a well-recognized scientific discipline around the world by the mid-1990s. In their ground-breaking book, Forman and Godron (1986) defined landscape ecology as the study of the structure, function, and change of landscapes of kilometers wide over which local ecosystems repeat themselves (also see Forman 1995). Landscape structure refers to "the spatial relationships among the distinctive ecosystems"; function refers to "the flows of energy, materials, and species among the component ecosystems"; and change refers to "the alteration in the structure and function of the ecological mosaic over time" (Forman and Godron 1986). This definition of landscape ecology is consistent with Troll's original definition in that both aim to integrate the spatial pattern of landscapes with ecological processes within them. However, Forman and Godron (1981, 1986) provided the first systematic conceptual framework for studying landscape pattern and processes, signified by the patch–corridor–matrix model. As a convenient spatial language, this model has played an important role in promoting the development of landscape ecology worldwide since the 1980s.

Several other definitions of landscape ecology have been developed in North America (see Table 15.1). In particular, the report by Risser *et al.* (1984) was an important landmark publication because it reflected the collective view by North American ecologists on what landscape ecology should be and because it has served as a blueprint for the development of landscape ecology in North America in the past decades. The document is a synthesis of a workshop on landscape ecology held in the USA in April 1983, with 25 participants many of whom were leading ecologists and geographers (23 from the USA, 1 from Canada, and 1 from France). Risser *et al.* (1984) defined landscape ecology as the study of the development, management, and ecological consequences of spatial heterogeneity, or "the relationship between spatial pattern and ecological processes [that] is not restricted to a particular scale." They further identified four "representative questions" in landscape ecology: (1) How does landscape heterogeneity interact with fluxes of organisms, material, and energy? (2) What formative processes, both historical and present, are responsible for the existing pattern in a landscape? (3) How does landscape heterogeneity affect the spread of disturbances (e.g., pest outbreaks, diseases, fires)? (4) How can natural resource management be enhanced by a landscape approach? These earlier ideas of landscape ecology in North American were significantly influenced by the theory of island biogeography (MacArthur and Wilson 1967, Wu and Vankat 1995) and patch dynamics (Levin and Paine 1974, Pickett and White 1985, Wu and Loucks 1995).

In line with Risser *et al.* (1984), the different definitions developed in North America all have considered spatial heterogeneity as the cornerstone of landscape ecology. Of course, this does not mean that all North American landscape ecologists hold the same view on landscape ecology. Their major differences seem to hinge on how a landscape is perceived. In the seminal work of Forman and Godron (1981, 1986), a landscape is a kilometers-wide land area with repeated patterns of local ecosystems (also see Forman 1995). But most landscape ecologists consider landscape simply as a spatially heterogeneous area whose spatial extent varies depending on the organisms or processes of interest (Wiens and Milne 1989, Wu and Levin 1994, Pickett and Cadenasso 1995, Turner *et al.* 2001). In this case, landscape is an "ecological criterion" whose essence is not its absolute spatial scale, but rather its heterogeneity relevant to a particular research question (Allen and Hoekstra 1992, Pickett and Cadenasso 1995).

As such, the idea of "landscape" is also applicable to aquatic systems (Steele 1989, Turner *et al.* 2001, Poole 2002, Wiens 2002, Turner 2005). This multiple-scale or hierarchical concept of landscape is more appropriate because it is consistent with the scale multiplicity of patterns and processes occurring in real landscapes, and because it facilitates theoretical and methodological developments by recognizing the importance of micro-, meso-, macro-, and cross-scale approaches. Today, the most widely used definition of landscape ecology in North America, and arguably worldwide, is simply the study of the relationship between spatial pattern and ecological processes over a range of scales (Pickett and Cadenasso 1995, Turner *et al.* 2001, Turner 2005). Reflective of this dominant ecological paradigm in contemporary landscape ecology are several chapters in this volume, addressing a series of key issues focusing on the interrelationship among spatial pattern, ecological processes, and scale (see Chapters 2 to 9, this volume).

15.3   The elusive goal of a unified landscape ecology

It is evident that the European and North American approaches to landscape ecology have differed historically. On the one hand, the European approach is characterized by a holistic and society-centered view of landscapes, the focus on user-inspired and solution-driven research, and the combination of qualitative empirical methods with surveying and mapping techniques. On the other hand, the North American approach is dominated by an analytical and biological ecology-centered view of landscapes, the focus on basic science-oriented and question-driven studies, and the emphasis on the use of quantitative methods (particularly spatial pattern analysis and modeling). This dichotomy, of course, is an oversimplification of the reality because neither of the two approaches is internally homogeneous in perspectives and because both have been changing as an inevitable consequence of increasing communications and collaborations among landscape ecologists worldwide.

Both European and North American approaches can be traced back to the original definition of landscape ecology by Carl Troll (1939, 1968, 1971). The focus of the North American approach on the interrelationship between spatial pattern and ecological processes is not only consistent with Troll's original definition, but also represents a significant advance in implementing Troll's proposal to integrate the geographical and structural approach with the ecological and functional approach. Also, as noted earlier, the emphasis on large geographic areas, spatial patterns, and scale multiplicity that characterizes the

North American approach was evident in Troll's earlier writings. One may argue that Carl Troll was inspired as much by landscape patterns revealed in aerial photos in the 1930s as contemporary landscape ecologists are by those displayed in GIS. Indeed, it was the resurgence of interest in linking ecological processes with spatial pattern in the 1980s that led to a revitalization of the entire field of landscape ecology. Studies of spatial heterogeneity have laid an important foundation for landscape ecology as a scientific enterprise. On the other hand, landscape ecological studies in Europe have epitomized the ideas of landscapes as human-dominated *gestalt* systems, which were also evident in the early works of Troll and other holistic landscape ecologists (Troll 1971, Naveh and Lieberman 1984, Bastian and Steinhardt 2002). They have promoted the development of interdisciplinary and transdisciplinary approaches that transcend natural and social sciences. Undoubtedly, these studies provide valuable methods and exemplary solution strategies for dealing with various complex landscape issues, which must also be considered as an integral part of landscape ecology.

The simplistic dichotomy of landscape ecology approaches also obscures the fact that North American landscape ecology has recognized the important role that humans may have in shaping landscapes from its very beginning. In most cases, humans have been treated as "one of the factors creating and responding to spatial heterogeneity" (Turner *et al.* 2001, Turner and Cardille, Chapter 4, this volume), but perspectives from landscape architecture and planning are quite prominent in other instances (e.g., Nassauer 1997, Ahern 1999, Vos *et al.*, Chapter 13, this volume). In contrast, human society becomes the focus in European landscape ecology as presented by Naveh and Lieberman (1984, 1994). While advocating this holistic landscape ecology perspective, Naveh (1991) claimed that North American landscape ecology was merely "a ramification and spatial expansion of population, community, and ecosystem ecology," and that Risser *et al.*'s (1984) vision of landscape ecology as "the synthetic intersection of many related disciplines which focus on spatial and temporal pattern of the landscape" was inadequate. However, although the North American approach does not always consider landscapes in "their totality as ordered ecological geographical and cultural wholes," even the most ardent holists cannot deny that studies using this approach "are important and of great theoretical and epistemological value to the science of landscape ecology" (Naveh 1991). On the other hand, few would doubt that a holistic landscape ecology approach is essential for resolving problems of biodiversity conservation and ecosystem management.

During the past decade, there have been an increasing number of books and articles attempting to unite the two primary approaches to landscape ecology (Farina 1998, Wiens 1999, Bastian 2001, Wu and Hobbs 2002, Burel and

Baudry 2003). While landscape ecologists converge on the desire for a unified landscape ecology, they differ significantly as to how to achieve the goal. How can different perspectives be unified? There is no simple way to add them up to form a coherent scientific core of landscape ecology even if such a "core" exists. One common approach that many ecologists have adopted is to include humans and their activities as factors influencing and responding to landscape heterogeneity. In this case, landscape ecology is viewed as a branch of ecology, and issues of land use, biodiversity conservation, ecosystem management, and landscape planning and design belong to the domain of practical applications of landscape ecology, or "applied landscape ecology" (Turner *et al.* 2001).

Others do not seem to agree. For example, Naveh (1991) asserted that "landscape ecologists cannot restrict themselves merely to the study of the ecology and/or geography or history of landscapes, projected according to the definition of Forman and Godron (1986)," and that "landscape ecological studies have to be carried out along multidimensional, spatio-temporal, functional, conceptual and perceptional scales by multidisciplinary teams, using innovative interdisciplinary methods and having a common systems approach and transdisciplinary conception of landscape ecology." We agree that interdisciplinarity and transdisciplinarity are critically important to landscape ecology (Wu and Hobbs 2002), and this point has been made clear and loud in most of the chapters of this volume (e.g., Hof and Flather, Chapter 8, Mackey *et al.*, Chapter 11, Bowman, Chapter 12, Fry *et al.*, Chapter 14). However, we do not believe that each and every landscape ecological study has to be done "along multidimensional, spatio-temporal, functional, conceptual and perceptional scales by multidisciplinary teams." Interdisciplinarity and transdisciplinarity are not monolithic, but hierarchical. Thus, we argue that the unification of landscape ecology needs a complementary framework that clearly recognizes and takes advantage of the hierarchical structure in cross-disciplinarity.

15.4   A hierarchical and pluralistic framework for landscape ecology

When a group of leading scientists from around the world was asked about the future of landscape ecology, they unanimously agreed that the field is characterized, most prominently, by its interdisciplinarity or transdisciplinarity (see Wu and Hobbs 2002). It is logical, then, to take this consensus as a point of departure for exploring the possibility of unifying different landscape ecology perspectives. However, we need to understand what landscape ecologists mean by the terms interdisciplinarity and transdisciplinarity because they have been used rather ambiguously in the literature. Particularly, transdisciplinarity sometimes sounds like "a mystic supra-paradigm" that can hardly be understood in practical terms, much less implemented (Tress *et al.* 2005). Thus, we believe that clearly defined terms for cross-disciplinary interactions are a prerequisite for effective discussions on the possible unification of landscape ecology approaches.

Based on an extensive review of the literature, Tress *et al.* (2005) and Fry *et al.* (Chapter 14, this volume) have provided a much needed clarification on four frequently used terms with increasing degrees of cross-disciplinary integrations: disciplinarity, multidisciplinarity, interdisciplinarity, and transdisciplinarity. Disciplinary research operates within the boundary of a single academic discipline with no interactions with other disciplines, thus producing disciplinary knowledge; multidisciplinary research involves two or more disciplines with loose between-disciplinary interactions and a shared goal but parallel disciplinary objectives, thus producing "additive" rather than "integrative" knowledge; interdisciplinary research involves multiple disciplines that have close cross-boundary interactions to achieve a common goal based on a concerted framework, thus producing integrative knowledge that cannot be obtained from disciplinary studies; and transdisciplinary research involves both cross-disciplinary interactions and participation from nonacademic stakeholders or governmental agencies guided by a common goal, thus producing integrative new knowledge and uniting science with society (Tress *et al.* 2005, Fry *et al.*, Chapter 14, this volume). According to these authors, both interdisciplinary and transdisciplinary, but not multidisciplinary, studies are "integrative" research, and transdisciplinarity is essentially interdisciplinarity plus nonacademic involvement. Of course, disciplines or sub-disciplines are relative and dynamic terms that depend necessarily on the classification criteria used. Thus, it is important to recognize that cross-disciplinarity (i.e., multi-, inter-, and transdisciplinarity) may be discussed in different domains, such as within biological sciences, among natural sciences, or across natural and social sciences.

Before we discuss our cross-disciplinary framework for landscape ecology, let's make some general observations of the science of ecology first. Ecology has often been described as an interdisciplinary science because the relationship between organisms and their environment involves a myriad of biological, physiochemical, and geospatial processes. Thus, ecological concepts, theories, and methods come from a number of different disciplines, including botany, zoology, evolutionary biology, genetics, physiology, soil science, physics, chemistry, geography, geology, meteorology, climatology, and remote sensing. Without a common ecological context, some of these disciplines may seem rather unrelated. Various interactions among these disciplines characterize different ecological sub-disciplines (e.g., molecular ecology, chemical ecology, physiological ecology, ecosystem ecology, geographical ecology, etc.). Arguably, the most popular way of classifying ecological sub-disciplines, at

least among bio-ecologists, has been based on the hierarchical levels of biological organization from the organism to population, community, ecosystem,
landscape, and the biosphere. Although this is not a nested hierarchy (meaning
that the levels do not always correspond to spatial and temporal scales in a consistent order), some general patterns of cross-disciplinarity emerge along the
hierarchy.

Moving up the hierarchy of biological organization from physiological ecology at the level of individual organisms to global ecology that focuses on
the entire Earth system, research questions and methodologies, in general,
become increasingly multidisciplinary and interdisciplinary, spatial and temporal scales characterizing each field tend to increase, and mechanistic details
of phenomena under study tend to get increasingly coarse-grained. The need
and actual frequency of explicitly considering human activities in research also
tend to increase. For example, interdisciplinary studies that involve both natural and social sciences are much more frequently encountered in ecological
studies at the landscape and global levels than those focusing on individual
organisms and local biological communities. As different ecological disciplines
provide different perspectives and approaches to the study of nature, they all
contribute crucial knowledge to understanding how nature works in the multiscaled and diversely complex world. Generally, studies at lower levels of the
ecological hierarchy provide the mechanisms for patterns observed at higher
levels, whereas higher-level studies provide the context and significance for
lower-level processes. For instance, it is impossible to understand how terrestrial biomes respond to global climate change without invoking the knowledge of plant ecophysiology and ecosystem ecology. On the other hand, global
climate change has provided tremendous impetus and new directions for physiological and ecosystem ecology.

The above general patterns suggest that the interdisciplinarity of ecology is
quite heterogeneous. We argue that landscape ecology has similar disciplinary
characteristics in that landscape ecology involves essentially all the levels of
ecological organization and as diverse disciplines as ecology itself. Although
the landscape sometimes is considered as a level of ecological organization, it
is fundamentally a hierarchical concept that is operational on a wide range of
scales in space and time. Different from the traditional ecological disciplines,
landscape ecology focuses explicitly on the relationship between spatial pattern and ecological processes on the one hand and nature–society interactions
on the other, with the human landscape as arguably the most common scale of
research activities.

To promote synergies and unification in the extremely heterogeneous field
of landscape ecology, we argue that interdisciplinarity and transdisciplinarity should be interpreted in a hierarchical and pluralistic view (Fig. 15.1).

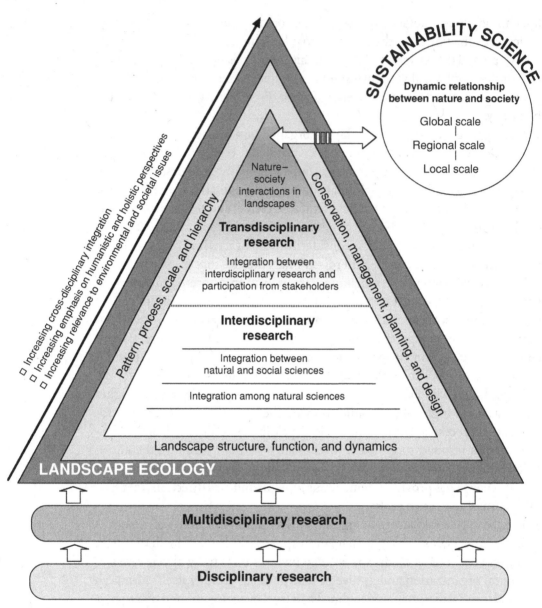

FIGURE 15.1

A hierarchical and pluralistic view of landscape ecology as an interdisciplinary and transdisciplinary science (Wu 2006). Landscape ecology is composed of research with various degrees of cross-disciplinary integration from interdisciplinary studies involving multiple natural sciences (e.g., bio-ecology and physical geography) to transdisciplinary studies that include natural and social sciences as well as active participation by stakeholders. Relevant multidisciplinary and disciplinary studies can also provide important contributions to the science of landscape ecology. The definitions of cross-disciplinarities are based on Tress *et al.* (2005)

"Hierarchical" here refers to the multiplicity of organizational levels, spatiotemporal scales, and degrees of cross-disciplinary interactions as well as the relativity of the definition of discipline. As a whole, landscape ecology is an integrative science that consists of studies with different degrees of interdisciplinary and transdisciplinary integration. This basic cross-disciplinary structure is not only reflective of what landscape ecology has been, but also germane to its future development. For example, it seems consistent with the general theme emerging from a list of major research directions and challenges suggested by a group of leading landscape ecologists (Wu and Hobbs 2002), as well as the chapters in this volume. In addition, it is hard to imagine how a credible transdisciplinary science can be developed without resorting to interdisciplinary and multidisciplinary efforts as well as solid disciplinary bases. "Pluralistic" here indicates the necessity to recognize the values of different perspectives and place them in a proper context characterized by a hierarchical cross-disciplinarity. This is indispensable for landscape ecology because of its diverse origins and objectives.

In this hierarchical and pluralistic framework, various approaches and perspectives correspond to different levels in the pyramid of cross-disciplinary integration (Fig. 15.1). In reality, landscape ecological studies usually have varying degrees of cross-disciplinary integration that are determined by specific research goals and questions. Many influential landscape ecological studies have involved different degrees of interdisciplinarity concerning primarily natural sciences, such as biological, ecological, physical, and geographical disciplines. The research topics include the effects of landscape pattern on animal behavior or "behavioral landscape ecology," metapopulation dynamics, spread of disturbance across landscapes, spatial ecosystem processes, patch dynamics, and neutral landscape models (e.g., Turner 1989, Farina 1998, Burel and Baudry 2003, Turner and Cardille, Chapter 4, Fahrig, Chapter 5, Gardner *et al.*, Chapter 6, this volume). In general, moving from the bottom to the top of the cross-disciplinarity pyramid in Fig. 15.1, landscape ecology increases the degree of integration among disciplines, prominence on humanistic and holistic perspectives, and relevance to environmental and societal issues (e.g., Hof and Flather, Chapter 8, Ludwig, Chapter 9, Mackey *et al.*, Chapter 11, Bowman, Chapter 12, Vos *et al.*, Chapter 13, this volume). Correspondingly, human–environment interactions increasingly become the focus of landscape ecology towards the transdisciplinarity end. There are outstanding examples from Europe and elsewhere in which natural and social sciences are successfully integrated with direct involvement of stakeholders, policy-makers, and governmental agencies (see Fry *et al.*, Chapter 14, this volume). Such transdisciplinary research ultimately unites science with society, and is an indispensable part of landscape ecology. In this case, landscape ecology is a critical part of the emerging sustainability science that focuses on the dynamic interactions between nature and society from the local to global scale through place-based and problem-driven projects (Kates *et al.* 2001, Clark and Dickson 2003).

15.5 Discussion and conclusions

Landscape ecology is the science and art of studying and influencing the relationship between spatial pattern and ecological processes across hierarchical levels of biological organization and different scales in space and time. The relationship among pattern, process, and scale is as essential in human-dominated landscapes as in natural landscapes, and is as important in theory as in practice. The "science" of landscape ecology focuses on understanding the dynamics of spatial heterogeneity and the relationship among pattern, process, and scale in natural as well as human-dominated landscapes. The "art" of landscape ecology emphasizes the necessary use of humanistic and holistic perspectives for integrating biophysical with socioeconomic and cultural components in general, and design, planning, and management in particular.

As we discussed earlier, two salient approaches have evolved, both of which can be traced back to the original definition of landscape ecology by Carl Troll (1939, 1968, 1971). The pattern–process–scale perspective that characterizes the North American approach is a continuation and indeed a breakthrough of realizing Troll's aspiration to integrate the geographical (structural) and ecological (functional) approaches. On the other hand, inspired and constrained by the close interactions between land and human society, scientists particularly in European and the Mediterranean countries have transformed the early holistic ideas into a transdisciplinary vision for landscape ecology. Differences in perspectives have apparently caused some landscape ecologists to worry about an identity crisis for landscape ecology (e.g., Moss 1999, Wiens 1999), and others have increasingly called for a unification of different approaches to landscape ecology (Wiens and Moss 1999, Bastian 2001, Wu and Hobbs 2002). Nonetheless, landscape ecology has been maturing as a science in recent years as it has apparently become more quantitative and precise with increasing use of modeling and statistical approaches, more concentration on methodology, and more concerted efforts to bring together different perspectives (Hobbs 1997, Wu and Hobbs 2002).

We believe that the diversity, but not divergence, of perspectives is an essential characteristic and strength of landscape ecology. The hierarchical and pluralistic framework proposed in this chapter help unite the different approaches to landscape ecology and allows for the continuing development of diverse perspectives and approaches. Unification is not to make certain views more prominent by diminishing others, but rather to join different perspectives complementarily in order to produce a whole that is larger than the sum of its parts. This is especially true for broadly interdisciplinary and transdisciplinary sciences such as landscape ecology that cut across natural and social sciences. Landscape ecology may never have a monolithic disciplinary core, and it should not in view of its diverse origins and goals. As a science of spatial heterogeneity, landscape ecology can benefit from its disciplinary heterogeneity. On the one hand, landscape ecology will continue to improve our understanding of the relationship among pattern, process, and scale; and on the other hand, it should play an increasingly important role in sustainability science in years to come.

Acknowledgments

JW's research in landscape ecology has been supported by the US Environmental Protection Agency, the US National Science Foundation, and the National Natural Science Foundation of China.

References

Ahern, J. 1999. Integration of landscape ecology and landscape design: an evolutionary process. Pages 119–23 in J. A. Wiens and M. R. Moss (eds.). *Issues in Landscape Ecology*. Snowmass Village: International Association for Landscape Ecology.

Allen, T. F. H. and T. W. Hoekstra. 1992. *Toward a Unified Ecology*. New York: Columbia University Press.

Bastian, O. 2001. Landscape ecology: towards a unified discipline? *Landscape Ecology* **16**, 757–66.

Bastian, O. and U. Steinhardt (eds.). 2002. *Development and Perspectives in Landscape Ecology*. Dordrecht: Kluwer.

Burel, F. and J. Baudry. 2003. *Landscape Ecology: Concepts, Methods and Applications*. Enfield, NH: Science Publishers, Inc.

Clark, W. C. and N. M. Dickson. 2003. Sustainability science: the emerging research program. *Proceedings of the National Academy of Sciences (USA)* **100**, 8059–61.

Farina, A. 1998. *Principles and Methods in Landscape Ecology*. London: Chapman & Hall.

Forman, R. T. T. 1981. Interaction among landscape elements: a core of landscape ecology. Pages 35–48 in S. P. Tjallingii and A. A. de Veer (eds.). *Perspectives in Landscape Ecology: Contributions to Research, Planning and Management of Our Environment*. Wageningen: Pudoc.

Forman, R. T. T. 1990. The beginnings of landscape ecology in America. Pages 35–41 in I. S. Zonneveld and R. T. T. Forman (eds.). *Changing Landscapes: An Ecological Perspective*. New York: Springer-Verlag.

Forman, R. T. T. 1995. *Land Mosaics: The Ecology of Landscapes and Regions*. Cambridge: Cambridge University Press.

Forman, R. T. T. and M. Godron. 1981. Patches and structural components for a landscape ecology. *Bioscience* **31**, 733–40.

Forman, R. T. T. and M. Godron. 1986. *Landscape Ecology*. New York: John Wiley & Sons, Inc.

Hobbs, R. J. 1997. Future landscapes and the future of landscape ecology. *Landscape and Urban Planning* **37**, 1–9.

Kates, R. W., W. C. Clark, R. Corell, *et al.* 2001. Sustainability Science. *Science* **292**, 641–2.

Levin, S. A. and R. T. Paine. 1974. Disturbance, patch formation and community structure. *Proceedings of the National Academy of Sciences (USA)* **71**, 2744–7.

MacArthur, R. H. and E. O. Wilson. 1967. *The Theory of Island Biogeography*. Princeton: Princeton University Press.

Moss, M. R. 1999. Fostering academic and institutional activities in landscape ecology. Pages 138–44 in J. A. Wiens and M. R. Moss (eds.). *Issues in Landscape Ecology*. Snowmass Village: International Association for Landscape Ecology.

Nassauer, J. I. 1997. Culture and landscape ecology: insights for action. Pages 1–11 in J. I. Nassauer (ed.). *Placing Nature: Culture and Landscape Ecology*. Washington, DC: Island Press.

Naveh, Z. 1991. Some remarks on recent developments in landscape ecology as a transdisciplinary ecological and geographical science. *Landscape Ecology* **5**, 65–73.

Naveh, Z. 2000. What is holistic landscape ecology? A conceptual introduction. *Landscape and Urban Planning* **50**, 7–26.

Naveh, Z. and A. S. Lieberman. 1984. *Landscape Ecology: Theory and Application*. New York: Springer-Verlag.

Naveh, Z. and A. S. Lieberman. 1994. *Landscape Ecology: Theory and Application*, 2nd edn. New York: Springer-Verlag.

Pickett, S. T. A. and M. L. Cadenasso. 1995. Landscape ecology: spatial heterogeneity in ecological systems. *Science* **269**, 331–4.

Pickett, S. T. A. and P. S. White. 1985. *The Ecology of Natural Disturbance and Patch Dynamics*. Orlando: Academic Press.

Poole, G. C. 2002. Fluvial landscape ecology: addressing uniqueness within the river discontinuum. *Freshwater Biology* **47**, 641–60.

Risser, P. G., J. R. Karr, and R. T. T. Forman. 1984. *Landscape Ecology: Directions and Approaches*. Special Publication 2. Champaign: Illinois Natural History Survey.

Schreiber, K.-F. 1990. The history of landscape ecology in Europe. Pages 21–33 in I. S. Zonneveld and R. T. Forman (eds.). *Changing Landscapes: An Ecological Perspective*. New York: Springer-Verlag.

Steele, J. H. 1989. The ocean "landscape". *Landscape Ecology* **3**, 185–92.

Tansley, A. G. 1935. The use and abuse of vegetational concepts and terms. *Ecology* **16**, 284–307.

Tress, G., B. Tress, and G. Fry. 2005. Clarifying integrative research concepts in landscape ecology. *Landscape Ecology* **20**, 479–93.

Troll, C. 1939. Luftbildplan and okologische bodenforschung. *Zeitschraft der Gesellschaft fur Erdkunde Zu Berlin*, 241–98.

Troll, C. 1968. Landschaftsokologie. Pages 1–21 in *Pflanzensoziologie und Landschaftsokologie – Symposium Stolzenau*. Junk: The Hague.

Troll, C. 1971. Landscape ecology (geoecology) and biogeocenology – a terminological study. *Geoforum* **8**, 43–6.

Turner, M. G. 1989. Landscape ecology: the effect of pattern on process. *Annual Review of Ecology and Systematics* **20**, 171–97.

Turner, M. G. 2005. Landscape ecology in North America: past, present, and future. *Ecology* **86**, 1967–74.

Turner, M. G., R. H. Gardner, and R. V. O'Neill. 2001. *Landscape Ecology in Theory and Practice: Pattern and Process*. New York: Springer-Verlag.

Wiens, J. A. 1992. What is landscape ecology, really? *Landscape Ecology* **7**, 149–50.

Wiens, J. A. 1999. Toward a unified landscape ecology. Pages 148–51 in J. A. Wiens and M. R. Moss (eds.). *Issues in Landscape Ecology*. Snowmass Village: International Association for Landscape Ecology.

Wiens, J. A. 2002. Riverine landscapes: taking landscape ecology into the water. *Freshwater Biology* **47**, 501–15.

Wiens, J. A. and B. T. Milne. 1989. Scaling of "landscape" in landscape ecology, or, landscape ecology from a beetle's perspective. *Landscape Ecology* **3**, 87–96.

Wiens, J. A. and M. R. Moss (eds.). 1999. *Issues in Landscape Ecology*. Snowmass Village: International Association for Landscape Ecology.

Wu, J. 2006. Cross-disciplinarity, landscape ecology, and sustainability science. *Landscape Ecology* **21**, 1–4.

Wu, J. and R. Hobbs. 2002. Key issues and research priorities in landscape ecology: an idiosyncratic synthesis. *Landscape Ecology* **17**, 355–65.

Wu, J. and S. A. Levin. 1994. A spatial patch dynamic modeling approach to pattern and process in an annual grassland. *Ecological Monographs* **64**(4), 447–64.

Wu, J. and O. L. Loucks. 1995. From balance-of-nature to hierarchical patch dynamics: a paradigm shift in ecology. *Quarterly Review of Biology* **70**, 439–66.

Wu, J. and J. L. Vankat. 1995. Island biogeography: theory and applications. Pages 371–9 in W. A. Nierenberg (ed.). *Encyclopedia of Environmental Biology*. San Diego: Academic Press.

Zonneveld, I. S. 1972. *Land Evaluation and Land(scape) Science*. Enschede, The Netherlands: International Institute for Aerial Survey and Earth Sciences.

Index

Page entries for **tables** appear in bold type.
Page entries for main headings which have subheadings refer only to general/introductory aspects of that topic.

288

Printed in the United States
by Baker & Taylor Publisher Services